T0192082

Springer Biographies

More information about this series at http://www.springer.com/series/13617

Leone Montagnini

Harmonies of Disorder

Norbert Wiener: A Mathematician-
Philosopher of Our Time

 Springer

Leone Montagnini
Ufficio Biblioscienze
Biblioteche di Roma
Rome
Italy

Translation from the Italian language edition: Le armonie del disordine by Leone Montagnini
Copyright © Istituto Veneto di Scienze, Lettere ed Arti All Rights Reserved.

ISSN 2365-0613   ISSN 2365-0621   (electronic)
Springer Biographies
ISBN 978-3-319-84455-8   ISBN 978-3-319-50657-9   (eBook)
DOI 10.1007/978-3-319-50657-9

Cover illustration: AGF/Science Photo Library with permission

Printed on acid-free paper

This Springer imprint is published by Springer Nature
The registered company is Springer International Publishing AG
The registered company address is: Gewerbestrasse 11, 6330 Cham, Switzerland

Let me here pay the greatest tribute to those older scientists [...] who look sympathetically on these struggles of their younger colleagues; and who, without considering every adventurous ugly duckling to be a young swan, are alert to foster power and originality.

Norbert Wiener, *Science and Society* (61b, 52)

Foreword

A Baedeker for our Challenging and Troubled Times

Norbert Wiener is an old mathematician, passed away more than fifty years ago and practically unknown outside the specialists. One can add that, among the few who have heard his name, suddenly the association is done with cybernetics, a ghost discipline, we could say, presently almost disappeared from the horizon of frontier investigations and circulating, instead, as a sort of characterizing prefix of many different everyday words (cybersociety, cyberbully, cyberpunk, just to provide a few examples). Moreover, in this book, he is presented as "a mathematician-philosopher", a strange qualification, indeed. Almost all that is written in the subtitle seems to refer to very old things: precious, maybe, but with a flavor of dusty erudition. An exception is provided by the last words of the subtitle "of our time". What can be the interest for "our time" of such an old men should be what is indicated, although mysteriously, by the title which, instead, seem to point out to some very contemporary notion. The one of disorder. In a mysterious way, we wrote, since what is promised and prospected are the harmonies which the disorder can produce. Well, Leone Montagnini was able to introduce all the complexity of the themes he will develop in these pages just through title and subtitle.

Leone has been able to afford and conclude a sort of impossible mission, the one of presenting in its entirety and complexity this unusual figure of scientist, integrating the various different aspects of his life and his work. Integration is really attained here at a level never obtained before. A book, as is written in the introduction, that is not "another biography" but "an essay about his thought". One could ask where is the novelty. Good reconstructions of the scientific life of any scholar are—always—also essays on thought. Yes, of course, but in Wiener's case there is one additional problem. He was really many things at the same time. Certainly a great mathematician, no doubt about that. A towering figure as is stressed in a lot of papers that can be found in the numerous volumes that the American Mathematical Society has published for celebrating him. But he was also many other things; his Ph.D. in philosophy—after the desire of studying medicine and intertwined interests in biology and mathematics—was not something accidental. He had studied with great philosophers of his time, not only Bertrand Russell, as is carefully documented here. He had planned to do philosophy when "the door was slammed in his face" by the Philosophy Department of Harvard University (§ 2.6 below). We could also suspect—in light of all his various interests —that he was, in fact, undecided about his future and "Harvard's door effect" only helped him to find his way. Writing "an essay on his thought" in Wiener's case means, then, doing a careful study of the modalities according to which different aspects of his intellectual life interacted. One could say that his various interests gave rise to results of different value, and due to his total engagement in mathematics starting around 1920, the most precious gems are to be found here. This is true, but what is at stake here is whether his way of doing mathematics is separable from the other loci of his mind. And beware: how to classify his invention of cybernetics? There is a lot of (his) mathematics here but also many other things. And what you have now in your hands shows also why he decided to write the book *Cybernetics*, where and when to publish it. The different aspects of his personality were really always strongly interacting. Also ethical questions are an essential part of the way in which scientific results are found, presented and published. The reader is invited to follow carefully Montagnini when he describes Wiener's doubts and dilemmas, his "acute attack of conscience", fearing the possibility that his "inventions go to the wrong hands" (§ 8.7, below).

And more. How to classify his work with neurophysiologists? And his lucid analysis of the problems posed by the new technological advances? A real fan, we could say, of technological developments (besides being one of the few great characters in the new play of information revolution), whose second book is a warning just from the title (*The human use of human beings*: the reader should have noticed the care in choosing words and adjectives). A few lines above, we used the word integration for describing the method used by Leone for presenting the different aspects. We must take the word seriously. An occasional reader of some of Wiener contributions could have simply concluded that he was a person with many intellectual interests as well as a superficial storytelling of his activities could resemble Pirandello's novel *One, No One and One Hundred Thousand* or Kurosawa's movie "Rashomon". Both are far from providing a good picture of Wiener's thought: in him the different facets always interact. The results in one of his activities must be seen—in order to fully appreciate their novelty—in light of what had happened or was happening in the other ones. That means that each point of view cannot be fully understood in isolation. Interdisciplinarity one would suggest as a keyword, although Wiener hardly used, if ever, this term. Yes, but only if we interpret the term in a correct way. He was all his avatars. We do not understand Wiener if we look at him as (only) a scholar doing interdisciplinary work. He, as a person, was "interdisciplinary"—if we can use this image, a very odd one, indeed. Both his work and his life have been guided by problems and questions, not by the compliance to formal rules or social duties. Problems are the very essence of our quest, as humans as well as scientists. Problems, however, do not obey disciplinary separations. These are useful for affording the former with refined and stronger tools, but should at all costs not be elevated at the status of absolutes.

This is true for Wiener not only for what we could call "nearby" disciplines, but for everything. If an ethical question was crucial, this interacted with the technical problems at which he was working and, in case of a contrast between them, he would have changed attitude, working at the technical problems in a different way. Neither abandon them nor renounce to his ethical commitments. He really is "one and many", a heavy burden he carried on, along all his life, apparently with ease and the lighthearted attitude of a child. But who knows the price that, as a human being, he payed for this burden that was the key of his originality.

Let's go back to the title. Although, at a first reading, it seems to point to technical (mathematical) findings by Wiener, it can be read also as a resumé of the crucial role that information technologies were (and are) coming to play in our world since they appear much more "disordered" than the traditional ones to which we were (and are) accustomed. From a social point of view this means that they must be understood and harnessed (and understood in order to be correctly harnessed). We must find their "harmonies". Glimpsing the complexity of Wiener's thought can help us in understanding new questions of our present society. He could appear also as a prophet. This is not satisfactory for a scientist even if we interpret the word "prophet" in the plain sense of a person able "to see things that others are unable to see". We should understand why he was able to do this. The answer may simply consist in the fact that he assumed "the right point of observation". For the understanding of our society as well, then, he is a very needed man. His message, till now, has not been known as it deserves, perhaps for the simple reason that his thought has not been seen in a unitary way. Montagnini has accomplished the superb goal of showing the intricacy of the connections among his life and (all) his thought(s), suggesting that for capturing his message not only in its entirety, but in his integrity, we must consider the complexity of the ways in which Wiener was able to do what he did.

How Leone was able to succeed in this task may appear a mystery. Easily explainable, indeed, by reading the book. It is simply the strategy—of Wienerian nature—of crossing the disciplinary frontiers and the capacity—after the crossing—to acquire all and only the tools necessary for moving in the new territories. There is no need of being a universal savant or another Wiener (both impossible tasks). This is Montagnini's lesson. What is required—and what he did—is to understand the kernel of his implicit epistemology, sharing with Wiener himself the capability of looking for links among distant things. What's at stake here, in fact, is not erudition but the ability of picking up fragments in which the complexity of his thought, of his work and his life opens out in a transparent way, allowing us to see the back-office of his mind.

When starting a journey, some guide is needed. This is true both in the case in which one simply plans to visit a new Country and in the case (nowadays, very rare, for what regards geographical places in our Earth) in which one dares to venture in still unexplored territories. In the first case what is needed is only a Baedeker, as one would have said a few decades ago and, when in Italy, the famous "Guide rosse" of Italian Touring Club would be a must. In the case of an unexplored territory no such useful help is available, one cannot but use the few elements reported by the first explorers, checking the most reliable.

The present book is useful for both types of journeys. It is a Baedeker for knowing and understanding Wiener, his thought in all the intricacy of the connections and interactions of its different layers. But playing this role it is useful also for exploring the still unknown territory of a society moulded and shaped by information technology. Our "mathematician-philosopher" still remains one of the most reliable and acute interpreter of our age, notwithstanding the fact that half century has elapsed from his death. So, we must thank Leone Montagnini who, providing a detailed guide to Wiener's thought in all his intricacy and complexity, has given us the possibility of using his suggestions and indications for moving in the still uncharted territory of information society.

Palermo, Italy   Settimo Termini

Contents

Foreword.   vii

Abbreviations for Archives.   xv

Introduction.   xvii

Part I   (1894–1918)

1   **Formative Years**.   3
  1.1   Family Roots   3
  1.2   Clay of New England.   7
  1.3   Child Prodigy.   8
  1.4   From Evolutionary Positivism to Pragmatism.   10
  1.5   In the Temple of American Philosophy   14
  1.6   The Positivistic Idealism of Santayana.   16
  1.7   The Idealism of Royce.   19
  1.8   Royce's Seminar on Scientific Method.   21

2   **The Young Philosopher of Mathematics**.   23
  2.1   The Choice of the Philosophy of Mathematics.   23
  2.2   Bertrand Russell.   25
  2.3   The Term Under Husserl   29
  2.4   Philosophy of Relativism   30
  2.5   A Future Mathematician with a Future Poet: Wiener
        and Eliot   32
  2.6   Wiener's Logical Research and the "No" from Harvard.   34
  2.7   The University of Maine   39
  2.8   Collaboration with the *Encyclopedia Americana*   40

Part II   (1919–1939)

3   Interwar Scientific Research 47
  3.1   An Unorthodox Mathematical Education 47
  3.2   From Harvard University to Massachusetts Institute
          of Technology 49
  3.3   Wiener's First Steps in Mathematical Research 51
  3.4   Wiener's New Esthetic Sense in Science 54
  3.5   The Influence of Peirce via Royce on Wiener's Science 57
  3.6   Wiener's Leibnizian Framework 62
  3.7   Pragmatic Factors in the Fertility of Wiener's Science....... 64
  3.8   Researches Alongside Engineers 67
  3.9   Participation in Projects to Build Technical Equipments 69
  3.10   Research Alongside Physicists 70

4   Reflections on Science and Technology 73
  4.1   Mathematics as a Fine Art 73
  4.2   The Relation with the Natural Philosophy
          of J.B.S. Haldane 80
  4.3   Arturo Rosenblueth 82
  4.4   John von Neumann 85
  4.5   "The Role of the Observer" 88
  4.6   "The Shattered Nerves of Europe" 93
  4.7   Thinking Technology: Matter and Energy 94

Part III   (1940–1945)

5   Wiener and the Computer. Act One 103
  5.1   Wiener's Principles for a High Speed Computing Machine ... 103
  5.2   Wiener's Numerical Method 104
  5.3   The Choice of a Digital and High Speed Architecture 105
  5.4   The PDE Project Packed Away in Camphor Balls.......... 108

6   The Research Project on "Anti-aircraft Directors" 111
  6.1   Anti-aircraft Directors: A Strategic Project............... 111
  6.2   An Eccentric Idea.................................. 112
  6.3   Carrying out Research 114
  6.4   The First Synthesis: Inclusion of Control in Communication
          Engineering 117
  6.5   The Second Synthesis: Inclusion of Communication in
          Statistics... 120
  6.6   The Role of Living Organisms in this Research............ 123
  6.7   "Behavior, Purpose, and Teleology"................... 125
  6.8   The Epilogue of the Research on A. A. Directors 128
  6.9   The Paternity of Discoveries and Inventions 131

7   Wiener and Computers. Act 2 133
    7.1   From the NDRC to the OSRD: The Time of Designers
                and Developers. 133
    7.2   McCulloch and Pitts. 136
    7.3   "A Logical Calculus" and Turing Machines. 138
    7.4   The Collaboration Between Pitts and Wiener. 139
    7.5   The Manhattan Project and the Los Alamos Laboratory. 141
    7.6   Von Neumann's Search for Computing Resources. 144
    7.7   Was Wiener Involved in Work for the Manhattan Project? ... 146
    7.8   A First Phase of Cooperation: January–July 1944 148
    7.9   The ENIAC Before von Neumann 151
    7.10   The ENIAC from von Neumann's Arrival 155
    7.11   Meanwhile, Wiener with Pitts, Bigelow,
                and von Neumann 158

8   From the Princeton Meeting to Hiroshima 161
    8.1   The Preliminary Phase of the Princeton Meeting 161
    8.2   The Princeton Meeting. 163
    8.3   The Work of the Four Groups 166
    8.4   The "Dark Side" of Cybernetics. 169
    8.5   Wiener and Rosenblueth Study Cardiac Conduction 171
    8.6   At the Height of Enthusiasm 172
    8.7   The Nuclear Bombing of Japan and Wiener's Crisis 174

Part IV   (1946–1964)

9   "1946" 179
    9.1   The MIT Rockefeller Electronic Digital Computer. 182
    9.2   The Preparatory Phase of the Macy Conferences on
                Cybernetics 184
    9.3   A "Duet" Between Wiener and von Neumann 189
    9.4   The Prevalence of Physiology 192
    9.5   Social Sciences. 193
    9.6   The Discussions on Circular Causal Mechanisms. 196
    9.7   Von Neumann's Doubts. 198
    9.8   A Scientist Rebels 202

10   Cybernetics 205
    10.1   The genesis of the book Cybernetics 205
    10.2   Clocks and Clouds. 208
    10.3   Clouds and Communication. 210
    10.4   The Evolution of Natural and Artificial Automata 212
    10.5   The Philosophy of *Cybernetics*. 216

10.6   The New Industrial Revolution 219
10.7   The Information Society............................. 222
10.8   Criticism of Social Engineering 224
10.9   Wiener and the USSR 227

11   **After *Cybernetics***.. 231
11.1   Wiener and von Neumann in the Cold War.............. 231
11.2   Cybernetics as a "Grammar of the Cold War"............. 234
11.3   An Anthropology for the Information Society 239
11.4   The Golem and Its Creator........................... 242
11.5   The Butterfly Effect and Limits to Growth............... 244
11.6   Fertility of Science................................ 245
11.7   The Tempter..................................... 249

Bibliography.. 253

Index.. 301

Abbreviations for Archives

DTIC   US Defense Technical Information Center
            http://www.dtic.mil/dtic
FDRPL   Franklin Delano Roosevelt Presidential Library and Museum di
            Hyde Park, New York. http://www.fdrlibrary.marist.edu Accessed
            10 January 2002
HGAP   Herman Goldstine Archive, Library of the American Philosophical
            Society of Philadelphia
HSRC   Honeywell-Sperry Rand Litigation Records, Charles Babbage
            Institute Archives, Minneapolis
LANLA   Los Alamos National Laboratory Archives. www.lanl.gov/history/
            atomicbomb/computers.shtml. Accessed 21 July 2010
LATFAS   Los Alamos Technical Reports. On line Archive of Federation of
            American Scientists [FAS], www.fas.org/sgp/othergov/doe/lanl.
            Accessed 21–27 March 2017
MCAPS   McCulloch Archive, Library of the American Philosophical
            Society of Philadelphia
MMLC   Margaret Mead Archivio, Library of Congress
NARA   US National Archives and Records Administration
NARS   US National Archives Record Records Services, Washington,
            D.C. Now NARA
SUAP   Stanislaw Ulam Papers, American Philosophical Society Library,
            Philadelphia
UAWL   University Archives of the Widener Library, Harvard University
UPA   University of Pennsylvania Archives
USNA-AMP   US National Archives. Boxes AMP (Applied Mathematics Panel),
            session OSRD (Office of Scientific Research and Development)
VNLC   John Von Neumann Archive, Library of Congress
WAMIT   Wiener Archive, MIT
WAUM   Wiener Archive, University of Maine

Introduction

The best philosopher today is perhaps a specialized scholar who, so to speak, stands with his feet in one discipline but actually seeks comprehensive connections of knowledge - always concretely - and stands in interaction with reality as it is actually present. It could be that [...] a political economist, classical philologist, historian, mathematician more than any other deserves to be called a philosopher.
Karl Jaspers, Psychology of the worldviews (1919, 2).

According to Norbert Wiener his identity as a mathematician had such an importance that he gave the second volume of his autobiography the title *I am a Mathematician*. And that he was in the most exquisite manner. He held a professorship in the Department of Mathematics at the Massachusetts Institute of Technology (MIT) for over forty years, from 1919 to 1960, and his death was commemorated by the journal of the American Mathematical Society (AMS) with a voluminous special edition, "in recognition of his towering stature in American and world mathematics, his remarkably manysided genius, and the originality and depth of his pioneering contributions to science" (BAMS 1966).

The importance today of his legacy as a mathematician was witnessed in 1994, on the thirtieth anniversary of his death and the centenary of his birth (1894–1964), by two important congresses, lasting one week each, one organized by the MIT and the other by the AMS, together with some cybernetics organizations (see Jerison, Singer, and Strook 1997; Mandrekar and Masani 1997). More recently, the IEEE Conference on Norbert Wiener in the twenty-first Century took place in Boston from 24 until 26 June 2014 (Cf. 21CW).

But Wiener was actually a mathematician-philosopher. A sort of Leibniz fallen from the sky in our time, worthy of being studied and debated by philosophers and sociologists, to understand the entangled and confused times in which we live today. And Wiener had a very precocious and lucid awareness of the "dangerous problems of the new machine age", which began at the end of the Second World War. An epoch in which, as he wrote in 1950, "we are proceeding on our course on the basis of charts of the idea of progress which do not mark the threatening shoals" (50j, 214).

In reconstructing his intellectual itinerary I paid attention to his philosophical background, without limiting myself to the logic and the philosophy of mathematics: I am convinced that we cannot understand Wiener by looking only at his relation with Russell; we must also take into account William James, Royce, and Santayana. In Wiener's scientific style there are clear footprints inherited by the probabilism of Peirce, which influenced him mainly through the Roycean mediation.

Being a philosopher was one of the "dark sides" of Wiener, saddened by the memory of his youthful anguish as a child prodigy, deprived of his childhood, who received his Bachelor of Science, majoring in mathematics, at the age of fourteen, followed by his Ph.D.—in philosophy—at eighteen, and a postdoctoral period of two years under Russell, Husserl, and Dewey.

On Christmas Eve of 1914, this finely educated American youth had a sad conversation in London with his friend Thomas S. Eliot—the author of *The Waste Land*, a masterpiece of American poetry. They discussed their personal destinies, and the doom threatening the world, while just across the English Channel, hundreds of thousands of other young men, Europeans, were slaughtering each other in the trenches. Soon these two men would both have left philosophy, one of them opting for math, the other for literature.

He returned to Boston for a year as assistant and lecturer in the philosophy department of Harvard University. But then he was met with a categorical refusal to go on. The fact that the door was slammed in his face was for him very painful, but it actually gave a very special gift to humankind. In fact, in 1919, the man who had been trained at the school of most of the greatest philosophers of the early twentieth century, Anglo-American but not only, was hired by the Massachusetts Institute of Technology, working up to 1960 in one of the major forges of technology in the world, and not even in a department of human sciences, but teaching mathematics alongside engineers and physicists.

So this is how the world gained this "towering" mathematician. Because the entire work that Wiener carried on from 1919 to his death appears as a sort of scientific and mathematical embodiment of the philosophical dreams of the earlier years of the twentieth century, and a prolonged effort to answer, with the fallible tools afforded by science, dualisms and questions that had been raised in that period. So, through Wiener, the early American Pragmatism became science, and in particular mathematics.

But that door slamming in Wiener's face also brought us a very sophisticated intellectual who was able to express, throughout his career, a philosophy of science and technology written by an insider, a vision of technology which, even when critical, is never technophobe. As he explained in an address to American intellectuals and beyond:

> I do not blame the American intellectual for a hostile attitude to science and the machine age. [...] I do blame him for a lack of interest in the machine age. [...] He shows a willingness to accept the trends of the day as disagreeable but inevitable. In fact, he reminds one of the refined creatures in a fable of Lord Dunsany. These delicate and refined beings have become so used to being consumed by a grosser and more brutal race that they accept their fate as natural and proper, and welcome the axe which takes their heads off. (50j, 163)

There are still many little-known or poorly investigated aspects of this intellectual figure, which are indeed worth a historiographical investigation. Regarding the period between the wars, I would emphasize the central role played on his thought by the experience of the Great Depression and the relationship with the current of left-wing British scientists, Haldane primarily, with whom Wiener could not fully identify, first of all because he always maintained a certain distance from Marxism. With his new appointment, Wiener gave up every commitment to philosophy as a profession. And yet philosophy nevertheless went on working within him, as a deep background of beliefs which guided his scientific research.

We cannot fully understand the nature and tone of the works written by Wiener after World War II without taking into account a second transition, which marked his life as much as the transition from the world of philosophers to that of MIT. That is the 'narrow' path of the war years, when Wiener devoted himself to intensive research to defeat the Nazis, fighting from the trench provided by his own laboratory. I have devoted over twenty years of archaeological work to this theme, reflecting and constantly seeking out new records. The results are included mainly in the third part of this book, now completely rewritten, where I try to shed light on two features: on the one hand, the prominent role played by Wiener in the development of the digital computer (and, therefore, in von Neumann's research on the plutonium bomb, for which digital computers were of crucial importance); and on the other, a reconstruction in phases of the genesis of Wiener's idea of cybernetics, clarifying, among other things—and I hope once and for all—the true role played in his thought by controversial aspects like feedback.

The way in which the war ended—the atomic bombing of Japan—produced in Wiener an acute crisis of conscience, which of course led him to a drastic change, raising at the heart of his research the need for a more ethical and responsible science.

The following chapters are mainly devoted to the various philosophical and sociological themes of his postwar books, from *Cybernetics* (48f1) to *God & Golem, inc.* (64e), the posthumous *Invention* (93), and the novel *The Tempter* (59e). Their interpretation becomes clearer when we are aware of Wiener's entire intellectual and human itinerary.

I wish to make it clear that this book has not been written as yet another Wiener biography. This is an essay about his thought. But to do this it was really impossible to separate Wiener the scientist from Wiener the philosopher, or to neglect the events of his life which are crucial to understand his intellectual itinerary. Events that are often an essential part of a hard historical study, but in many cases never discussed. And this particular historical study was often made difficult by the atmosphere of secrecy in which Wiener had to operate, or by his deliberate or unconscious self-censorship. This is why a chronological order seemed the most appropriate.

To begin with, if one wants to know about Wiener's life, there is his autobiography published in two volumes, which has also been a very important source for me here: *Ex-Prodigy* (53h) and *I am a Mathematician* (56g); and then there are other autobiographical accounts (especially 58f). After the Second World War, Wiener became a popular figure, wrapped up in a halo of brilliant but unreliable genius for practical and political affairs. After his death, oblivion came quickly.

We had to wait until 1980 to witness the beginnings of a historiographically rigorous recovery, accompanied by the events of early cybernetics. This historical and conceptual work was inaugurated in the book *John von Neumann and Norbert*

Wiener by the physicist Steve J. Heims (1980), followed ten years later by *Norbert Wiener 1894–1964,* written by Pesi Rustom Masani (1990), a very well documented and rigorous study by a mathematician who was one of Wiener's collaborators, and which is also a good reference for Wiener's mathematical work. In 1991 Heims published *The Cybernetics Group,* a fundamental book about the early development of cybernetics, and in particular about the role of socio-human scientists during the *Macy Conferences on Cybernetics.*

In addition, one cannot help but mention here the four volumes of *Wiener's Collected Works,* the last of which appeared in 1985, focusing not only on the mathematician. This volume collects his philosophical articles published in now difficult to access locations, as well as some unpublished documents such as the memorandum on digital computers of 1940. Since then a number of studies, often very well documented, have appeared, throwing light on yet more aspects of Wiener's life and thought, and the environment surrounding his intellectual itinerary.

The first edition of the present book originally appeared in Italian as *Le Armonie del Disordine* in September 2005.[1] I later published several other essays on the theme[2], whose results have been merged here. *Harmonies of Disorder,* although it partially preserves the content of *Le Armonie del Disordine,* can be considered in many respects a new book.

---

[1] However, the full Italian manuscript of Montagnini 2005 was presented almost two years earlier to Istituto Veneto di Scienze, Lettere ed Arti for acceptance (given during the academic session of 24 April 2004), according to the custom of this distinguished academy, which subsequently published it (cf. Ibid. II). That book contains a series of my essays on various aspects of Wiener, the man, and early cybernetics, written before 2003 (see Montagnini, Leone 1996, 1999–2000, 2000, 2000–2001, 2001–2002a, 2001–2002b), more or less partially merged, integrated, and expanded organically.

[2] See in particular, Montagnini, Leone 2007, 2008, 2010a, 2010b, 2012, 2013, 2015a, 2015b, 2016, 2017a, b, and c, and Montagnini, Tabacchi, and Termini 2015.

Part I
(1894–1918)

The moods of the waters of the river were always
delightful to watch. To me, as a mathematician and a
physicist, they had another meaning as well. How
could one bring to a mathematical regularity the study
of the mass of ever shifting ripples and waves, for was
not the highest of mathematics the discovery of order
among disorder?
Norbert Wiener, *I am a Mathematician* (64g [56g], 33)

I sat upon the shore
Fishing, with the arid plain behind me
Shall I at least set my lands in order?
T. S. Eliot, *The Waste Land* (1922, 48)

Chapter 1
Formative Years

1.1 Family Roots

Norbert Wiener was born on 26 November 1894 in the US, in the town of Columbia, Missouri, where his father, Leo, a Russian Jew, had arrived 13 years earlier. His father taught modern languages at the local university, while his mother, Bertha Kahn, was a wealthy Missourian, also of Jewish descent, but with a bit of Lutheran blood in her veins. She came from a family which had been settled in the US for three generations. She always took care of the house and their four children, of which Norbert was the eldest. In 1896 his father obtained a chair of Slavic Languages and Literature at Harvard University, which he held until his retirement in 1930. Hence, the whole family moved close to Boston, which became Norbert's homeland.

Norbert Wiener's education took place under the constant supervision of his father. He would confess: "among all the influences that operated on me, during my childhood and adolescence, the most important was that of my father" (64g [56g], 33). This prompted critics to dig into the history and culture of Leo Wiener. So we shall do the same, without forgetting that the young man soon began to express independent views, knowingly encouraged by his father, who became equally early —according to another evocative expression of the son—his "closest mentor" indeed, but also his "dearest antagonist" (50j).[1]

Leo Wiener was born in 1862 to a Jewish family in Bialystok, a city in a territory where the political boundaries of Protestant Prussia, Orthodox Russia, and the very Catholic Poland had moved up and down for centuries like waves on the sea shore (Cf. Anon 1949). This produced the multicultural and multilingual land par

---

[1]In the dedication (50j): "To the memory of my father LEO WIENER, formerly Professor of Slavic languages at Harvard University. My closest mentor and dearest antagonist". The other dedications in Wiener's books are: "To Arturo ROSENBLUETH, for many years my companion in science", (48f1) and "TO MY WIFE. Under whose gentle tutelage I first knew freedom", (53h); "To those inventors who have preferred the claims of truth to the gifts of fortune" (59e).

© Springer International Publishing AG 2017
L. Montagnini, *Harmonies of Disorder*, Springer Biographies,
DOI 10.1007/978-3-319-50657-9_1

excellence, from which it was no accident that a personality like that of Ludwik Lejzer Zamenhof might spring; the father of Esperanto, created to be a universal language (Boulton 1960). On today's maps, the town must be sought within the borders of Poland, with the name Białystok, situated halfway between the nearby Warsaw and Kaliningrad-Königsberg, Kant's homeland. But when Leo Wiener was born it was located within the borders of White Russia under the Tsarist court. About the middle of the nineteenth century, the city had undergone rapid industrialization by welcoming, for customs reasons, numerous textile factories from western Poland, and it was also an important railway junction: from west to east, it joined Paris to Saint Petersburg via Berlin, and from north to south, Königsberg to Ukraine. There were significant Russian, Bulgarian, Hungarian, Baltic, and Romanian minorities, while the two main ethnic groups were made up of Poles and Ashkenazi Jews. The latter gave to the variegated universe of Bialystok a color that can now only be witnessed through the black and white pictures collected in *A Vanished World* by Roman Vishniac (1983), which luckily reached us in spite of the systematic attempts by the Nazis to exterminate men and memories.

Norbert Wiener did not mind the legendary family tradition of which his father was so proud, with its lineage from Maimonides, the Jewish physician and philosopher of Arab Spain. But his judgment of Akiba Eger, another illustrious ancestor, was somewhat different. He writes in his autobiography:

> The next outstanding figure in our ancestry is a much more certain one, even though I find him far less attractive. He is Aqiba Eger, Gran Rabbi of Posen from 1815 to 1837. Like Maimonides, he was recognized as one of the greatest Talmudic authorities, but unlike Maimonides, he was opposed to secular learning, which was coming into Judaism through such men as Mendelssohn. On the whole, I feel quite content that I did not live in his times and that he does not live in mine (64h [53h], 10).

The end of the passage is a measure of the critical relationship that Norbert Wiener had with the Orthodox Judaism that his father had left behind. The Mendelssohn evoked here, Moses Mendelssohn (1729–1786), the composer's grandfather, had become a symbol for the Jews who aspired to emancipation. The son of a humble Jewish scribe of the Duchy of Anhalt, while remaining deep in his Jewish faith, had studied Latin and modern philosophy, especially Locke, Leibniz, and Christian Wolff. After stepping around a series of anti-Jewish restrictions and mastering the German language, he managed to get permission to stay in Berlin, where he was admitted into the philosophical entourage of the German Enlightenment, as a philosopher and man of culture. Finally, the enlightened ruler Frederick the Great granted him legal residence in Berlin. Moses Mendelssohn promoted the replacement of German with Yiddish, also translating the Pentateuch into German and sparking the ire of the most hard-line rabbis, who saw in this act a betrayal of the Jewish community, as Yiddish, the language of Ashkenazi Jews, was considered thoroughly Jewish. This polemical situation lasted throughout the next century, even in the case of the ancestor Akiba Eger, Chief Rabbi of the Prussian city of Posen, now the Polish Poznan (Cf. 64h [53h], 10).

The Wieners, at least since Norbert's grandfather Solomon, who had adopted German in the family and had sent his son Leo to a Lutheran school, had set out decisively on the trail of Mendelssohn (Cf. 64h [53h], 12–13). Leo Wiener had a restless youth: from Bialystok he had gone to school in Minsk, then Warsaw. He studied medicine at the University of Warsaw, then engineering at the Berlin Technical University, without completing his studies. Meanwhile, he soaked up socialist utopian ideals, and in 1881 at the age of nineteen, he decided to emigrate to America, chasing the dream of founding a commune (Cf. 64h [53h], 15).

But in the US, Leo's community project soon foundered. He carried out several activities until he got into language teaching: he was a polyglot, knowing at least a dozen modern languages, as well as Greek and Latin. He also dabbled in mathematics. And as you can read on the website of the Department of Slavic Languages and Literatures at Harvard University, their current courses in Russian, Polish, and Slavic liturgy were introduced by Leo Wiener in the academic year 1896–97; a few years later he also started the Czech course. His presence in the university was invaluable, thanks to the linguistic and cultural bridge he represented with the European, especially German and Russian, world. He had long-lasting and intense relationships with many colleagues, including the philosopher William James: Norbert Wiener tells us that he was "one of the heroes of my father" (64h [53h], 109).

In his first few years of research and translation, Leo's efforts were largely directed towards the Yiddish language and literature: he wrote articles on Yiddish language elements that are found scattered in Polish, German, Ukrainian, and Belarusian [*Wiener, Leo* in *EJ*]. In 1898 he went to Europe to collect the material that would be used in the following year for a *History of Yiddish literature in the Nineteenth century* (Wiener, L. 1899) and *The popular poetry of the Russian Jews* (Wiener, L. 1898). During the trip he was encouraged to study the Yiddish literature of Isaac Leib Peretz, novelist, poet, and translator of scientific papers in Yiddish, motivated by the desire to raise the cultural level of his co-religionists. On the same occasion, Leo Wiener met the librarian of the Asian Museum in Saint Petersburg, which provided him with a supply of about a thousand books in Yiddish. These would subsequently form the nucleus of the Yiddish collection in Harvard (Cf. Wiener, Leo, in [EJ], v. 16, col. 499). Leo was also the first to introduce to the United States the poet and songwriter Morris Rosenfeld (1862–1923), whom he translated in 1898, with the title of *Songs from the Ghetto*, the *Lider Bukh* (1897), a collection of Yiddish folk songs, in which Rosenfeld sang the misery of a Jewish proletariat without a country, subjected to harsh industrial exploitation (Cf. Miller 2007; *Rosenfeld, Morris* s.d; 64h [53h], 146).

With the turn of the century these interests quite suddenly made way for Russian literature in general, and especially Tolstoy. Leo edited an *Anthology of Russian literature from the earliest period to the present time* (Wiener, L. 1902–1903), which was followed in the years 1904–5 by an impressive translation of the *Complete works of Count Tolstoy*, in 24 volumes. This translation, made when Tolstoy (1828–1910) was still alive, also includes works expressing the evangelical, universalist, pacifist, and non-violent radicalism which would have so much influence on Mahatma Gandhi, and which Tolstoy poured into novels such as

Resurrection (Tolstoy 1904) and essays such as *The Kingdom of God is Within You* (Tolstoy 1905).

However, Leo Wiener's Tolstoyism seems to be the result of a very personal interpretation. In fact, he had accepted vegetarianism—just as Norbert would do—and gave up alcohol, but was never a pacifist in the absolute sense of the later Tolstoy. The incontrovertible proof of this is the fact that, during the First World War, Leo Wiener did everything he could to support his son's attempts to join the army. Norbert Wiener was not even a pacifist in principle; despite his pronounced myopia, he tried by all the means at his disposal to enroll in the First World War and fought during the Second as an active scientist, working on strategically important scientific projects.

After the translation of Tolstoy, Leo Wiener applied himself to philological studies characterized by a marked anti-conformism, poorly welcomed by his colleagues, and especially the Germans, although in recent years these studies have received unexpected appreciation (Cf. Wiener, L. 1915, 1917, 1922).

The switch from Yiddish to Tolstoy took place when Norbert Wiener was only 6 years old and the reasons behind it were never entirely clear to his son. In the autobiography, Norbert hints at disagreements with some US Jewish organizations on whether to give priority to the universal interests of humanity or to the national interests of the Jewish people (Cf. 64h [53h], 146).[2]

As a matter of fact, Norbert Wiener was never educated into religious Judaism. On the contrary, the family adopted a life style guided by a complete assimilation into the American melting pot, to the point that they even decided to hide the Jewish origins of their children, and the young man only found out in a roundabout way at the age of fifteen.[3] As time went by, Norbert Wiener developed a keen and original sense of his own Jewishness, as he confessed in his memoirs by writing about the poet Heine, another example of an emancipated intellectual Jew:

> I came to love the heart-rending cries of Heine, in which not one word is missing or in excess to obscure his love and his venom. I know, as did my father, almost every word of his *Hebraische Melodien* [Heine 1851], and there are no poems that can move the Jew in me to greater pride or agony (64h [53h], 169).

André Neher (1987), a contemporary philosopher who has studied the culture of Judaism, especially Ashkenazi Judaism, and its influences on Western thought, has seen in Wiener's book *Cybernetics* (48f1), whether or not the author was "conscious" of it, a key link with the Hebrew mythography of Job and the Golem, which is later made explicit in his last book, *God & Golem, Inc.* (64e).

---

[2]Heims (1984 [1980], note 13, 418) notes that, in an earlier unpublished version of the autobiography, called *A Bent Twig*, "Leo Wiener's rejection of Judaism and of the Jewish community is given greater prominence".

[3]Moreover his mother used to repeat phrases disrespectful to Jews as other minorities, according hackneyed stereotypes: "Scarcely a day went by in which we did not hear some remark about the gluttony of the Jews or the bigotry of the Irish or the laziness of the Negroes" (64h [53h], 146); see also 64h [53h], 20).

Whatever the reason, a strong trait in the culture of Leo Wiener, who undoubtedly passed it on to his son, was the trusting acceptance of the spirit of emancipation, with romantic, emotional colorings. A spirit that was then acclimated in the American democratic environment and which sought, through scrupulous scientific research, both passionate and intellectually honest, even at the cost of great personal sacrifice, the very purpose of life (Cf. 64h [53h], 288–298). The truest worship by both father and son was for modern science, and their beloved "patron saint", so to speak, was Galileo who, locked in prison kept on saying "and yet it does move". On the other hand, this is also a character rooted in a secular way of interpreting Jewishness, as is explained for example by Albert Einstein:

> The second characteristic trait of Jewish tradition is the high regard in which it holds every form of intellectual aspiration and spiritual effort. I am convinced that this great respect for intellectual striving is solely responsible for the contributions that the Jews have made toward the progress of knowledge, in the broadest sense of the term. [...] I am convinced that this is not due to any special wealth of endowment, but to the fact that the esteem in which intellectual accomplishment is held among the Jews, creates an atmosphere particularly favorable to development of any talents. At the same time a strong critical spirit prevents a blind obeisance to any mortal [sic] authority (Einstein 1938, 10 and 38).

1.2   Clay of New England

There is another cultural river from which the young Norbert Wiener drank—perhaps less karst, and which has so far remained a somewhat neglected part of the autobiographical tale: namely, the American environment. This has not escaped the notice of such a careful scholar as Steve J. Heims, himself of Jewish origins, and an immigrant from Europe: the fact that, "however widely he traveled, Wiener was and remained a New Englander" (Heims 1984 [1980], 168). Rather than a "Cold War Americanism", as one Italian philosopher wrote in the late 1970s in a rare book on Wiener from that period, Wiener was simply an American, if anything just specified by *Old world traits transplanted,* to use the title of the sociological study by Park and Miller (1921) on immigration to America from Eastern Europe.

In his work, when Norbert Wiener uses the word "we", it is usually an American who speaks to Americans. In his phraseology, images of American landscapes appear constantly. It could not be otherwise if we remember that his childhood and adolescence were spent in close contact with the famous names of Harvard's academic world. Later he spent more than 40 years at the Massachusetts Institute of Technology. The American land has thus made an essential contribution to the clay by means of which the personality and culture of this man was shaped, a man who would throughout his life be a devoted hiker of the Appalachian Mountain Club, a man who would always live with his wife and children in Boston, spending holidays in their summer residence in Sandwich, New Hampshire. Politically, Norbert Wiener shared "the American civil religion", summarized in the Constitution and in the Jeffersonian preamble, with a predilection for the plight of the disadvantaged classes, which never gets to be confused with Marxism, and which is reminiscent of radical democrats such as the sociologist Charles Wright Mills (Cf. Ferrarotti 1977, XXXI).

Richard J. Bernstein, another American philosopher, has pinpointed the following elements working under the early American Pragmatism:

> Anti-foundationalism, fallibilism, and the nurturing of critical communities, […] the awareness and sensitivity to radical contingency and chance that mark the universe, our inquiries, our lives. […]. For the pragmatists, contingency and chance are not merely signs of human ignorance, they are ineradicable and pervasive features of the universe" (Bernstein 2007 [1991], 328; see also Bernstein 1992).

Anyone who has read the works of Norbert Wiener and knows his scientific style cannot but agree that these words are almost a portrait. This is not so surprising. Fisch lists the protagonists of that philosophical era: Peirce, Royce, W. James, Santayana, Dewey (Cf. Fisch 1996, 1).[4] Well, Norbert Wiener was educated by all these men, and he knew almost all of them very closely: William James was a friend of the family, under Royce he received his Ph.D. in philosophy, and he followed the university courses of Santayana and Dewey. He was not personally acquainted with Peirce, but Peirce led his life in a secluded manner: he influenced the thought of all the others, especially through his writings, and that was probably the way he influenced Norbert Wiener.

The miracle of the American Hellas, which had found its agora in the Harvard University philosophy department, was destined to end abruptly, between 1910 and 1916. Wiener was born in 1894 and the contingency of this particular birth date would not have allowed him to eat all the sweetest fruit that the tree of American philosophy ever produced, had he not had the precocious educational history that he himself told us about in *Ex-Prodigy*, the first volume of his autobiography (Cf. 64h [53h], 146).

1.3 Child Prodigy

Among Leo Wiener's translations there is an early nineteenth century German work (Witte 1914) which tells about the education of a certain Karl Witte, son of a minister of the city of Lochau who, dissatisfied with the methods used in the local school, had decided to take care of his son's elementary education by himself. He

---

[4]Fisch also cites Whitehead. But I prefer to omit this name, even though Wiener had a close relationship with him too, because Whitehead's philosophy of the American period, focusing on the definition of process, belongs to a stage subsequent to the early Pragmatism which Wiener was so deeply influenced by.

thus taught him to read when he was only 3 years old and gave him books without restriction as his cognitive abilities and his eagerness to learn progressed. Nine years later Karl had already learned five languages, and at fourteen he graduated from the University of Leipzig.

Leo Wiener did the same experiment with his firstborn. Norbert tells us:

> I learned my algebra and geometry at so early an age that they have grown into a part of me. My Latin, my Greek, my German, and my English became a library impressed on my memory. Wherever I may be, I can call on them for use. These great benefits I acquired at an age when most boys are learning trivialities. Thus my energies were released for later serious work at a time when others were learning the very grammar of their professions (64h [53h], 290).

As I discussed more widely in *The Mathematical Art of Norbert Wiener* (Montagnini 2015b), the effect of this very precocious education during that period, between the ages of 0 and 6 when a child's mind is so much more absorbent (Montessori 1949), should not be neglected when we try to understand the almost instinctive way he would later do mathematics ("algebra and geometry [...] have grown into a part of me" (64h [53h], 290).

When Wiener was just 3 years old, he could read, and soon, seconded by his father, he became a voracious reader (Cf. 64h [53h], 18 and 34). At six he was enrolled in the third grade of the Peabody Elementary School, passing immediately to the fourth. His father decided, however, that it would be more appropriate to educate him at home, and became his tutor until 1903 when, at the age of nine, Norbert was enrolled in the first year of Ayer High School. In 1906, eleven and a half years old, he was enrolled at Tufts College, where in the summer of 1909 he obtained a Bachelor of Sciences. At Tufts, in particular, Wiener attended courses in biological sciences, mathematics, and philosophy; on his graduation he majored in math (Cf. 64h [53h], 112). In the autumn of 1909 he was enrolled at the Harvard Graduate School: the original wish of the young man was to study medicine, but because of his tender age, his educators preferred him to study zoology. However, at the end of the first semester, even this choice seemed unsuitable, given the disastrous results obtained in the laboratory, due to his still immature manual dexterity. It was decided to advise the young man to take up philosophical studies and so he enrolled at the Sage School of Philosophy at Cornell University. In the autumn of 1911 Wiener went back to the Harvard Graduate School, where he received his Master of Arts degree in philosophy in 1912 and a Ph.D., still in philosophy, in 1913. In this itinerary, made especially tortuous by the fact that the father tended to test and as far as possible maximize the educational outcomes of the child, three fields of study became intertwined—biology, mathematics, and philosophy—all addressed at university level. Among the three, it was philosophy that would prevail.

1.4   From Evolutionary Positivism to Pragmatism

Let's look now at the contents of Norbert Wiener's philosophical studies and the
way he experienced them subjectively. His philosophical beginnings were closely
linked to positivism, under the influence of his father, immersed as he was in the
nineteenth-century admiration for science. Leo Wiener liked to recall that his son

> by six was acquainted with a number of excellent books, including works by Darwin,
> Ribot, and other scientists, which I had put in his hands in order to instill in him something
> of the scientific spirit.[5]

Théodule Armand Ribot was a comparative and experimental French psychologist, author of *The diseases of memory* (Ribot 1882 [1881]), a book considered as
one of the earliest attempts to study memory disorders in an experimental way.
Wiener may also have read *Essay on the creative imagination* (Ribot 1906), published in French in 1900. However, just the mention of Ribot testifies to a really
precocious introduction of the young man to the questions related to mind and
brain, which would become one of the main focuses of cybernetics.

In the letter he sent for his doctoral submission, Norbert explained

> I first became seriously interested in philosophy when as a child of 9 or 10 I read Haeckel's
> 'The riddle of the universe' [Haeckel 1899], although before this I had read some brief
> essays of a scientific character by Spencer, Huxley, etc., and Darwin's 'The Descent of
> Man' and 'The origin of Species'.[6]

Thomas H. Huxley was a British biologist who had been one of Darwin's
leading champions in the contemporary diatribes on evolution, and Herbert Spencer
was the leader of English evolutionary positivism. Norbert Wiener's autobiography
adds: "I have since found Herbert Spencer to be one of the most colossal bores of
the nineteenth century, but at that time I held him in esteem" (64h [53h], 103).

The riddles of the Universe by Ernst Haeckel (1899) also falls into this full
immersion in nineteenth-century positivism which, at the age of 9 or 10, as reported
in the letter for doctoral submission, aroused in Wiener his first serious interest in
philosophy. A few years earlier, the physiologist Emil Du Bois-Reymond had
argued in the speech on *The seven riddles of the universe* (1880) that it was
impossible for science to come to understand the essence of matter and energy, the
origin of movement and life, the proper goal of nature, and the way feelings,
consciousness, thought, and language arise.

Ernst Haeckel, German philosopher and evolutionary zoologist, known for
propounding the law that "ontogeny recapitulates phylogeny", published his *Riddle
of the universe* (1900 [1899]), with the explicit intention of countering the arguments of Du Bois–Reymond, earning the admiration of the public, whence 400,000
copies of the book were sold.

---

[5]Leo Wiener interviewed by Bruce (1911). Cit. by Heims 1984 [1980], 6.
[6]Letter for doctoral submission, cit. by Grattan-Guiness 1975, 106.

Haeckel opposed Du Bois-Reymond's skepticism with a monistic metaphysics of one eternal substance, 'matter/energy', which he claimed could be based on the two experimental conservation principles that emerged in nineteenth century science: the law of conservation of matter in chemistry and the law of conservation of energy in physics. He thought that all phenomena observed in the universe would be merely transformations of this unique substance, produced by blind, deterministic, evolutionary mechanisms, and that this consideration would be sufficient to explain the origin of both life and language. He denied that in these processes any purpose could be traced, also ruling out the idea that the freedom of the will was a mystery to be explained, since in his view it was nothing more than a mere illusion.[7]

The very young Norbert Wiener soon distanced himself from his father's stance. In the letter for doctoral submission, he adds:

> From Haeckel's position I first turned towards scepticism, then towards an agnosticism [sic]. About this time I began to study philosophy at Tufts College under Professor Cushman. Among other books we read Høffding's 'Problems of Philosophy' [Høffding 1903]. At the same time I read James' 'Pragmatism' [James 1907]. The combined influence of these two books made me try to combine in some manner the logical standards of formal consistency and practicality.[8]

This suggests that during high school Wiener started out from the materialistic monism of Haeckel, moved on to a skepticism akin to that of Du Bois-Reymond, and finally achieved a more moderate form of skepticism, namely, the "agnosticism" expressed by Thomas H. Huxley, which dates back to the minting of the same English term "agnosticism", which Huxley used to describe the position of those who, like him, thought that God and any other ultimate reality is unknown and probably unknowable (Cf. Huxley, Thomas Henry in *EB* 1997). It was this very agnosticism which must have found expression in that never published first essay entitled *The theory of ignorance*, "a philosophical demonstration of the incompleteness of all knowledge", (64h [53h], 96) of which Wiener says in his memoirs:

> It is no coincidence that my first childish essay into philosophy, written when I was in high school and not yet 11 years old, was called 'The Theory of Ignorance'. Even at that time I was struck with the impossibility of originating a perfectly tight theory with the aid of so loose a mechanism as the human mind (64g [56g], 324).

This psychological fallibility must have gained strength through his contact with the Neo-Kantianism of Harald Høffding and the Pragmatism of William James (1842–1910), philosophies he became familiar with at Tufts College. There, during his second year, aged thirteen, he attended several courses on philosophy and psychology held by Professor Cushman (Cf. 64h [53h], 109), who had adopted as a manual the English translation of the *The problems of philosophy* by Høffding 1905 [1903], a Danish philosopher with neo-Kantian tendencies whose works on the history of philosophy had a large circulation at that time and in which he stressed

---

[7]Here and elsewhere, for the profiles of the philosophers, when not expressly indicated, I am indebted mainly to Fisch 1996, Abbagnano 1995, Copleston 1996.

[8]Letter for doctoral submission, cit. by Grattan-Guiness 1975, 106.

the importance of the theory of knowledge in the history of philosophy. Høffding's epistemology was centered on the Kantian notion of synthesis, which, however, was no longer read in a Kantian way, as an epistemological condition for the possibility of human knowledge, but as an essentially psychological concept (see Brandt 1967).

At the same time Norbert Wiener discovered the philosophy of James, one of America's leading thinkers, who had an unquestionable influence on him. Wiener was struck by the philosophical difference in depth between James and the courses in college:

> The rather dilute material of my courses in philosophy and psychology showed up poorly in comparison with the outside reading and in particular with the great books of Professor William James, which I devoured almost as much as literary tidbits as for their serious content (64h [53h], 109).

One of these books was *Pragmatism* by James (1907). The young Wiener was invited by William James to visit him at home for a conversation (Cf. 64h [53h], 109). Later James invited him to attend his Lowell Lectures on Pragmatism, held in Boston at the end of 1906. Just these lectures constituted the content of *Pragmatism*, a book given by James as a gift to Wiener, but without letting him know (see 64h [53h], 109–110).

In order to understand the philosophy of Pragmatism, and the atmosphere that Wiener lived and breathed that year, listening to James, then frequenting the philosophers of Harvard University for 2 years, is useful to read a description by Richard Bernstein, who explains that Pragmatists abandoned

> the presupposition of much of modern philosophy that the rationality and legitimacy of knowledge require necessary foundations. Inquiry neither has nor needs any such foundations. The pragmatists did not think that abandoning all foundational claims and metaphors leads to skepticism (or relativism). They stressed the fallibility of all inquiry [...]. The classical pragmatist shares a cosmological vision of an open universe in which there is irreducible novelty, chance, and contingency (Bernstein 1992, 813–4).

James' Pragmatism was part of the rising tide which, in the early years of the twentieth century, was making a clean sweep of the nineteenth century infatuation for science, a revolt which saw Henri Bergson becoming a sort of standard bearer. It is likely that Wiener had been familiar with Bergson's philosophy since the days of his first acquaintance with the thinking of James, the latter having greeted the publication of *Creative evolution* (Bergson 1911 [1907]) as a "divine apparition", with the recommendation:

> But open Bergson, and new horizons loom on every page you read. It is like the breath of the morning and the song of birds. It tells of reality itself, instead of merely reiterating what dusty-minded professors have written about what other previous professors have thought. Nothing in Bergson is shop-worn or at second hand (James 1909, 265).[9]

---

[9]Bergson also maintained a dialogue with James' thinking. See Bergson (1934), and in particular the chapter "Sur le Pragmatisme de William James". An interesting connection is also to be found between Pragmatism and Boutroux' contingentism [see Boutroux 1911].

In *Creative evolution,* Bergson criticized the mechanistic evolutionism of Darwin, Spencer, and Haeckel, and in general all attempts to reduce biological and psychological phenomena to physico–chemical phenomena. His philosophy had started from a critique of the concept of time as used by physical science, a time which, when translated into mathematics, gets "spatialized", becoming an abstract set of instants that do not adequately represent real time, as experienced in real life, grasped by an intuition of the flow of our consciousness.

Bergson believed that the main enemy of real knowledge is the intellect, which he understood as that cognitive activity which expresses science and which constructs abstract patterns by means of analysis, producing an artificial, mechanical, deterministic "geometrical order". Against the intellect Bergson erects "intuition" as a different kind of knowledge, allowing us to grasp the states of consciousness, not as juxtaposed, but as interpenetrating each other; we ourselves are not like the blind gear of a physical–chemical determinism, but free spirits, finally, able to see biological and cosmic evolution as the result of a free spirit, a "vital impetus" or "vital force" [French: "élan vitale"], which operates creatively, producing an unpredictable and constant "stream of novelty".

James had received Bergson's ideas with joy, but without sharing the attack against science, which he reinterpreted in a fallible and non-deterministic key, in line with a typical approach already held by Charles Sanders Peirce. Accepting the ideas of Peirce, first of all that the truth content of a proposition lies in its practical effects (Cf. Peirce 1878), the James' Pragmatism was opposed to the idea of the universe as being made up of perfect, strictly ordered units,—whether it was a materialistic monism like that of Haeckel, or an idealistic monism—that of a "pluralistic" universe consisting of a plurality of things, only relatively independent of each other, and forming an imperfect unit which tends to correct its imperfections. On this view, James also established an ethic of perfectibility, the so-called Meliorism, somewhere in-between pessimism and optimism, according to which individuals, finite and free, have the opportunity to contribute to shaping their own destiny.

Those themes would recur in the Wiener's philosophy when he began to express himself autonomously, and would flow back into his own science. But there was one thing that left the youth unsatisfied, even though he was attracted by the depth of James' analysis, a deficiency that Wiener confesses politely in his memoirs:

> There was more than the style of the novelist in William James, and perhaps less of the philosopher than one might have thought, for his ability to evoke the concrete was to my mind many times greater than his ability to organize it in a cogent logical form (64h [53h], 110).

Wiener's staunch Pragmatism had to deal constantly with a requirement of almost instinctive logical rigor, which led him in his years at Tufts to admire Spinoza's *Ethica, more geometrico demonstrata* (Ethic demonstrated in geometrical order, 1883 [1677]), as well as the logical ideas of Leibniz (Cf. 64h [53h], 109).

We will find the same attitude toward Dewey, whose lectures at Columbia University he would attend during his postdoctoral studies, in 1915. About Dewey, he writes: "As a very young man I appreciated the help and discipline of a rigid logic and a mathematical symbolism" (64h [53h], 222–3). His need to reconcile, almost without compromise, an unbending logic with the complexity of practice is well expressed in the aforementioned letter for doctoral submission, where he says: "The combined influence of these two books [Høffding 1903 and James 1907] made me try to combine in some manner the logical standards of formal consistency and practicality".[10]

One can imagine that he hoped to create a sort of "*Pragmatismus Ordine Geometrico Demonstratum*", to paraphrase Spinoza's famous book on *Ethica*. A goal that appears intrinsically paradoxical: is it possible to put in a logical form a philosophy denying the need for, or better, the very possibility of a logical foundation of everything?

During the years at Tufts, the results for the young Wiener were disappointing: "finding little success in my undertaking", he states, "my view again shifted towards agnosticism, and my main interests turned towards the natural sciences".[11]

Therefore from failure arose a rejection of philosophy and the decision to seek refuge in science: a dialectic within which he would struggle for a long time. At Tufts he preferred to abandon abstract speculation and throw himself into the field of the biological sciences. For his graduation, as we mentioned before, he preferred to major in mathematics (Cf. 64h [53h], 112).

1.5   In the Temple of American Philosophy

In the fall of 1909, Wiener was enrolled in the zoology courses at Harvard Graduate School and at the end of the first semester, mainly due to his clumsiness in laboratory activities, his advisors decided he had better quit those courses and enroll in the Sage School of Philosophy at Cornell University, to study philosophy. There he followed mainly historical–philosophical courses on Plato's *Republic* and the English philosophers of the seventeenth and eighteenth centuries (Cf. 64h [53h], 149–150). Two years later he would comment negatively on the year at the Sage School, mainly because he was not allowed to do any creative work.[12] Such a judgment, seen from a different perspective, reveals the change that had taken place in the following 2 years at the Harvard philosophy department, which by contrast he considered extremely valuable and creative.

Although Wiener gave the title "Philosopher in spite of myself" to the chapter of his autobiography dedicated to his philosophical studies at Harvard, the two academic years (1911/12 and 1912/13) in which he studied there were certainly "two positive and successful years," as Norman Levinson (1966, 5), his former student, colleague and friend, would observe.

---

[10]Letter for doctoral submission, cit. by Grattan-Guiness 1975, 106.
[11]Ibid.
[12]Ibid.

When in 1911 the young Wiener arrived at the Department of Philosophy in Harvard, he found the American philosophical Eden on the wane, and he was just in time to savor the sweetest fruit that remained before everything changed. William James had died the year before and the most significant names remained the old and sick Josiah Royce and George Santayana. Let us read again the letter for doctoral submission:

> I came last year [1911] to Harvard again, where my work was turned towards epistemology by Prof. Perry, and my views were profoundly modified by Prof. Santayana. Among the particular problems which interested me, that of the place of terms and relations stands out most prominently. I took Prof. Royce's course in symbolic logic last year, and this line of work, owing to my early mathematical training, fascinated me greatly. My subsequent contact with Prof. Schmidt has greatly fostered my interest in this subject, in which I am now writing my thesis.[13]

In those years the overall atmosphere was marked by the controversy that surrounded the so-called neo-realists, a group of six professors (Ralph Barton Perry, Edwin B. Holt, Walter T. Marwin, William Pepperell Montague, Walter B. Pitkin, and Edmond Gleason Spaulding), who in 1910 had signed the manifesto *The Program and First Platform of Six Realists* (Holt 1910) and in 1912 gave to the press the collective book *The New Realism. Cooperative Studies in Philosophy* (Holt 1912). Arguing against idealism, they gave birth to a movement that would prove to be short-lived, with the explicit aim of rejuvenating and letting fresh air into the faculty, opening up to the English realism of G.E. Moore and Bertrand Russell (Cf. 64h [53h], 165).

Neo-realists agreed on the fundamental realistic assumption, meaning, as Pitkin put it, that, regarding their existence or their behavior, known things cannot be considered as the product of the cognitive relation, and nor can we think that they depend in an essential way on this relationship: they exist as they are independently of the knower (Cf. Copleston 1996, 380–401). They disagreed with each other when it came to describing the cognitive process itself in more detail, and the relationship between mental states and things. The two neo-realists whose courses Wiener followed, Edwin B. Holt (1873–1946) and Ralph Barton Perry (1876–1957), responded to this problem by proposing the "neutral monism" theory introduced by William James, and shared by Holt and Perry, which tended, in Wiener's words

> to regard the elements entering into the combinations of the material world and the components of mental states as not necessarily different. Mind and matter, in a philosophy of this type, are complexes of the same subject matter, but are viewed from different aspects or are ordered in a different manner (18h, 367).

The neo-realists represented the new wave of the department and Wiener would have been attracted into their orbits. In fact, Perry become his main point of reference, remaining his main contact in Harvard during his postdoctoral studies in Cambridge.

---

[13]Letter for doctoral submission, cit. by Grattan-Guiness 1975, 106.

However, Wiener was never a person to be easily captured by the song of the sirens of the moment. He appreciated the liberal political ideas of the neo-realists, but they gave him the impression "of an intolerable shoddiness and brashness" (64h [53h], 165). Instead he was attracted by the outstanding personalities of two men, but at that point upstream, at least inside the small circles of a university department: Royce and Santayana. And note that Wiener got to logic through Royce, the main target of the *Manifesto on New realism*.

In this preference for Royce, the patriarch of American idealism, as well as for that curious mix of positivism and idealism incarnated by Santayana, it is possible to grasp the true originality of Wiener's personality.

1.6   The Positivistic Idealism of Santayana

"Those who cannot remember the past are condemned to repeat it". This well known aphorism is indeed due to George Santayana (Santayana 1906, 284), the Harvard philosopher who had "profoundly modified" Wiener's views.[14] In his autobiography Wiener stated:

> In my first year I took a course with Santayana. I remember very little of its content but a considerable amount of its atmosphere. The feeling of a continuity with an old culture and the feeling that philosophy was an intrinsic part of life, or art, and of the spirit, gave me a great deal of satisfaction; and yet, after the passage of all these years, I cannot put my finger on any definite idea which that course has given me (64h [53h], 164).

And Santayana's footprint appears clearly in the following passage written in 1951, during Cold War, commenting on the poor sensitivity to future catastrophe shown by people in those days. Wiener explained:

> The future is terribly unreal to the majority of us. And how shall it not be, when the past is equally unreal to us?

> The continuity with the past which belongs to the older settled regions of Europe and to a lesser but a very real extent to the older settled regions of the US, is dependent not only on an acquaintance with written history, but with the continual presence of the houses, the roads, the farms and the cities established by past generations, and with a familiarity with the mode of life which developed them. [...]

> However, in the great cities, or in the esoteric hot house civilization of Southern California, where a man's parents lie in the soil of Iowa or Nebraska, and a man's neighbors are pursuing their purposes in life without any reference to his own, it is useless to ask that he should regard his own grandchildren as anything but slightly modified sorts of strangers. The span of social memory which is needed for the homeostatic action of an historical sense is too great for transients and squatters. Yet without this span, the regions which have been built out of the desert and the burnt hillsides in a magnificent defiance of nature, represent far too much of a defiance of nature to possess the elements necessary for their continued existence. *To respect the future, we must be aware of the past; and if the regions where that*

---

[14]Ibid.

awareness of the past is real have shrunken down to a pin point on our vast map, then so much the worse for us and our children, and our children's children. (51b, p. 68. Italics added).

The notion of homeostasis that is used here would be introduced many years later by the physiologist Cannon (1932), yet it is hard to deny that here Wiener's reasoning has also many similarities with Royce's "teleological community", which we will discuss in the next section.

Not to be confused with Giorgio De Santillana (1902–1974), a friend of Wiener's, Italian professor of history of science at MIT, George Santayana (1863–1952) was born in Madrid and settled in Boston at the age of nine. He taught at Harvard from 1889 right up to 1912, when he left teaching and returned to Europe, first to England, then Italy, and Rome in particular, where he lived for the second half of his life.

Santayana established the first chair of aesthetic philosophy of Harvard University. In his first work, *The Sense of Beauty* (Santayana 1896), one can discover the thread of his thought. Here he set the foundation of the sentiment of beauty on psychophysical mechanisms, and also on organic and environmental aspects (Cf. Bosco 1987, 31–7). In general, he sought to found the world of idealism on the basis of evolutionary positivism. This is a trend deepened in *Interpretations of poetry and religion* (1900), dedicated to the affinity between beauty and moral qualities (Cf. Bosco 1987, 38), and which sees its triumphs in *The life of reason* (1905–1922), comprising five volumes, all of them aimed "to correct Plato by means of Spencer", and describing the "phases of human progress" in their various aspects: common sense, politics, religion, fine art, and science (Cf. Bosco 1987, 41–77). During the period in which Wiener attended his lectures, Santayana wrote *The Intellectual temper of the age*, where we read:

The present age is a critical one and interesting to live in. The civilisation characteristic of Christendom has not disappeared, yet another civilisation has begun to take its place. [...] We still love monarchy and aristocracy, together with that picturesque and dutiful order which rested on local institutions, class privileges, and the authority of the family. [...] On the other hand the shell of Christendom is broken. The unconquerable mind of the East, the pagan past, the industrial socialistic future confront it with their equal authority. Our whole life and mind is saturated with the slow upward filtration of a new spirit that of an emancipated, atheistic, international democracy (Santayana 1913, 1).

It is interesting to note that in all Wiener's thought, even when it was focused on mathematics or technology, the concept of "spirit of the time" would be central. In particular, during the years 1911 and 1913, Santayana saw the emergence of a new spirit in the "life of the mind", the spirit of an industrial and socialist future, together with the return of Hume's skepticism, especially through Bergson and William James. It is a spirit that penetrates deep into science. Santayana writes:

For science, too, which had promised to supply a new and solid foundation for philosophy, has allowed philosophy rather to undermine its foundation, and is seen eating its own words, through the mouths of some of its accredited spokesmen, and reducing itself to something utterly conventional and insecure (Santayana 1913, 4).

It is the observation of the crisis of nineteenth-century certainties that shook science at the beginning of the twentieth century, a crisis that hopelessly calls into question the philosophy of Santayana itself, anchored as it was in an evolutionary materialism. This provoked disappointment in him, and he sought salvation in the attempts of philosopher–mathematicians like Bertrand Russell. In *The problems of philosophy* (Russell 1912), Russell had written: "The true spirit of delight, the exaltation, the sense of being more than man, which is the touchstone of the highest excellence, is to be found in mathematics as surely as in poetry." (Russell 1912 cit. by Santayana 1913c, 118)

Santayana argued: "Mathematics seems to have a value for Mr. Russell akin to that of religion. It affords a sanctuary in which to flee from the world, a heaven suffused with a serene radiance and full of a peculiar sweetness and consolation" (Santayana 1913c, 117). On the other hand Santayana deplores the attitude of Bergson. In *The philosophy of Henri Bergson* he asks himself: "Why, for instance, has M. Bergson such a horror of mechanical physics?", and considers it nothing more than "a black art, dealing in unholy abstractions, and rather dangerous to salvation" (Santayana 1913c, 64). For Bergson mathematics has only the function "to keep our accounts straight in this business world". On the contrary, for Santayana, mathematics' "inherent use is emancipating and Platonic, in that it shows us the possibility of other worlds, less contingent and perturbed than this one" (Santayana 1913c, 66).

In the article "Aesthetics" in *Enyclopedia Americana*, Wiener presents a brief discussion of Santayana's aesthetics in contrast to that of Bergson. According to Bergson, Wiener explains,

> art exhibits the intention of life, as opposed to science, which presents its analysis. Art is the sphere of intuition, science that of reason. As Bergson considers the knowledge gained by intuition as deeper and truer than that gained by reason, he assigns a higher place to art in his philosophy than to science. He names sympathy as one of the cardinal marks of art. (18b, 201)

Unlike Bergson—Wiener argues—Santayana does not oppose art and science:

> Bergson's antithesis between art and science is foreign to the philosophy of Santayana, who regards art as plastic instinct which is conscious of its aims. The instinct which constitutes art in Bergson's philosophy need not possess this consciousness. For Santayana art is reason propagating itself (18b, 201).

For his part, Wiener, would never share the idea of a Platonic foundation underlying the interpretation of mathematics and science, of the kind emerging from these words by Santayana, and which served as a compass for Russell's logicist project to build an absolute foundation for mathematics. However, Wiener's mathematics would never even be a way "to keep our accounts straight in this business world." He would remain for life convinced, like Santayana, that there are no walls between science and art. Indeed, he thought of mathematics as a fine art in all respects. And he would never stop questioning the spirit of his time. In the way of a true, but nowadays rare, philosopher, he would do science by looking for the *Grund*, the essence, of scientific and technological problems.

1.7 The Idealism of Royce

Another Harvard philosopher who was very influential for Wiener's thought was Josiah Royce, maybe the Harvard personality that Wiener spent most time with. Wiener says:

> Royce was a many-sided man, coming at a critical period in the intellectual world when the old religious springs of philosophical thought were drying up and new scientific impulses were bursting into life (64h [53h], 165).

Wiener considered Royce almost as an icon of that "critical period" designed by Santayana, and this idea was also shared by another of Royce's—and Santayana's —pupils, namely the future poet Thomas S. Eliot, according to whom the idealism defended by Royce represented the now moldy legacy of a time when in the American universities philosophy was taught by maverick ex-ministers "who had studied in Germany" (Skaff 1986, 17). Yet, for Royce as a person, Eliot kept the utmost respect, as an extraordinary patriarch of American philosophy, and this must have been what Wiener felt, too.

Royce's philosophy can be seen as an American variant of English contemporary idealism, which in those years was called "absolutism". England, the home of empiricism and evolutionism, has continued through the centuries to welcome a variously Platonizing line of thought, which aims to counter those trends and to give speculative support to traditional religious thought, as in the emblematic case of bishop Berkeley, who at the beginning of the eighteenth century denied the existence of things, reducing them to mere appearance ("esse est percipi") given by God to men for their usefulness.[15] In the last quarter of the nineteenth century, coinciding with Spencer's overflowing positivism, there arose a group of neo-idealists in England—including Thomas Green (1836–1882), John McTaggart (1866–1925), and especially Francis Herbert Bradley (1846–1924)—who reproduced with neo-platonic accents concepts such as the Kantian transcendental apperception and the Hegelian absolute, making them coincide with God.

In *Appearance and reality* (1893) Bradley, starting from the Hegelian affirmation that "the true is the whole", came to a radical skepticism regarding the experience of the world as it is investigated by science, through an analysis of the concept of relationship, both in terms of the relationships between things, and as regards the relations between things and our conscious experience of them. According to Bradley, each finite object caught in experience is never something that is in itself self-sufficient, because its properties depend on the relationships it has with other things. It is thus different if it is present in our consciousness or if it is not. The result is a world made up of a network of relationships among things that do or do not exist depending on how we look at them, and this for Bradley was inconceivable and contradictory; in the end these circumstances reveal the mere

---

[15]For this rather convincing judgment, it seems to me, I am indebted to the Italian neo-idealist Guido De Ruggiero (1921 [1912], 261 ff.).

appearance of the world, while only the absolute consciousness is real, i.e., God, conceived of as a whole, devoid of relationships and therefore not contradictory.

Josiah Royce (1855–1916)—a professor at Harvard from 1882, who had begun a favorable dialectic relationship with James which lasted throughout his life—was less radical than Bradley in demolishing individual experience. It is also true that, in Royce's philosophy, reality coincides with the absolute, while at the same time it is constituted by a multiplicity of individuals. This idea had already appeared in his first major work, *The religious aspect of philosophy* (1885), where, with reference to the ethics of Kant and the Kantian kingdom of ends, it is claimed that the life of a man in solitude is an insignificant fragment, but that it acquires meaning if framed in the light of relations with others, in a community of purposes, which Royce sees exemplified in the scientific community and in the Christian spirit.

Royce's magnum opus, *The world and the individual* (1900–1901), represents a supreme effort to settle the idea of the absolute as unity and multiplicity, obtaining a marriage between absolutism and pragmatism, the so-called "absolute pragmatism" (Cf. Mahowald 1972). Ideas are conceived by Royce, not so much as cognitive entities, but rather as volitional entities, and the meaning of an idea is seen to reside in its intentional character. In a pragmatist way he believes that ideas are true if they work, but in turn they only work because they are in harmony with the purposes of the absolute. Generally speaking, individuals, with their individual plans, but also any element of nature like a stick or a stone, although unaware, constitute a network of teleological relationships, sometimes on purpose, within a divine cosmic plan.

The idea of the "teleological" community as a communion of purposes is also taken up in *The problem of Christianity* (Royce 1913), a book that came out while Wiener was following his course, and in which the conditions of existence of Roycean communities are identified in the integration of the individuals through communication and sharing of a past (community of memory) and a future (community of hope).

In his autobiography, Wiener is nevertheless keen to stress that Royce was for him primarily the motivation for beginning to work on formal logic, although he was not a leading figure in logic. Royce always had a keen interest in mathematics and logic, fueled by the need to establish with them their metaphysical instances. This interest emerges with particular acuteness in the long *Supplementary essay. The one, the many and the infinite,* placed in appendix in *The world and the individual.* In the essay he seeks to confute the alleged denial of multiplicity by Bradley, resorting to Cantor's notion of an infinite set that contains within itself its own representation. Royce shows how the absolute is one, and can contain within itself an infinity, as would a perfect map of England drawn somewhere in England, which would represent the whole of England and therefore itself and so on ad infinitum. In later years we still find Royce struggling with math and logic. In particular in *The relations of the principles of logic to the foundations of geometry* (1905), he adopted a constructivist approach to logic that would later be taken up by Whitehead, and which would form Wiener's main commitment to logic (Cf. 64g [56g], 52–54).

1.8   Royce's Seminar on Scientific Method

In his autobiography, Wiener also tells us of "Royce's seminar on scientific method, which I attended for 2 years, and *which gave me some of the most valuable training I have ever had*" (64h [53h], 165–166. Italics added). Given the reluctance to speak positively of their philosophical past, this assertion must be emphasized. Wiener would also refer to it in the introduction to *Cybernetics* (48f1), where he would write that, even before attending Rosenblueth's neurological dinners in the thirties, "I had been interested in the scientific method for a long time and had, in fact, been a participant in Josiah Royce's Harvard seminar on the subject in 1911–1913" (61c [48f1], 1–2).

It will be useful to say a few words about this seminar, not only because of its importance in Wiener's intellectual itinerary, but also because it is rather poorly known for something of such value. It started as a regular seminar, during AY 1888–89. Initially known as the "Metaphysical Seminar", it focused first on Kant, then on Hegel, and after that the methods of psychology. It then changed its name to the "Logic Seminar" or "Seminar on Logic". From 1908–1909 the focus moved to the methods of the different sciences. From 1910 up to Royce's death, hence also during the two AY (1911–1912 and 1912–1913) when Norbert Wiener attended it, its denomination was: "A Comparative Study of Various Types of Scientific Method" revealing a definite interest for interdisciplinarity and investigations into the nature of the various scientific methods used in science (Cf. Smith 1963, 1).

Wiener's autobiography lists among the participants a Hawaiian volcanologist; the eugenicist Frederic Adams Woods, author of *Heredity in Royalty*; Southard, the psychiatrist and director of the Psychopathic Hospital in Boston; as well as two men of a certain caliber, whose influence on Wiener we cannot overlook: Bridgman and Henderson. We also know that, on 17 October 1911, during the first year of Norbert Wiener's attendance, his father Leo had addressed a philological paper, on the history of a word (Cf. Royce 1963, n. 6, p. 30]. Wiener likely preferred not to mention the name of his inconvenient father.

Regarding the physicist Percy Williams Bridgman (1882–1961), Wiener informs us that

[he] was even then beginning to be skeptical about the elements contained in experiment and in observation, and who understood the influence on physics of James's pragmatism, was definitely veering toward the operational position which he later assumed (64h [53h], 166).

Bridgman's operationalism developed considerations that can already be found in Mach's Phenomenalism as well as in Pragmatism, and were applied in Einstein's theory of relativity and in the Copenhagen interpretation of quantum mechanics. Bridgman insisted on the fact that—as stated in his famous book of 1927—"we mean by any concept nothing more than a set of operations; the concept is synonymous with the corresponding set of operations. If the concept is physical, as of length, the operations are actual physical operations, namely, those by which length is measured" (Bridgman 1927, 5). In essence, e.g. the temperature is meaningful for Bridgman only in terms of thermometers and other devices for measuring this quantity. In the same sense, in special relativity, Einstein had considered time as having no meaning outside of clocks. Wiener still considered valid the mindset in Bridgman's operationalism, intended primarily as a critical stance with regard to the actual contents of scientific knowledge. Wiener would apply the lesson of Operationalism on various occasions: during his stay in Göttingen when he studies problems relating to quantum mechanics, before the introduction of the uncertainty principle by Heisenberg; during the Second World War when he studied prediction theory; and even when he introduced the idea of an embodied logic, limited by the restrictions imposed by the computer or the brain that uses it.

We cannot even ignore the possibility that in that seminar Wiener might have made contact for the first time with systemic ideas, which would play such an important role in his cybernetic research. This happened through the physiologist Lawrence J. Henderson, to whom Wiener refers in his autobiography:

> There was also Professor Lawrence J. Henderson, the physiologist, who combined some really brilliant ideas about the fitness of the environment with what seemed to me a distressing inability to place them in any philosophical structure (64h [53h], 166).

Lawrence J. Henderson (1878–1942), after graduating in medicine at the Harvard Medical School, studied physical chemistry in Strasbourg, and in 1904 started teaching organic chemistry at the Harvard Medical School, where he remained until his retirement. Between 1904 and 1912 he studied the acid-base balance in solutions known as "buffer solutions", i.e., solutions which, exploiting the changes in the rate of dissociation of particular substances, are able to maintain a constant level of acidity when one adds extra amounts of an acid or a base. After this research he devoted himself to studying blood as a "buffer solution" and came to the idea that, among all the possible buffers, the body was using the most suitable in relation to all other body processes. In 1913 he published a book entitled *The fitness of the environment*, in which he argued that the organic functions cannot be studied in isolation, and that the main characteristic of living beings is to regulate many processes that guarantee their functionality. In general, he insisted on the concept of *milieu interieur*, introduced in the 19th century by the French physiologist Claude Bernard (1813–1878), who had hit upon the key theoretical point of physiology as the ability of the living organism to maintain constant internal physicochemical conditions, while also maintaining a continuous exchange with the external environment (Cf. Bernard 1878, 45 ff.).

There was another aspect of this seminar, concerning the statistical method in science and Brownian motion in particular, which I prefer to consider when discussing Wiener's mathematical style from the time he was appointed at MIT.

Chapter 2
The Young Philosopher of Mathematics

2.1 The Choice of the Philosophy of Mathematics

Wiener's doctoral thesis was very technical. He was eager to make an original contribution to formal logic, and this was what had prompted him to ask to do his doctoral dissertation in this discipline under Royce. The latter would have accepted to be his supervisor, but his state of health had worsened. Wiener was therefore entrusted to the care of Karl Schmidt at Tufts College. The topic chosen was strictly technical: *A comparison between the treatment of the algebra of relatives by Schröder and that by Whitehead and Russell.*[1]

The "algebra of relatives" by Ernst Schröder (1841–1902) constituted a systematization of the work on the logic of relations by C.S. Peirce. The logic of relations is a fundamental chapter of formal logic, which was developed by both Peirce and Schröder in the form of Boolean algebra. The logic of relations exposed in Russell and Whitehead's *Principia*, although largely modeled on the one by Peirce, had adopted the propositional formalism introduced by Peano. However, it is very likely that the choice of focusing on the logic of relations was also correlated with metaphysical questions, considering that the issue of relations was one of the hottest spots in contemporary epistemological debates, already discussed by Bradley and Royce. And it was also one of the main aspects of the neo-realist controversy.

The dissertation set out to examine the state of the art in the logic of relations. This research called for a strong formal commitment, as it set out to test by demonstrative procedures that the two formulations, algebraic and propositional, were equivalent (Cf. Mangione and Bozzi 1995). His teachers thus urged Wiener to study abstract algebra.

After discussion of the dissertation, his philosophical orientation seemed clear, and it was thought that he should follow the Ph.D. with 2 years of postdoctoral

---

[1]Letter for doctoral submission, cit. by Grattan-Guiness 1975, 132.

© Springer International Publishing AG 2017
L. Montagnini, *Harmonies of Disorder*, Springer Biographies,
DOI 10.1007/978-3-319-50657-9_2

studies in one of the major centers for the study of logic in Europe. As the first hypothesis of studying with Peano was abandoned on the advice of his supervisors, it was considered that the ideal place to continue those studies would have been Russell's school in Cambridge. As Leo Wiener wrote to Russell in 1913: "His [Norbert's] predilection is entirely for Modern Logic, and he wishes during his one or 2 years' stay in Europe to be benefited from those who have done distinguished work in that direction".[2]

Norbert Wiener arrived in Cambridge in September 1913. He spent the second half of the first year at the University of Göttingen, with Edmund Husserl, as had already been planned for some time, since Russell was due to be absent, because teaching in Harvard. The second AY (1914–1915) was disrupted by the war and, on the advice of Russell, it was finished at Columbia University in New York, under John Dewey.

Although half way between philosophy and mathematics, logic was then even more than today a philosophical discipline, so Wiener went to Europe as a young philosopher to study with philosophers, in the perspective of a philosophical academic career. This point should be stressed because the autobiography suffers from a kind of amnesia in this respect, dwelling rather on complementary experiences in math or physics, rather than on the central part occupied by philosophical studies. At Trinity College he followed the courses of some of the most prominent philosophers of the day: McTaggart, Moore, and Russell.

"I can't imagine what on earth you are doing with McTaggart unless you are reading Hegel or drinking whiskey",[3] Eliot wrote jokingly to his friend Wiener. McTaggart was a neo-idealist who, more than Bradley, had remained faithful to the philosophy of Hegel, of whom he was a translator and commentator. He differed from Hegel as he denied that the real was rational even as it appears in the present. In his view, current reality is contingent and imperfect, as it appears to be sensitive to knowledge and it is only through the dialectical process that it tends towards its future realization as rationality and perfection. This reassessment of contingency may have influenced Wiener, resonating with the ideas already acquired from James, Royce, and Bergson. In this sense Wiener may have felt somewhat more "at home" listening to McTaggart than to George E. Moore (1873–1958), the realist discourse of which had arisen in direct contrast to neo-idealism, especially in the form that it had taken with Bradley, and was explicitly expressed in the Scottish tradition which defended the beliefs of common sense. The first belief to defend according to Moore was the one concerning the existence of material objects which, in his opinion, remained as they were, whether or not there was a subject to perceive them. The relationships between objects were also exclusively external and not likely to affect their properties.

Moore's main interest was in ethics, thought of as an objective science that deals with realities we call "goods". In his opinion, just as anyone knows what the color

---

[2]Leo Wiener to Bertrand Russell, 15 June 1913, cit. by Russell 1967, 223–224.
[3]Eliot to Wiener, end of year 1914 (WAMIT). Cit. by Masani 1990, 60.

yellow is, so everybody knows what "good" is: the purpose of ethics is to analyze the nature of goodness in general, and to determine the direction in which an action can be said to be good.

2.2 Bertrand Russell

Bertrand Russell held positions similar to the realism of G.E. Moore. He had been initiated into the philosophy of Hegel, sharing the stance of Bradley, but later converting to realism. Russell writes:

> [G.E. Moore] also had a Hegelian period [...] He took the lead in rebellion, and I followed, with a sense of emancipation. Bradley argued that everything common sense believes in is mere appearance; we reverted to the opposite extreme, and thought that *everything* is real that common sense, uninfluenced by philosophy or theology, supposes real. With a sense of escaping from prison, we allowed ourselves to think that grass is green, that the sun and the stars would exist if no one was aware of them, and also that there is a pluralistic timeless world of Platonic ideas. The world, which had been thin and logical, suddenly became rich and varied and solid. Mathematics could be *quite* true, and not merely a stage in dialectic. Something of this point of view appeared in my Philosophy of Leibniz (Russell 2009, 15–16)

Russell's realism tended to turn mainly toward problems in the foundation of mathematics and hence in the direction of logic and the theory of knowledge. Russell had initially studied the problems in the foundations of geometry, dealing with the Kantian perspective, then with the logic of Leibniz (Cf. Russell 1900). In 1900, at the International Congress of Philosophy in Paris, he discovered the rigor that characterized Peano's symbolic logic and decided to adopt its symbolism to address the foundational problem unambiguously.

Russell's philosophy of mathematics has two characteristics. On the one hand, there was logicism, i.e., the belief that mathematics can find a rigorous and absolute foundation in logic. To prove this he undertook with Alfred Whitehead the project realised in the three volumes of the *Principia Mathematica* (1910–1913). On the other hand, there was Platonic realism: just like Frege, Russell believed that mathematics was just like India for Columbus, meaning that the truths it identifies are really a kind of discovery, rather than inventions.

Once he had finished with the Herculean effort involved in writing the *Principia*, whose third volume appeared in 1913, but which was concluded in 1910, Russell's reflection continued with *The problems of philosophy* (1912), which dealt with the theory of knowledge, and proposed the theory of *sense data*: factual knowledge is the only true knowledge. He subdivided this into "knowledge by acquaintance", that is, direct knowledge which is accessed "in person" as sense data, and "knowledge by description", a kind of knowledge derived from the first (Cf. *The problems of philosophy* by Russell 1912, and in parallel the *Philosophical studies* by Moore 1922).

When that odd student of philosophy who was Wiener arrived at Trinity College in the autumn of 1913, destined to spend his life as a mathematician alongside the engineers, Ludwig Wittgenstein, another young man with an unusual character, had just taken leave of Russell. Wittgenstein, an Austrian, had arrived in the United Kingdom to study engineering, but had dropped those studies when he became fascinated by the logic of Russell. Powered by a kind of philosophical fire that had rarely been seen before, he had become in those 2 years—from 1911 to 1913—first Russell's student, and later a colleague and a kind of adviser. Wittgenstein had just disappeared to Norway and shortly thereafter would enlist in the Austrian army at the outbreak of the war, during which he wrote the *Tractatus Logico-Philosophicus* (1921), which would become a sort of Bible of logical positivism (Cf. Monk 1990, 284).

Wiener attended two courses given by Russell. One of them was on the theory of knowledge. Among other things, Russell presented his theory of sense data. It was, Wiener tells us, "an extremely elegant presentation of his views on sense data as the raw material for experience" (64h [53h], 191). But Wiener disagreed with his teacher. He argued:

> I have always considered sense data as constructs, negative constructs, indeed, in a direction diametrically opposite to that of the Platonic ideas, but equally constructs that are far removed from unworked-on raw sense experience. (64h [53h], 191)

Apart from this disagreement, he "found the course new and tremendously stimulating". Russell introduced his students "to Einstein's relativity, and to the new emphasis on the observer" (64h [53h], 191). He "also saw the present and future significance of electron theory, and he urged me to study it, even though it was very difficult for me at that time, in view of my inadequate preparation in physics" (64h [53h], 194).

Russell seems to have been influenced by Wiener's ideas. In fact, he assumed a constructivist stance in his book *Our knowledge of the external world as a field for scientific method in philosophy* (1914) (Cf. Russell 1918–1919). Among other things it uses the Whitehead constructivist approach, already mentioned when discussing Royce, and which allows us to interpret geometric points and instants of time as a series of events of finite dimensions, and in general to consider external things as a "logical construction" that starts from and uses sense data (Cf. Restaino 1978, 98).

The other course was "a reading course on the *Principia Mathematica*". It was the same course that Wittgenstein had attended up until the previous year, and, as then, there were only three students.[4] "His [Russell's] presentation of the *Principia* was delightfully clear; and our small class was able to get the most out of it" (64h [53h], 193–194).

It is of paramount importance to pursue the matter of the relationship between the already illustrious teacher and the student who was just nineteen. Right at the beginning, Russell had a totally negative impression. In October 1913, he had told a friend that a Harvard professor, Leo Wiener, had barged into his study, and in a few

---

[4]Cf. Wiener to Perry, October 1913, cit. by Grattan-Guiness 1975, 106.

minutes had briefed him on his eclecticism and his adventurous life, and had introduced his son, about whom Russell had concluded: "The youth has been flattered, and thinks himself God Almighty - there is a perpetual contest between him and me as to which is to do the teaching".[5]

Wiener felt Russell's hostility towards him and wrote to his father in the same period:

> I have a great dislike for Russell; I cannot explain it completely, but I feel detestation for the man. As far as any sympathy with me, or with anyone else, I believe [sic], he is an iceberg. His mind impresses one as a keen, cold, narrow logical machine.[6]

This mutual coldness, however, soon melted away and this seems to have happened especially after Russell had read the doctoral thesis. Wiener wrote in a letter to Perry, his teacher at Harvard:

> I have shown my thesis to Mr. Russell, and he has read it. Before he read it, his attitude to me (as well as that of Mr. Hardy, whom I also met) seemed rather cold and indifferent, and they seemed inclined to doubt my mathematical ability, but after Mr. Russell had read my thesis, he warmed up considerably towards me. He praised my thesis, saying that it was a very good technical piece of work, and even went so far as to give me a copy of volume III of the *Principia*.[7]

In this change of attitude towards Wiener there is a strong analogy with what happened to the young Wittgenstein soon after Russell had read his manuscript. About Wittgenstein, Russell had written at the time that it was something "very good, much better than my English pupils do. I shall certainly encourage him. Perhaps he will do great things".[8]

Wiener received the same encouragement; "[Russell] said some very complimentary things to me",[9] and, as in the case of Wittgenstein, Russell became a real mentor, something that was not obvious. Wiener writes in the memoirs,

> My chief teacher and mentor was Bertrand Russell, with whom I studied mathematical logic and a good many more general matters concerning the philosophy of science and mathematics (64g [56g], 21).

Russell was always convinced that philosophy cannot be fruitful if it is detached from science, and it is no coincidence that in his course he introduced the students to the most recent views in physics, in particular Einstein's theory of relativity, as well as the nascent atomic physics (Cf. 64h [53h], 191, 193–194) and 64g [56g], 25). Here, Wiener came up against his "inadequate preparation in physics" (64h [53h], 194), a matter which had remained quite foreign to his curriculum until then. Russell also advised him to follow some courses in mathematics in the strict sense. A year earlier Russell had confided:

---

[5]Russell to Lucy Donnelly, 19 October 1913, cit. by Grattan-Guiness 1975, 105.

[6]Norbert Wiener to Leo Wiener, cit. by Grattan-Guiness 1975, 104.

[7]Wiener to Perry, October 1913, cit. by Grattan-Guiness 1975, 106.

[8]Russell to Ottoline Morrell, 23 January 1912. Cit. by Monk 1990, 41.

[9]Norbert Wiener to Leo Wiener, February 1914. Cit. by Grattan-Guiness 1975, 104.

It has been one of my dreams to found a great school of mathematically-minded philosophers, but I don't know whether I shall ever get it accomplished. I had hopes of Norton, but he has not the physique, Broad is all right, but has no fundamental originality. Wittgenstein of course is exactly my dream.[10]

Writing these things, Russell was putting his hopes on Wittgenstein in particular, but we could probably apply them also to Wiener and to the handful of young people who attended his seminar on *Principia*. This dream was after all that, a decade later, the plan to create, through symbolic logic (just that of *Principia*), a genuinely "scientific" philosophy would have triumphed with logical positivism. It was a dream that, given the logical tools adopted, could not meet Wiener's endorsement, considering he had long been attracted by the idea of a strict philosophy in formal terms. Nevertheless he did not share Russell's foundational demands all the way. So Wiener wrote to his father, at the end of October 1913:

His [Russell's] type of mathematical analysis he applies as a sort of Procrustean bed to the facts, and those that contain more than his system provides for, he lops short, and those that contain less, he draws out. He is, nevertheless, within his limitations, a wonderfully accurate thinker.[11]

During the seminar on the *Principia*, Wiener became fully aware of the theory of logical types and the important philosophical considerations included in it, as well as the shortcomings of his doctoral thesis in the light of the question of paradoxes. On the other hand, despite getting over the early disagreements, he had an incurable difficulty in fully accepting the way his master operated. Wiener did not accept and would never accept the logicist project, that is, to reduce the whole of mathematics to a closed set of logical axioms. As he would say in his autobiography:

As for myself, I already then felt that an attempt to state all the assumptions of a logical system, including the assumptions by which these could be put together to produce new conclusions, was bound to be incomplete. It appeared to me that any attempt to form a complete logic had to fall back on unstated but real human bits of manipulation. To attempt to embalm such a system in a completely adequate phraseology seemed to me to raise the paradoxes of type in their worst possible form. I believe I said something to this effect in a philosophical paper which later appeared in the *Journal of Philosophy, Psychology and Scientific Method*. Bertrand Russell and the other philosophers of the time used to term this journal "the Whited Sepulchre," an allusion to the simple white paper cover in which it appeared. (64h [53h], 192–3)

Here Wiener appears too modest regarding his philosophical past since, albeit without the complex techniques of Hilbert's meta-mathematics used by Gödel, he had tried at that time to work out a genuine philosophical system based on those issues, which he expressed in at least four publications in the so-called "Whited Sepulchre".

---

[10]Bertrand Russell to Ottoline Morrell, 29 December 1912 cit. by Schwartz 2012, 46.
[11]Norbert Wiener to Leo Wiener, 25 October 1913 (WAMIT). Cit. by Heims (1984 [1980], 18–19).

2.3 The Term Under Husserl

During the second semester of the first year abroad, Wiener went to Göttingen to
follow "three courses with Husserl, one on Kant's ethical writings, one on the
principles of ethics, and the seminar on phenomenology".[12] However, he did not
like Husserl's way of doing philosophy. As he confessed to Russell:

> The intellectual contortions through which one must go before one finds oneself in the true
> phenomenological attitude are utterly beyond me. The applications of phenomenology to
> mathematics, and the claims of Husserl that no adequate account can be given of the
> foundations of mathematics without starting out from phenomenology seem to me absurd.[13]

He also attended a course on abstract algebra given by Landau and one on
calculus by Hilbert. He wrote to Russell:

> At present I am studying here in Gottingen, following your advice. I am hearing a course on
> the theory of groups with Landau, a course on differential equations with Hilbert (*I know it
> has precious little to do with philosophy but I wanted to hear Hilbert*).[14]

The phrase in brackets testifies on the one hand to Wiener's instinctive attraction
to mathematics, while on the other it further confirms that the purpose of the visit
was primarily to listen to Husserl and understand how mathematics was included
there in terms of philosophy alone.

The Göttingen philosophy department also included the mathematics department, and Wiener had the opportunity to frequent both communities, the philosophers and the mathematicians (Cf. Rowe 1986 and 1989). An argument which took
place with a philosophy student who had asked him for clarification about Bertrand
Russell's work would remain emblematic for Wiener throughout his life. When he
heard the reply, Wiener quickly retorted: "But he [Russell] doesn't belong to any
school" (64h [53h], 208). The idea of belonging to a movement of thought would
always be judged by Wiener as "intellectual gregariousness". The incident, however, is especially symptomatic of the situation of confusion in which he found
himself. He became interested in the meetings of the Mathematische Gesellschaft
chaired by Hilbert, where he befriended two young mathematicians: Felix
Bernstein, who had done a remarkable job on Cantor's theory, and Otto Szasz, who
became his close friend and protector (Cf. 64h [53h], 211). However, he did not feel
at home in either environment. He wrote to Russell:

> Symbolic logic stands in little favor in Göttingen. As usual, the mathematicians will have
> nothing to do with anything so philosophical as logic, while the philosophers will have
> nothing to do with anything so mathematical as symbols. For this reason, I have not done
> much original work this term: it is disheartening to try to do original work where you know
> that not a person with whom you talk about it will understand a word you say.[15]

---

[12]Wiener to Russell, June or July 1914, cit. by Russell 1968, 41.
[13]Ibid.
[14]Ibid. Italics added.
[15]Ibid.

The only person he really found himself at ease with was Frege, whom Wiener went to see at Brunnshaupten in Mecklenburg. Of this meeting he wrote to Russell: "I had several interesting talks with him about your work" (see Footnote 12). This was a time when logical positivism had not yet made its big impact. Russell's research, like Frege's, were still something esoteric for most.

2.4 Philosophy of Relativism

Against the claim of an ultimate foundation for mathematics and, in general, for knowledge, Wiener wrote in that period a wide-ranging essay on *Relativism* (14d). This put forward a line of thought, "relativism" in fact, which excluded the possibility of absolute certainty in all areas of knowledge and extolled systematic doubt, but without falling into a skepticism that completely denied the possibility of knowing. *Relativism* begins by confronting the views of new realism and new idealism. Both, Wiener argued, admit knowledge only when it is self-sufficient, that is, only when it refers to an object for which the fact of being in relationship with another is not essential to know it.

According to Moore these are the particular objects, or according to Bradley the absolute. On the basis of similar arguments to those used by neo-idealists, Wiener concluded that "in no significant sense can we assert the existence of self-sufficient knowledge", because there is nothing that is devoid of essential relationships with others, and our consciousness forever modifies the observed object. From this observation he comes to the most radical conclusion: "But if no knowledge is self-sufficient, none is absolutely certain" (14d, 566). To deny certainty is not to say, in his opinion, denying the possibility of knowledge.

The 'agnosticism' which first appeared in Wiener's high school days now seemed to have swallowed and digested the lessons of pragmatism, intuitionism, and neo-idealism, offering a new and rich image of science; one which, defending cognitive ability, avoids both the deterministic outcome of the old nineteenth-century materialism and the Platonism of the realistic epistemology of Moore and Russell.

The fallibilist epistemology of *Relativism* is associated with a sort of fallibilist ethics, outlined in his article *The highest good* (14c), also written in Göttingen. Here he took a position opposite to that of Moore, arguing that "there is no highest good" (14c, 520), and denying the possibility of an objective morality, owing to the mutability of the ideals of morality, both geographically and in time. In *The highest good*, one could perceive the influence of James' meliorism, while *Relativism* frankly acknowledges the debt contracted by pragmatism, as well as by Bergson's philosophy itself. However, Wiener suggests, on the one hand, that pragmatism ought not to be considered as the ultimate philosophy and, on the other, that intuitionism should be intuitionist to the very end, avoiding, where science is concerned, those clear distinctions it criticizes in science itself. Wiener claims in *Relativism*:

> But, all things considered, relativism is far nearer to pragmatism than to Bergsonianism. Relativism only objects to pragmatism in so far as it seems to claim to have said the last word in philosophy: a relativistic pragmatism is quite possible. But Bergsonianism contains elements which are essentially non-relativistic. Bergson postulates gulfs which can not be bridged between homogeneous duration and mathematical time, between purposes and mechanism, between life and matter, between language and thought, between that intuitive thought which allows the mutual interpenetration of idea with idea, and intellectual thought, —that thought which deals in absolutely hard-and-fast concepts and clear-cut distinctions. The world is for Bergson divided by a set of fundamental dichotomies, which are made with absolute sharpness. Though he believes that the opposing sides of these dichotomies are found everywhere intertwined and interrelated with one another, their opposition is for him a fundamental and irreducible fact. Now, to suppose the existence of absolutely sharp distinctions runs directly counter to the spirit of relativism, and, I believe, of Bergsonianism itself. (14d, 570)

Wiener sought a "Hegelian synthesis", some would say, taking up an expression dear to him, which would fill the "abysses" postulated by Bergson, a need felt strongly by both Santayana and Royce. But the latter philosophers both retained a nostalgia for a Platonically understood science, and both viewed the absolute certainties of logic and mathematics as a life raft at sea in the storm of skepticism, whereas the showdown between Wiener and Bergson takes place right on the ground of logic and mathematics, where he launched his final attack in *Relativism*, evocatively writing:

> Bergson believes that the physical sciences and mathematics deal with notions that are absolutely rigid. Though the world of space and matter is for him but a surface-world, but the external manifestation of the true world of time and life, it is a world of pure space and pure matter and *pure forms*, uncontaminated by any taint of time or of life or of the "mutual interpenetration" of idea with idea. [...] But we have seen that such a world is a mere nonentity; that natural science, like every other intellectual discipline, must deal with imperfectly defined concepts, and hence must permit a certain amount of the interpenetration of idea with idea. Even in the case of mathematics, the most abstract and most formal of all disciplines, we have seen that no assignable set of rules will ever exhaust the conditions of the validity of a single deduction; we have seen how the very use of a symbolism is conditioned by our thinking according to the *spirit of the symbolism*, which can never itself be exhaustively and adequately symbolized. No! Bergson's dualism is a false one: pure formal thought exists only as a misinterpretation of mathematics by Bergson and certain formalistic philosophers of mathematics. [...]

> Since Bergson regards mathematics and the allied sciences as purely formal disciplines, and puts them in a world by themselves, he is forced to consider the realm of the mutual interpenetration of idea with idea as free from all taint of mathematics. In our true insight into the world, he believes, we cast aside the shackles of formal reasoning, and with a sort of a systematical intuition perceive immediately the inmost nature of reality. [...] This mysticism is the necessary result of a belief in the purely formal character of mathematics and physical science. But, if we do not believe that mathematics and physical science are purely formal, [...] then there is no ground for thinking that they, too, do not play their part in our true insight into the universe. [...] *Bergson sets up a windmill, calls it physical science, and then charges it most valiantly. But it is only because it is a windmill, and not true science, that he attacks, that he comes off victorious.* (14d, 570–1. Italics added)

The paper reaches its climax with the consideration that no set of rules can be assigned that could be considered exhaustive of the conditions of validity of a single deduction, as the actual use of any symbolism requires the reference to unwritten rules, the so-called "spirit of symbolism", which in turn cannot be properly and comprehensively symbolized. One had to conclude that the complete closure of any axiomatic system was impossible, whence absolute certainty could not even be granted either for mathematics or for logic. These concepts were further developed in *Is mathematical certainty absolute?* [15b]; cf. also *Mr. Lewis and implication* (16a). Following this path he had come very close to the consequences of Gödel's theorems (1967 [1931]). Wiener writes in his autobiography:

> When I studied with Bertrand Russell, I could not bring myself to believe in the existence of a closed set of postulates for all logic, leaving no room for any arbitrariness in the system defined by them. Here, without the justification of their superb technique, I foresaw something of the critique of Russell which was later to be carried out by Gödel and his followers, who have given real grounds for the denial of the existence of any single closed logic following in a closed and rigid way from a body of stated rules. (64g [56g], 324; see also 64h [53h], 193).

In pushing its operation to the level of logic and mathematics, Wiener, however, was going to find himself truely alone. As the philosopher Ernst Nagel, an heir to Viennese logical positivism, acknowledged in an all too short comment on *Relativism*:

> In expressing his doubts concerning the possibility of a completely inclusive axiomatization of formal logic, [Wiener] was challenging what was perhaps the dominant conviction of the foremost students of the subject at that time (Nagel 1985, 67).

As for logic and mathematics, the strong ideal of science would continue to reign supreme for many years yet, an ideal that would be crystallised by logical positivism and in the *Tractatus Logico-Philosophicus* by Wittgenstein. The book ends with the sentence: "What we cannot speak about we must pass over in silence" (Wittgenstein 1963 [1921], 151).

And for the *Tractatus* one "can speak" only about what is verifiable with logical algorithms (regarding the propositions of logic and mathematics) or by observational and experimental procedures (for factual propositions of empirical science). This principle, called the "principle of verification", was used by the logical positivists like a sharp, clear-cut, line to separate "real science", the territory of verifiable propositions, from "non-science", the field of unverifiable meaningless pseudo-sentences: metaphysics, art, religion, etc. This was a way of seeing things that Wiener would criticize throughout his life.

2.5   A Future Mathematician with a Future Poet: Wiener and Eliot

Pesi Masani has rightly emphasized the appreciation for *Relativism* coming from a single but very significant person among his contemporaries, Thomas S. Eliot, who expressed his "hearty agreement" with the views in the essay.

The scion of an old New England family who had moved to St. Louis, Mo., Eliot had studied at Harvard between 1906 and 1909, obtaining a Bachelor of Arts degree in 1909. At Harvard he was greatly influenced by George Santayana and the literary critic Irving Babbitt. In the academic year 1909–1910 he was an assistant in philosophy at Harvard. Later, he spent the academic year 1910–1911 in France to listen to the lectures of Henri Bergson at the Sorbonne. He returned to Harvard as a candidate for the Ph.D. from autumn 1911 to June 1914, where he studied the poetry of Dante as well as British and French writers, but also Indian philosophy and Sanskrit. During the 1913–1914 academic year, when he was already in England, he attended the seminar on the scientific method of Royce (Cf. Skaff 1986, 16). In 1913 he read *Appearance and Reality* by Bradley and in the autumn of 1914 left for Oxford with the intention of writing a doctoral thesis on Bradley. This was completed in 1916 with the title *Knowledge and Experience in the Philosophy of F.H. Bradley* (published many years after as Eliot 1964). He sent it to Harvard where it met with the appreciation of Royce, who considered it "the work of an expert". (Cf. Lowe, s.d.). But Eliot never went to the US to discuss it, partly because submarine warfare had made transatlantic crossings unsafe. Meanwhile, in 1914, he met the poet Ezra Pound, who was as great a poet as a mentor of literary talents. It was a meeting that definitely influenced his choice to move from the philosophical career to a literary one.

Wiener and Eliot had previously spent time with each other at Harvard in the 1911–1913 biennium. They had in common their interest in the philosophies of Santayana and Royce. They met again in London to spend Christmas together in 1914. Eliot had come to England in the autumn of that year to study Bradley's philosophy at Oxford, and Wiener was at the end of the first quarter of his second year with Russell in Cambridge (Cf. 64h [53h], 220).

On 6 January 1915, in a long letter to Wiener, Eliot writes to say that he has read all the papers published by Wiener, and especially the one on *Relativism*. Eliot believed it contained a new doctrine that could be "officially promulgated". He also felt that it could "be worked out, under different hands, with an infinite variety of detail".[16] Echoing previous conversations, Eliot added:

> Of course one cannot avoid metaphysics altogether, because nowhere can a sharp line be drawn; to draw a sharp line between metaphysics and common sense would itself be metaphysical and not common sense. Any relationship does I think suggest this recommendation: not to pursue any theory to a conclusion, and to avoid complete consistency. Now the world of natural science may be unsatisfying, but after all it is the most satisfactory that we know, so far as it goes. And it is the only one which we must all accept.(see Footnote 16)

After what looks like a sort of an ante litteram rejection of logical positivism, at least a decade prior to its advent, Eliot added, however, that "relativism, strictly interpreted, is not an antidote for the other systems. [...] Who is to be the referee?" The consequences that he drew on a personal level were anti-philosophical:

---

[16]Eliot to Wiener, 6 January 1915, cit. by CW4, 73–75, 74.

although we can never do without philosophy, it had to be abandoned, for him to move from the philosophy of art to art *tout court*, and for Wiener to move from the philosophy of science to real science. Eliot argued:

> I am quite ready to admit that the lesson of relativism is: to avoid philosophy and devote oneself to either *real* art or *real* science. (For philosophy is an unloved guest in either company.) Still, this would be to draw a sharp line, and relativism preaches compromise. For me and for Santayana philosophy is chiefly literary criticism and conversation about life; and you have the logic, which seems to me of great value. The only reason why relativism does not do away with philosophy altogether, after all, is that there is no such thing to abolish! There is art, and there is science. And there are works of art, and perhaps of science, which would soon have occurred had not many people been under the impressions that there was philosophy.[17]

We do not know what Wiener's response was. He took advantage of his holidays in London to visit Whitehead, some of whose ideas Wiener was using in that period to develop his own "synthetic logic", a direct consequence of his "relativistic" epistemology.

2.6   Wiener's Logical Research and the "No" from Harvard

Since the first months of his stay in England, Wiener felt an intense desire to work on a logical research in the strict sense. The result was an activity which ran alongside his reflections on relativism, as evidenced by the numerous papers he published during his European biennium. His first paper ever came out in 1913, and it was on mathematics: *On a method of rearranging the positive integers* (13a). Shortly afterwards his first work on logic came out, the fundamental and justly famous (Cf. Kyburg 1976, 33) paper on *A simplification of the logic of relations* (14a), in which Wiener reduced the logic of relations to the logic of classes, through the notion of ordered pairs, "the importance of which", according to the historians of logic Mangione and Bozzi, "it is difficult to overestimate in its simplicity" (Mangione and Bozzi 1995, 431).

The proposal by Wiener had left Russell doubtful and he wrote to him: "I do not think a relation ought to be regarded as a set of ordered couples".[18] To this, the 19 year old curtly replied: "It seems to me that what is possible in mathematics is legitimate".[19] As we can see, Wiener continued to behave in an extremely self-assured manner, even towards his mentor.

Wiener's itinerary of logical research continued during his stay in Göttingen, where he devoted himself to an entirely new field, which he referred to with the terms "synthetic or constructive logic". In his autobiography Wiener explains:

---

[17]Eliot to Wiener, 6 January 1915, cit by CW4, 73–75, 74–75. Italics in the original.

[18]Russell to Wiener (WAMIT) Folder MC cit. by Masani 1990, 55.

[19]Wiener to Russell, in WAMIT, Folder MC cit. by Masani 1990, 55.

Whitehead had been perhaps the chief English postulationalist, but he supplemented a pure postulationalism with the view that the objects of mathematics were logical constructions rather than simply the original concepts described in the postulates. For example, at times he regarded a point as the set of all convex regions which in our ordinary language might be able to contain this point. As a matter of fact, Huntington has formulated very similar ideas quite independently, and an important essay in this direction had been made by the philosopher Josiah Royce several years earlier. But the classical example of constructionalism in mathematics is the definition of the whole numbers which occurs in the *Principia Mathematica* of Whitehead and Russell (64g [56g], 52).

The "principle of extensive abstraction" maintains that the ideal entities necessary to science, such as the point, the number, etc., can be considered as non-postulated concepts, but as defined in terms of familiar notions suggested by perception. This is the same principle as the one used by Russell to define the number as a set of all the sets with the same numerosity; e.g., the number 3 is the set of all the sets containing 3 elements: 3 apples, 3 trees, 3 pens, etc. (Cf. Masani 1990, 54–77, in particular 63). Wiener was particularly attracted by this approach, which he felt could have become his own field of research, probably perceiving it as a way, in the spirit of *Relativism*, to allow the concrete to penetrate into the abstract, and almost as a way to give tangible examples of the vision of science he had drawn upon to reply to Bergson.

His first logical work along this line of research was *A contribution to the theory of relative position* (14b), completed just before *Relativism*, in which he dealt with the concept of 'total or complete relation'. He used this concept to develop a theory of space and time in which entities such as moments and points are considered as constructs derived from psychologically less remote entities such as temporal events characterized by duration and spatial extension (Cf. Masani 1990, 56). It was followed by *Studies in synthetic logic* (15a), written in Göttingen in the Summer of 1914 and dealing with a type of relationship which could be used to measure psychic sensations such as the intensity of a sound, which cannot be dealt with in the same way as extended quantities such as length and duration (Cf. 64h [53h], 201). Wiener says this in his autobiography:

> I had the idea that a method I had already used to obtain a series of higher logical type from an unspecified system could be used to establish something to replace the postulational treatment for a wide class of systems. The idea occurred to me to generalize the notions of transitivity and permutability, which had already been employed in the theory of series, to systems of a larger number of dimensions. I lived with this idea for a week, leaving my work only for an occasional bite of black bread and Tilsiter cheese, which I bought at a delicatessen store. I soon became aware that I had something good; but the unresolved ideas were a positive torture to me until I had finally written them down and got them out of my system. The resulting paper, which I entitled *Studies in Synthetic Logic*, was one of the best early pieces of research which I had done. It appeared later in the *Proceedings of the Cambridge Philosophical Society* and served as the basis for the Docent Lectures which I gave at Harvard about a year afterward (64h [53h], 211–2).

Before submarine warfare reached its climax with the sinking of the ocean liner Lusitania on 7 May 1915, making it extremely difficult for the young man to return to his homeland, Russell advised him to end the second year postdoctoral fellowship with Dewey at Columbia University in New York. We read in Wiener's autobiography:

Following the advice of Bertrand Russell, I studied with John Dewey. I also took courses with some of the other philosophers. In particular, I listened to lectures by one of the New Realists, but I was only able to confirm my impression of an undigested mass of the verbiage of mathematical logic, completely uncombined with any knowledge of what it was all about.

My term at Columbia was a makeshift at best and although I began to develop the intellectual consequences of my own ideas, I did not get much help from my professors. Indeed, the only one of them who was a great name comparable to those I had learned to appreciate at Cambridge and Göttingen, was John Dewey; and I do not think I got the best of John Dewey. He was always *word-minded* rather than *science-minded*: that is, his social dicta did not translate easily into the precise scientific terms and mathematical symbolism into which I had been inducted in England and Germany. As a very young man I appreciated the help and discipline of a rigid logic and a mathematical symbolism. (64h [53h], 222–3. Italics added).

Wiener had been strongly influenced by the pragmatist's Weltanschauung to introduce elements of contingency and fallibilism even into his own logic and philosophy of mathematics. By now his old dream "to combine in some manner the logical standards of formal consistency and practicality" had become a solid program that he was developing with pretty clear ideas: "relativism" in the theory of knowledge with "synthetic logic" as its logical aspect. At Columbia University, however, Wiener did not meet interlocutors who were able to understand his program or encourage the young rookie.

In the AY 1915–1916, he was granted a job lasting 1 year as an unpaid assistant and docent lecturer at Harvard University, which he was entitled to as a Ph.D. graduate. In this context he held two courses for undergraduate students, one on philosophy and one on logic in the strict sense (Cf. 64h [53h], 228), and a series of docent lectures, dedicated to the theme of his "synthetic logic and measure theory," in which he presented the content of the papers already published and other new ideas. In the last part of these lectures, he dealt with the philosophy of space, discussing the theses of Kant, and then, using the procedures of synthetic logic, he went on to speak about combinatorial topology, or what is now known as algebraic topology. The material of the course, once revised, was used in *The relation of space and geometry to experience* (22a), and in *A new theory of measurement. A study in the logic of mathematics* (21a). Russell wrote as follows about the latter paper to recommend its publication:

This is a paper of very considerable importance, since it establishes a completely valid method for the numerical measurement of various kinds of quantity which have hitherto not been amenable to measurement except by very faulty methods.

Although Dr. Wiener's principles can be applied (as he shows in the later portions of his paper) to quantities of any kind, their chief importance is in respect of such things as intensities, which cannot be increased indefinitely. Much experimental work in psychology, especially in connection with Weber's law, has been done with regard to intensities and their differences. But owing to lack of the required mathematical conceptions its results have often been needlessly vague and doubtful. So far as I am aware. Dr. Wiener is the first to consider, with the necessary apparatus of mathematical logic, the possibility of obtaining numerical measures of such quantities. His solution of the problem is, so far as I can see, complete and entirely satisfactory. His work displays abilities of high order, both technically and in general grasp of the problem; and I consider it in the highest degree desirable that it should be printed.[20]

As an aside, the Weber–Fechner law is a psychological law that establishes a relationship between how stimuli feel as distinct from the absolute magnitude of their intensity. Historically the law has been expressed in different analytical forms (cf. Galimberti 1992, 977). Coming to the letter, we can notice a very positive judgment indeed, which does not seem a circumstantial compliment. The Wiener's studies on constructive logic were really deep and seminal. They would later give rise to a vast field of research for the formalization of experimental psychology. As Peter Fishburn and Bernard Monjardet showed in an essay that does a retranslation in today's symbolic language of the research conducted by Wiener at that time,

> Wiener's contributions to measurement theory deserve to be remembered because they include important concepts that were rediscovered by others and now have a central place in the representational theory of measurement and in graph theory (Fishburn and Monjardet 1992, 165).

In spite of everything, in 1916, the young Wiener's performances were not considered sufficient to get him the yearned, stable commitment at the Harvard Department of Philosophy. Not only that, but the young man found the doors closed everywhere and wherever principals and heads of every departments of philosophy that were looking for staff (Cf. 64h [53h], 236–7).

Wiener never knew the reasons why Harvard did not want to hire him. Münsterberg, an applied psychologist, his former professor at Harvard, and the mathematician G.D. Birkhoff had been auditors of his docent lectures. Wiener assumed that problems might have arisen because of some disagreements, for personal reasons, between Münsterberg and his father; perhaps there had been also a negative opinion from Birkhoff, who had criticized some statements during the docent lectures.

In my opinion there is a contingent environmental aspect of great importance that one cannot be neglected. And that curiously nobody, Wiener first of all, has ever taken in consideration: the fact that in a very short time the miracle of the Pragmatist Hellas had mostly ended. James died in 1910, Peirce in 1914, Royce in 1916, while Santayana—just like Eliot 2 years after—had gone to Europe in 1912, to stay there for the rest of his life. Rightly so, the autobiography of Wiener has tones of regret for the fact that he did not draw the best from John Dewey, who remained the only big name in the American philosophical environment; and the choice of Russell to direct him to Columbia University, at Dewey, had been very wise. Moreover, on the other side of the ocean, in 1916, Russell had been removed

---

[20]Russell to Hardy for the publication of (Wiener 21a) by the London Mathematical Society, cit. by Grattan-Guiness 1975, 104.

from teaching at Trinity College because of his pacifist ideas. Because of that he was even sent for six months to prison in 1918 and for many years had no more academic assignments. There were only Hardy and Huntington left, they were the only in the academic world that might recognize the value of the young Wiener, and in fact they kept encouraging the young man. But they were mathematicians, not philosophers.

The only point of philosophical reference of Wiener remained Perry, which did not consider him "worthy of recommendation" (64h [53h], 236–7). And also this fact is not strange. The coolness between the two men was mutual, considering the expressions Wiener uses toward him in the autobiography. In addition, despite Perry's neo-realist interest for Russell, it is reasonable to doubt that he could easily understand and appreciate the young Wiener's philosophical character, which remained really unknown to him. The young age, but overall the peculiarity of Wiener's philosophy and logic, must have actually played a most important role in the exclusion judgment. He had learned in depth the logical techniques used by Russell and Whitehead, techniques that very little people were able to handle and appreciate in that period, as the Wiener's experience at Göttingen proved. Philosophically speaking Wiener was not a neo-realist like Perry; on the contrary he his Pragmatism was excessively mixed with the Roycean idealism, the same Perry had harshly criticized in his *Manifesto*. The choice of the auditors of his docent lectures seems to have been the best that environment could offer. To try to grasp Wiener's difficult research on logic applied to psychology was ask a distinguished experimental psychologist and an outstanding mathematician, Hugo Münsterberg— who by the way will die in 1916. But he was an experimental psychologist educated in the Germany in the second half of the Nineteenth century. How Münsterberg had could understand the perspectives inner in Wiener's measure theory rediscovered only almost a century later? Similarly, it is difficult to believe that George David Birkhoff (1884–1944), a great mathematician and mathematical physicist, but firmly anchored to the idea of science of that Poincaré who was a staunch opponent of Cantor's set theory, would welcome with open arms the completely based on logic Wiener's approach, who will no doubt must have shown some serious shortcoming in the mathematical training.

In his autobiography Wiener is, in my opinion, very close to the truth of things when he states:

> I was assured by Professor Perry that I was not good enough to merit much of a recommendation. I was not a very promising bet at that time, but I cannot help believing that some part of my department's coolness was based on my lack of years and on a conservative unwillingness to experiment with the unknown. (64h [53h], 236–7); cf. also (64g [56g], 27).

Anyway, Wiener's predicament does not actually appear so different from other lacks of understanding with which the history of science is littered. We may think, for example, of Laplace and Lagrange's refusal to acknowledge Fourier's heat theory: too bizarre for their scientific canons, and also containing mathematical gaps in its early stages (Cf. e.g. Narasimhan 1999).

2.7   The University of Maine

Faced with the impossibility for his child to get recruited by any department of philosophy, his father urged him to seek a position as a professor of mathematics, and suggested to do so through an employment agency for teachers (Cf. 64h [53h], 237). And that was how that in the academic year 1916–1917 Wiener was hired as a mathematics college instructor at the University of Maine in Orono. This period lasted only a year but was extremely significant because it marked the first attempt to move on to the teaching of mathematics. A well-documented article by Eisso J. Atzema (2003) adds new information to what we already knew about this from *Ex-prodigy*.The University of Maine, or rather the Maine State College of Agriculture and the Mechanical Arts, was an institution whose main purpose was to support agriculture, with a student population of rather undisciplined, mostly rural extraction (Atzema 2003, 13). For a 20 year old man who had been educated by the finest philosophical minds of the time, it was not the ideal place to have his first experience as a mathematics teacher, and this can help us to understand why he speaks in his autobiography about a cultural environment that seemed humanly poor and unambitious (64h [53h], 10). The Dean of the University, James Norris Hart (1861–1958), struck by the quantity and quality of Wiener's publications, was hesitant about hiring him, especially considering his young age and the general turbulence of the students.

Hart asked for references from William Fogg Osgood (1864–1943), head of the Harvard mathematics department, who replied that he could not judge Wiener's ability as a teacher, because he had not been involved in any activity of this kind in his department. On Wiener's own suggestion, Hart then turned to Edward Huntington, who had just gone to teach at Harvard. Here is the precious testimony of a mathematician and logician among those best qualified to speak of the young man:

> I am not surprised that you are rather skeptical as to Dr. Wiener's fitness for your work, and I think that this skepticism is in a measure justified, on account of the fact that Dr. Wiener has had, as yet, no experience of this kind of teaching. On the other hand, he is not at all the type of over-brilliant, unpractical scholar which one might naturally suppose him to be. We have had other 'precocious' students here, whom I would not for a moment recommend for your position. He is not of that type. His brilliancy consists simply in having a mind which works more smoothly and rapidly than is the case with most of us – the kind of mind that is likely to do well anything that he undertakes to do. He has a good sense and good humor, and infinite patience, and has been successful in private tutoring and in handling conference sections. While his appointment would certainly be in the nature of an experiment, I believe that the experiment would be worth making. I have talked with him a good deal about methods of teaching mathematics, and have been greatly impressed with the soundness of his ideas, which are progressive and at the same time thoroughly sensible.[21]

Wiener was granted a teaching term on trigonometry and algebra, one on elements of analysis, and one on solid geometry. In his letter of acceptance he also

---

[21]Huntington to Hart, 20 April 1916 (WAUM). Cit. by Atzema 2003, 10.

asked to give a course in logic, as this was his 'Fach' and he wished to "keep his hands in it".[22] So Hart talked to Craig Wallace, teacher of philosophy, and he was allowed to teach an elementary logic course. Hart also suggested that he teach a course on the history of mathematics.

The "experiment", in Huntington's words, was not successful as great difficulties emerged in maintaining discipline, and Wiener realized that the interests of the learners would not in any way have allowed him to go as deeply into the logic and the history of mathematics as he would have wished. Moreover, the human environment provided by his colleagues did not seem to meet his cultural expectations. The letters he sent to his mother and father show an almost complete despair. He also reported these difficulties and his intention to quit to the Dean.

It took the American entry into World War I with the declaration of war on Germany in April 1917 to break the deadlock. The general mobilization led the university to dismiss all instructors until the following year; new hires were only confirmed for those faculties which remained active for war needs. Wiener resigned in early May, before being officially discharged, in order to follow the training program of the Harvard Reserve Officers' Training Corps [R.O.T.C.]. In the end, however, he was not mobilised, for reasons of myopia and high blood pressure. He then managed to find a technical job in a factory in Lynn, Massachusetts. Despite the fact that Norbert Wiener felt at ease working alongside the engineers in the factory, his father found him a new job at the *Encyclopedia Americana* in New York, where shortly afterwards he was admitted to the editorial staff.

2.8 Collaboration with the *Encyclopedia Americana*

The articles that Wiener wrote between 1917 and 1918 for the *Encyclopedia Americana*, which we have often already referred to, were subject to strict instructions from the publisher to avoid personal judgment. In fact, they consist largely of historical treatments, although the strong personality of the author does not fail to shine; we cannot even rule out the possibility that they were sometimes subjected to cuts. In any case they allow us to draw a picture of the main philosophical interests of Wiener at the time in which he was about to enter the Massachusetts Institute of Technology.

In light of what we know so far about Wiener, it is not surprising to discover that these articles were all about strictly philosophical or logical subjects: *Æsthetics* (18b); *Metaphysics* (19f); *Soul* (20h); *Ecstasy* (18j); *Apperception* (18f); *Meaning* (19d); *Universals* (20j); *Category* (18g); *Substance* (20i); *Pessimism* (19g); *Duty* (18i); *Dualism* (18h); *Mechanism and Vitalism* (19e); *Induction, in logic* (1919b); *Infinity* (19c); *Postulates* (19h). Only two of them can be considered as dealing with mathematics in the strict sense, *Algebra, definitions and fundamental concepts* (18c)

---

[22]Cf. Ibid., 12.

and *Geometry, non-Euclidean* (19a), although they both also have logical and philosophical aspects.

The most discussed issues are those concerned with mind-body dualism and the vitalism/mechanism dichotomy: ubiquitous issues in the philosophical debate of that period. The articles often draw upon the new vigor given to the dualist thesis by Bergson, in contrast with the monistic solutions offered by the mechanistic materialism, absolutist idealism, and above all the neutral monism expounded by James, later taken up by Holt and Perry (cf. 18h). The article *Soul* (20h) also refers to this latter theory, and Wiener also includes a detailed account of Russell's arguments against it. Years later, when Wiener had ceased to be a professional philosopher, Russell would have changed his mind about those views. The article notes that:

> [the concept of neutral monism] may be expressed in Aristotelian languages by calling the mode of aggregation the entelechy of the stream of states. Like Aristotelianism, neutral monism finds the unity of consciousness in the embodiment of a form or structure or relation, but unlike Aristotelianism, it maintains that the matter shaped by this form is not the body, but the stream of consciousness itself (20h, 271).

Yet another reference to Aristotle's notion of form can be found in *Dualism*, which ends up by stating that—in addition to the classical mind/matter dualism—there is another dualism, perhaps less stressed but no less ancient, "between form and matter" (18h, 367), which had already been conjectured by Anaxagoras. Such a dualism, Wiener explains, had been developed by Plato and Aristotle. These seem to be the roots of the subsequent reflections by Wiener about information as a radically different reality from matter and energy.

In a sense, these articles are a portrait of a young philosopher, an expert of the issues debated in Europe and the United States at the time. They are clear and affirmed an anti-dualistic approach, which still gave a nod to Bergson. In the article *Dualism*, we may read that:

> [Bergson is] the most interesting form of latter-day dualism [...] based on a sharp contrast between the mental, possessing the continuity of memory which allows the present as it were to contain the past, and the material, the subject-matter of physical science, forming a kinematographic succession of spatial arrangements of particles. (18h, 367)

The adjective "interesting" used to refer to Bergson's ideas also comes back in the article *Soul*:

> Among the more interesting of the current views concerning the nature of the soul is that which assimilates the soul to the phenomena of life in general. While this is the tendency of all vitalists, it reaches the highest degree of metaphysical development in the philosophy of Henri Bergson (20h, 270).

Wiener sees a change in Bergson's thought as regards the dualistic distinction of earlier writings like *Matter and Memory* (1911 [1896]). In *Dualism* Wiener argues:

> In his earlier writings, the distinctness of these two worlds is emphatically asserted, but more recently he has come to regard matter as an arrested, atrophied manifestation of the same vital impulse that constitutes life and mind. Bergson thus forsakes dualism for monism (18h, 367).

We also find a short but incisive article entitled *Mechanism and vitalism* (19e). Mechanism is seen as "a tendency to reduce biology as well as chemistry, astronomy, optics, etc., to a mechanical basis and to explain all biological phenomena in terms of motions of particles" (19e, 527). The enthusiasm for this approach, in Wiener's opinion, is justified by the fact that "the Newtonian mechanics has long constituted an ideal for all the natural sciences on account of the elegance of its form and the clearness of its definitions" (19e, 527). On the other hand, Wiener remarks, living organisms "manifest a distinct and highly complicated structure", and as a consequence the mechanical explanations are of "an extremely sketchy nature" (19e, 527). On the contrary vitalists such as Driesch and Bergson deal with "consciousness", "purpose, desire, sensation", "indeterminism", as "not merely non-mechanical, but counter to the current of mechanism, involving either indeterminism, or determination through factors which have no mechanical correlates" (19e, 527–8). Wiener does think that:

> The methods of the vitalist are generally so crude and his definitions so vague that there is no great body of biological knowledge which has been gained from the vitalistic standpoint. The terminology of vitalism abounds in such expressions as élan vital, or 'entelechy', which are only defined *per ignotius* (19e, 528).

Therefore, for Wiener, up to that point in time, neither of the two approaches had really given significant results in biology. He concluded:

> In short, whether a complete mechanization of biology be possible or not, biological investigation has been fertile precisely in so far as it has subjected itself to the norms, if not to the concepts, of physical science. It would consequently seem that mechanism is methodologically correct, even if it be metaphysically wrong (19e, 528).

In this strong conclusion, also a little difficult to interpret, Wiener clearly demonstrates a lack of enthusiasm for mechanism. In his opinion "mechanism is metaphysically wrong". Why? Maybe because in Newtonian science there is no room for indeterminism, or indeed for consciousness or purposive behaviours. However, Wiener insists on his stance, already stated in *Relativism*: we have to work with the only science we have.

In *Mechanism and vitalism*, one sentence struck me as even more significant. Wiener states that: "biology as it exists is permeated through and through by *anthropomorphic concepts*" (19e 527–8). Wiener understood that the solution to the mechanism/vitalism dispute consisted in considering the need to reduce vital phenomena, not so much in Newtonian terms as "motions of particles", as in finding ways to "deanthropomorphize" phenomena related to life and mind, traditionally considered as an exclusively human field, bringing these phenomena back under the aegis of a unified science. This would be one of the main goals of Wiener's future cybernetics. And maybe here one could find a solution to myriad dualisms that still today hamper the advancement of science in certain fields.

One fingerprint left on Wiener by Royce concerns the way in which he tackles the theme of the infinite in the *Encyclopedia Americana*. More than half the article *Infinity* is dedicated to discussing Royce's "Complementary Essay". After supporting Royce's stand against Bradley, Wiener ends by saying:

The Royceian theory of the infinite is based on the analogy of cardinal infinitude, and presupposes that there is such a thing as a complete universe. Certain paradoxes discovered by Russell and Burali-Forti tell very strongly against the existence of a complete unity embracing all lesser unities. Royce's work possesses value rather as an account of the potential infinity of a universe capable of indefinite enlargement than as a description of a given complete infinite. Furthermore, it is clear that systems of much less extent than the Royceian infinite may possess the self-reflecting property of cardinal infinity (19c, 122).

Wiener just entering MIT therefore shows that he took Royce's metaphysical arguments very seriously, although he considers, in the light of his now mature knowledge of mathematical logic and of the research conducted alongside Russell, that the Roycean absolute—understood as one infinite manifold—cannot be considered other than as a "regulative idea", and that the self-imaging property of cardinal infinity can, beyond logic, be detected only in local areas of the real. This was an insight that shortly afterwards would lead him to discover the "Wiener process".

Part II
(1919–1939)

*It is the same with mathematics,which would certainly
not have come into existence if one had known from
the beginning that there was in nature no exactly
straight line, no real circle, no absolute magnitude.*
F.W. Nietzsche, *Human, all too human*
(2005 [1878], 16)

*I formed a new respect for the irregular and a new
concept of the essential irregularity of the universe.*
Norbert Wiener, *I am a Mathematician*
(64g [56g], 323)

Chapter 3
Interwar Scientific Research

3.1 An Unorthodox Mathematical Education

We need to backtrack a little, to ask ourselves what was at this time Wiener's mathematical training. In fact, if not self-taught or even "amateurish", as stated by Levinson (1966, 12), his training in this discipline had in fact been at least unorthodox. He had learned the rudiments of mathematics from his father (Cf. 64h [53h], 290). Later on, during the 3 years as an undergraduate student he had studied calculus, including differential equations, under a teacher with an engineering style and a poor opinion of abstraction, as Wiener complains in his autobiography (Cf. 64h [53h], 112). He was fascinated by mathematics, as evidenced by the fact that he chose this discipline for his B.S. examination—which would remain his only qualification in mathematics throughout his life. The rest of his mathematical studies were extra-curricular. Actually, they were more the result of a curious attraction than a convinced choice. The autobiography refers to courses taken in a manner well separated from his curriculum, often too advanced for his level, "escapades" almost, as it happened in 1910, at the Sage School of Philosophy in Cornell University, where he attended a course by Edward V. Huntington on the theory of functions of a complex variable, with little profit in Wiener's opinion (Cf. 64h [53h], 150).

The decision to devote himself to formal logic had matured while he was following the lectures on logic by Royce. As evidenced by the letter for doctoral submission for his doctorate,[1] these fascinated him, because there was a resonance with his earlier mathematical studies. Since his Ph.D. dissertation had been very technical, requiring mastery of the formalism, he had attended lessons in abstract algebra by Huntington (Cf. 64h [53h], 167–168, and also 181 and 232–233).

Huntington's influence cannot be minimized: "I learned—Wiener confessed—the mathematical aspect of my philosophy from Professor E. V. Huntington" (64h

---

[1]Letter for doctoral submission, cit. by Grattan-Guiness 1975, 106.

© Springer International Publishing AG 2017
L. Montagnini, *Harmonies of Disorder*, Springer Biographies,
DOI 10.1007/978-3-319-50657-9_3

[53h], 167; cf. also 64h [53h], 181 and 232–3). Along with Oswald Veblen and Benjamin A. Bernstein, Huntington was one of the leaders of a school in America known as the "postulate theorists", who introduced the axiomatic approach in mathematics; in Europe this trend would be taken up by Hilbert, some years later. Huntington had also conducted significant research on constructivist logic.

In Cambridge, Wiener began to study mathematics in the strict sense on Russell's advice: "Russell impressed upon me that to do competent work in the philosophy of mathematics I should know more than I did about mathematics itself" (64g [56g], 21–22); cf. also 64h [53h], 190). That is how it was that he enrolled in various courses in mathematics at Trinity College, attended "without immatriculation", as "a quasi undergraduate" (64g [56g], 156–157). Hardy's teaching in particular left a permanent imprint on him. Wiener faithfully followed this program, with the same contents as Huntington's course in Cornell, contents which Wiener finally felt to be within his reach (Cf. 64h [53h], 190) and 64g [56g], 22).

> Hardy's course, however, was a revelation to me. He proceeded from the first principles of mathematical logic, by way of the theory of assemblages, the theory of the integral, and the general theory of functions of a real variable, to the theorem of Cauchy and to an acceptable logical basis for the theory of functions of a complex variable. In content it covered much the same ground that I had already covered with Hutchison of Cornell, but with an attention to rigor which left me none of the doubts that had hindered my understanding of the earlier courses. (64h [53h], 190)

In addition to Hardy's course, he also attended those given by Baker, Littlewood, and Mercer (Cf. 64h [53h], 190). In his autobiography Wiener confesses several times that the mathematical courses into which he threw himself almost instinctively were outside the line prescribed by his philosophical studies, and too advanced for him, as would be the case of the courses taken during his stay in Göttingen in 1914. He recalls in his autobiography:

> Landau's group theory course was *a hard-driving plunge through a mass of detail with which I was not fully prepared to cope. I was able to follow Hilbert's course in differential equations only in parts*, but these parts left on me a tremendous impression of their scientific power and intelligence (64h [53h], 214–215. Italics added].

However, Wiener made the personal discovery "that mathematics was not only a subject to be done in the study but one to be discussed and lived with" (64h [53h], 215); a discover he made while attending the meetings of the "Mathematische Gesellschaft". It is no accident that, when he went back to Harvard, he began to attend regular meetings of the Harvard Mathematical Society.

In this way, his mathematical studies appear to be characterised by a sort of global education. And his autobiography makes no secret of the shortcomings that this path may have led to. He filled these gaps during the 3 years between his teaching at the University of Maine (1916–1917) and his employment at MIT as mathematics instructor in the fall of 1919.

The final step in this direction was given by the study of some books that his sister Constance had inherited from her boyfriend, G.M. Green, a brilliant mind in the Harvard Mathematics Department who had died prematurely due to the epidemic of Spanish flu that followed the First World War. The collection included some texts on integral equations, functional theory, and Lebesgue integration, by authors such as Volterra, Fréchet, Osgood, and Lebesgue. Norbert Wiener could triumphantly declare: "For the first time, I began to have a really good understanding of modern mathematics" (64h [53h], 265).

3.2 From Harvard University to Massachusetts Institute of Technology

As we can see, mathematics was gradually becoming Wiener's true field. In 1918 Wiener gave up his work at the *Encyclopedia Americana*. And while the philosophical commentaries written for it were still coming out—they appeared until 1920—he found war employment as a civilian at the Ballistics Research Laboratory of the US Army, at the Aberdeen Proving Ground, where he dealt with the calculation of ballistic tables.

It was probably this experience of collaboration with some of the younger and more brilliant American mathematicians that gave him the confidence and the awareness that he was destined to be a mathematician, a choice sealed in the fall of 1919 when he joined MIT. His arrival "at the Massachusetts Institute of Technology meant that I had come safely into port in the sense that I was no longer to be rushed by the problems of finding a job and knowing what to do with myself". (64h [53h], 276).

With the end of the war, Wiener had dreamed of being hired by Veblen, the head of the Aberdeen Proving Ground, at the Department of Mathematics of Princeton University, if not in the newly established Institute for Advanced Study, where he would have been able to carry out research in pure mathematics. Instead, Wiener had to settle for a job as instructor in the Department of Mathematics of the Massachusetts Institute of Technology, which he got through a recommendation by the mathematician W.F. Osgood, a friend of his father (Cf. 64h [53h], 270–1).

In the end, the stormy 3-year period 1916–1919 had meant for him the simple transition from Harvard to MIT, two institutions so close to each other as the crow flies that anyone might confuse them; and yet subjectively the leap was huge for the young man.

Harvard University, founded in 1636 as a theological school, then secularized with the American Revolution and the advent of Unitarianism in the early nineteenth century, was the oldest university in the US (Cf. EI 1949, v. 18, 387) and, like today, quite possibly the most prestigious one. Harvard was the apple of Boston's eye, a city made rich by the businesses surrounding its shipping and its industries, and upon which the university conferred a nice and somewhat less provincial face, while claiming to be the heart of American culture, or "The Hub of the Solar System" (Holmes 1858, 172). It was Holmes, a physiologist at Harvard and also a poet who created the neologism "Brahmins", which stood for a group of poets and historians active in the years following the Civil War, and who were in their day much more popular then their contemporaries Whitman or Melville. The Brahmins were people like Holmes or James Russell Lowell (1819–1891), the first director of *The Atlantic Monthly*, the Boston literary magazine, which immediately became the center of American literary life. "Brahmin" soon became a word to define the Harvard man. To be part of that scene, you needed to have studied and/or taught at Harvard. The anti-Bostonians called them "smug", accusing them of Europeanism, turning to the past, and despising the frontier (Cf. Cunliffe 1963 [1954], 134 ff.).

So let us take a look at the Massachusetts Institute of Technology as it was in 1919. It was very different from today's futuristic MIT and, by comparison with the temple of the Brahmins, must have seemed like some kind of big engineering workshop. Founded in the mid-nineteenth century by the geologist William Barton Rogers, MIT was born as an institution for training civil engineers to meet the needs of Boston's industrial development in the second half of the nineteenth century. And in 1880, Boston became the leading city in the United States for the number of patents. Essentially applied research was carried out at MIT. In its laboratories Alexander G. Bell did his first experiments on the telephone. But it was only from the beginning of the 1930s that MIT gradually became a center for study and research on the level of fundamental science, and this resulted mainly from the work of the generation of young teachers of which Wiener was a prime example (Cf. Fontana 1991).

Now, with a Ph.D. in philosophy from Harvard and being the son of a professor of that university, Norbert Wiener could rightly be considered a Brahmin, as much as his friend Eliot. It is not surprising, therefore, that the matter of the refusal by Harvard, together with his final destination at MIT, left in him a long trail of resentment against Harvard's cultural world, which had shown little support for him, especially when compared to the atmosphere he had known in Cambridge (UK), Göttingen, and other European cultural environments. This was a resentment which he could express through the words of other illustrious anti-Bostonians, from Edgar Allan Poe, who had given his home town the nickname of *Frogpondium*, to Irving, who considered it the home of "Preserved Fish" (Cunliffe 1963 [1954], 77).

In retrospect, it is clear that this was the best fate that could have befallen him, not only for his personal career, but also for us. Indeed, it is precisely this integration of the refined Brahmin into the world of the engineer with greasy hands that becomes the main feature of Wiener's intellectual biography, and makes him one of the most interesting scientific and philosophical figures of the twentieth century. On the one hand, this integration characterized his specific style of research in mathematics, connoted by a unique blend of theory and practice; and on the other it provided us with an intellectual able to observe and judge the technical and scientific developments of the twentieth century, learned first hand in one of the world's leading institutions for technological research; and not in a department of human sciences, but working closely alongside engineers and physicists.

3.3   Wiener's First Steps in Mathematical Research

In 1919, the MIT Mathematics Department was simply a subsidiary department providing mathematical services to future engineers; for that reason when he had been recruited and was on approval for the first year, the department director, professor Tyler, asked Wiener to focus on teaching and work on applied mathematics (Cf. 64h [53h], 271).

Wiener responded obediently by abandoning his philosophical research, including his investigations on logic. However, he aimed to deepen those aspects of mathematics most akin to logic, i.e., themes of pure and higher mathematics. He did this in two ways: working on them during his holiday periods, and finding pure aspects within the applied problems his MIT colleagues would put to him.

In the most recent years he had been investing more and more in constructivist logic, whose border with mathematics strictly speaking had become increasingly blurred, so much so that he had already arrived at the threshold of topology at Columbia University in the Spring-Summer of 1915. This research route, once he had opted for mathematics per se, resulted in a rather natural way in the decision taken in 1919 to elect functional analysis as his own specific field of research. Functional analysis was an emerging field of mathematics in those years. It studied functions whose arguments are in turn still functions. As a guide and collaborator he chose the French mathematician Maurice Fréchet, who had introduced a highly abstract style into functional analysis. Wiener would remark:

> One of the specific things which attracted me to Fréchet was that the spirit of his work was closely akin to the work I had tried to do at Columbia on topology. My training with Russell and my later contact with the work of Whitehead had sensitized me to the use of formal logical tools in mathematics, and there was much in Fréchet's work which was suited from the very beginning to be embodied in the peculiar and highly original mathematico-logical language which Whitehead and Russell had devised for the *Principia Mathematica* (64g [56g], 50–51).

In 1920, after teaching for several years at the University of Poitiers, Fréchet was appointed professor of higher analysis at the University of Strasbourg. Wiener joined him to study for a few months together before the start of the International Congress of Mathematics in Strasbourg, planned for the month of September of that year. During their collaboration, Wiener was able to build a complete and consistent set of axioms for the so-called vector spaces; it was a major achievement which, in the same month, in a definitely independent way, was also accomplished by Stefan Banach, a professor of mathematics at the Institute of Technology in Lvov at the time (64g [56g], 60).

The subject of this discovery was referred to as a Banach–Wiener space for a while, and since Wiener later tended to disregard the subject, it subsequently became known as a Banach space (64g [56g], 60). In the meantime, despite this success, Wiener's attention was already focused on another problem he had always been working on in the framework of functional analysis: a revolutionary mathematical model for Brownian motion, which would later become known as a "Wiener process" (64g [56g], 64).

In 1919, having made the decision to devote himself to functional analysis, Wiener had been looking for some current open problem to which to apply himself. Discussing with the mathematician Isaac Albert Barnett of the University of Cincinnati, who was visiting Harvard, he heard of the problem of integration in the space of functions, on which René Gâteaux (1913, 1919 and 1922) had written some papers in France, and Percy J. Daniell several papers between 1917 and 1921, in the USA and in the UK (for more about Daniell, see Aldrich 2007). Daniell attended the International Congress of Mathematicians of 1920 in Strasbourg, where Wiener was present. They both gave papers (Cf. Daniell 1921a and for Wiener [21f] and [21g]).

The problem consisted in modifying the Lebesgue integral—initially applied only to families of points—so that it could apply also to families of curves. From another point of view this meant extending the concept of probability so that it could be applied not only to points, as it was normally done, but to trajectories (64g [56g], 35).

Like the other mathematicians who had dealt with the problem before him, Wiener tried some variants of the Lebesgue integral which would be suitable for the purpose, and wrote a first paper (20f). So far the study fell within the same kind of pure research that had led him to the Banach spaces. But in this case, as Wiener would later tell us, the mathematical formalism obtained had left him feeling dissatisfied, and he wondered if there could be some counterpart in the real world (cf. 58f, 1). He dwelt mainly on two physical phenomena: turbulence and Brownian motion.

Wiener had always considered an article by Geoffrey I. Taylor, an expert on the theory of turbulence, as very significant for his research, mainly because Taylor in his own work had given a predominant role to the notion of self-correlation (Cf. 58f, 2). Later Wiener was struck by the recurrence of a similar notion, even in the way in which Jean Perrin had handled the physical phenomenon of Brownian motion in his book *Atoms* (Perrin, 1916 [1913]; (Cf. Wiener 58f, 2).

Brownian motion is an extremely disordered and random movement of very small fragments suspended in a liquid, which are visible only under the microscope. The motion is caused by the thermal agitation of the molecules of the liquid hitting the fragments. It is named after the botanist Robert Brown, who observed it for the first time in 1827. In the late nineteenth and early twentieth century, the hypothesis of whether matter was made of atoms and molecules, as had been claimed by the ancient atomists, or was constituted by a continuum, was still under discussion. Some physicists understood that, if the effect of Brownian motion was due to repeated blows by small atoms or molecules, then it provided a way to detect their actual reality and indirectly to measure certain characteristics, such as the number of molecules in a given volume of the substance (Avogadro's number), together with their size and their mass. In particular Einstein in 1905 and, shortly afterwards, Smoluchowski, proposed two probabilistic models for Brownian motion that proved adequate to describe the experimental observations, and allowed them to calculate various properties of molecules.

These two mathematical models were rather well known, especially the one by Einstein 1998 [1905], whose publication was one of his four *Annus Mirabilis* papers. However, in both cases the discussions had been about particles (both those visible in suspension, and the invisible ones corresponding to the molecules of the liquid). Hence, they were about the motion of points, while Wiener was interested in studying trajectories. Precisely this approach had been adopted in *Les Atomes*, a book by the French physicist Perrin,

> [...] where he said in effect that the very irregular curves followed by particles in the Brownian motion led one to think of the supposed continuous non-differentiable curves of the mathematicians (64g [56g], 39).

Speaking for the layman, a curve is continuous if it can be drawn without ever lifting the hand from the sheet of paper. Where a continuous curve has a peak or a cliff at a point, one cannot draw a tangent there. This means that, at that point, there is no derivative (the curve is non-differentiable at that point). Between the mid-nineteenth century and early twentieth century, mathematicians discovered various curves with the property of being continuous but never differentiable in the interval of validity. These curves are obtained through iterative, self-representational operations. An example is the curve suggested by the Swedish mathematician Helge von Koch, often referred to as the "Koch snowflake".

Looking at the trajectories followed by particles subjected to Brownian motion, not at the particles themselves, Wiener applied to this physical phenomenon the mathematical formalisms previously obtained by extending the Lebesgue integral from points to families of curves, as achieved by Gâteaux, Daniell, and Wiener himself. Wiener supposed that each curve was one of many possible trajectories that could be followed by a single particle subjected to Brownian motion, and that this particle was subject to the probabilistic laws of Einstein's model. The results, published in (21d), were surprising. In Wiener's own words:

> To my surprise and delight I found that the Brownian motion as thus conceived had a formal theory of a high degree of perfection and elegance. Under this theory I was able to confirm the conjecture of Perrin and to show that, except for a set of cases of probability 0, all the Brownian motions were continuous non-differentiable curves (64g [56g], 39).

The mathematical description of a Brownian trajectory that comes from Wiener's hypothesis, called the "Wiener process", is the typical case of a "pathological curve": just like the Koch curve, this trajectory has no derivative at any point and is obtained through a process of self-representation. Unlike the Koch curve, however, a Wiener process is the result of chance, to such a point of unpredictability that, even knowing where the particle is at any given moment, we are unable to determine where it will be at the next instant. The Lebesgue integral as specifically modified by Wiener provided a way to calculate how likely it was that a particle undergoing Brownian motion would follow such an "absurd" path (with a probability measure called the "Wiener measure") and the outcome is amazing: with probability 1, except for a set of cases of probability 0, that is, always. In other words, all possible trajectories that a particle undergoing Brownian motion describes are Wiener processes. Perrin's intuition was correct and Wiener had given a consistent mathematical model of it in accordance with the physical observations.[2]

3.4 Wiener's New Esthetic Sense in Science

This work by Wiener showed an approach full of novelties. First of all, mathematicians had initially thought of these curves as "pathological cases", as exceptions devoid of any practical applications (Cf. Kline 1972). Some had even expressed an instinctive disgust at the increasing number of similar new cases, like Charles Hermite who, writing to his colleague Thomas J. Stieltjes, confessed: "Je me détourne avec effroi et horreur de cette plaie lamentable des fonctions qui n'ont pas de dérivés" ("I turn with fear and horror from this lamentable plague of functions that have no derivatives". Cit. by Heims (1984 [1980], 70). However, completely reversing Hermite's opinion in his book *Les Atomes*, the physicist Jean Perrin remarked that:

> Though derived functions are the simplest and the easiest to deal with, they are nevertheless exceptional [...]. The contrary, however, is true [...]. Consider, for instance, one of the white flakes that are obtained by salting a soap solution. At a distance its contour may appear sharply defined, but as soon as we draw nearer its sharpness disappears. The eye no longer succeeds in drawing a tangent at any point on it [...]. So that if we were to take a steel ball as giving a useful illustration of classical continuity, our flake could just as logically be used to suggest the more general notion of a continuous underived function (Perrin 1916 [1913], IX).

The general opinion about the universality of smoothly flowing curves, without abrupt changes in direction, had merely been the result of a "comfortable simplification of reality", while curves which are not derivable are actually to be found everywhere. In particular, Perrin gave the example of the actual contour of the coast of Brittany and the trajectories of particles undergoing Brownian motion.

The continuous undifferentiable curves would later reappear in the form of the geometric figures called "fractals", developed by Benoît Mandelbrot, who confessed that Wiener's work had been his main source of inspiration. Mandelbrot noted that Perrin's considerations had gone largely unnoticed by most of his

---

[2]For a more technical account about Wiener research on Brownian motion, see Ito 1976.

contemporaries, while amazingly they had shocked the young Norbert Wiener, inspiring him to build his own probabilistic model of Brownian motion (Cf. Mandelbrot 1975).

The Italian physicist Marcello Cini focused in even further depth on a new sensitivity that characterized the young Wiener's way of doing mathematics. In Cini's opinion, the very way of looking at reality had changed in Wiener, compared to more traditional physicists and mathematicians. Cini draws attention to a passage in the autobiography where Wiener writes down the thoughts inspired by the Charles River, which flowed just below his window at MIT, shortly after his arrival, when he was working on Brownian motion:

> The moods of the waters of the river were always delightful to watch. To me, as a mathematician and a physicist, they had another meaning as well. How could one bring to a mathematical regularity the study of the mass of ever shifting ripples and waves, for was not the highest of mathematics the discovery of order among disorder? At one time the waves ran high, flecked with patches of foam, while at another they were barely noticeable ripples. Sometimes the lengths of the waves were to be measured in inches, and again they might be many yards long. What descriptive language could I use that would portray these clearly visible facts without involving me in the inextricable complexity of a complete description of the water surface? This problem of the waves was clearly one for averaging and statistics, and in this way was closely related to the Lebesgue integral, which I was studying at the time. Thus, I came to see that the mathematical tool for which I was seeking was one suitable to the description of nature, and I grew ever more aware that it was within nature itself that I must seek the language and the problems of my mathematical investigations (64g [56g], 33).

The observation of a river, Cini explains, would have posed but very few problems to a traditional physicist, to whom the classical hydrodynamic theory seemed finally to have solved all problems. At the end of the eighteenth century, the problem of extending the Newtonian mechanics of point-like bodies to ideal fluids had already been solved. For real fluids the solution had come in 1895, from Osborne Reynolds, who had demonstrated how, in the flow of the river, there were two types of regimes: a "laminar" one, when the flow is quiet, and a "turbulent" one when it is not. Reynolds had also established that the transition from one regime to the other depended on a single parameter, later called the Reynolds number, related to the properties of the fluid. And so the issue was definitively closed, at least for a science that aimed traditionally to formulate "the simplest law regulating the most general class of phenomena linked to a single physical principle" (Cini 1994, 111. Original in Italian). But Wiener, Cini observes, was interested in describing complexity, "to represent the mutability and variety of those waves", "to reproduce the unique characteristics of that 'process', inventing the appropriate formalism to provide an accurate and detailed description of how it takes place".[3] Evidently, Cini concludes, once again, the way we see reality has changed.

---

[3]When commenting Heims (1980), Marcello Cini (1985) had already intervened extensively on this topic and the different epistemological approaches motivating the research of von Neumann and Wiener.

I would agree with Cini's thesis. A new way to view reality appears with Wiener. In Wiener work, there is a new "influential metaphysics", to quote Popper and Watkins, something to consider as "metaphysics", as "non-scientific" ideas, according to the view of the logical positivists. Nevertheless, it was really "influential" because, in spite of the logical positivists, it led to the creation of new pieces of science; new compared with the preoccupations of a Newtonian scientist.

After the archaeological work regarding Wiener's philosophical background carried out in the previous two chapters, we have here a rare opportunity to grasp in detail how an influential metaphysics works; in other words, how the set of pre-theoretical assumptions in the mind of a scientist works.

In general, I like to think that the spirit that animated Wiener at MIT between 1919 and 1921 was the same that inspired Eliot when he produced *The Waste Land* (Eliot 1922), written in the same period. It was for me always striking that, when I was considering the title for the present book, I thought first of all of *Le Geometrie del Disordine* [Geometries of Disorder]. But I had to change my mind because Dario Calimani (1998) had already used the title for a book on the poetry of T.S. Eliot! Therefore I came up with *Armonie del Disordine* [Harmonies of Disorder], considering the fundamental role played by harmonic analysis in Wiener's mathematics, as we shall see later.

It is fascinating to think that, not long after their Christmas meeting in London in 1914, with Wiener choosing mathematics and Eliot literature, they went on, at least for a while, to have the same aesthetic spirit, expressed in two masterpieces, *The Waste Land* in literature and the *Wiener process* in science. Without reference, of course, to Wiener, the literary critic Dario Calimani remarks:

> It seems as though Eliot was looking for and found in his style the formal dimension in which to set and fix that mobility of the emotions that situations of disorder and chaos inspired in his sensitivity (Calimani, D. 1998, 21. Original in Italian).

The mathematician Pesi Rustom Masani writes in his biography of Wiener that he "attempts to trace the interaction between mathematical genius and history that has led to the conception of a stochastic cosmos" (Dedication in Masani 1990, 5). I remind the reader that "cosmos" in Greek refers to an "orderly structure".

Eliot, according to Calimani, looks for the "Modernist marriage between a form that aims to contain and control and a content that, by its very nature, rebels, refusing to communicate in spite of it, any rational and certain meaning" (Calimani, D. 1998, 21).

Wiener himself, in *Mathematics and art. Fundamental identities in the emotional aspects of each* (29h), claiming that similarities between artistic styles and ways of doing mathematics can be observed in every age, asserts that mathematics in his epoch took part in the modernist style:

> The modernist says, 'This idea looks interesting to me. Let me see where it leads me, even though I may give it no ultimate approval.' This is the genesis of futurism in painting, of cubism and such bizarreries, and of such modern literary movements as expressionism. The mathematical counterpart of all these is the interest in postulational systems irrespective of

the assigning of any ultimate validity of the initial postulates. [...]. It is not an accident that the period of the bizarre physical theories of Einstein is the period of bizarre music, of bizarre architecture, of bizarre literature, and of a bizarre stage (29h, 160).

Wiener here seems to attribute to others what is his innermost spirit. Considering Einstein's theories as "bizarre" could be in fact questioned. Einstein maintained always that "God does not play dice with the universe". And notwithstanding the fact that he had changed Newtonian mechanics, introducing the notions of relative time and space, he was in fact looking for a perfect architectural vision of the universe. On the other hand, how could anyone deny that the trajectories of the Wiener process, as much as the poetic visions of *The Waste Land*, are not bizarre, or "freaky" as Heims says?

3.5 The Influence of Peirce via Royce on Wiener's Science

Actually, Wiener's new approach to Brownian motion might have appeared strange to his contemporaries working in mathematical, physical, or engineering fields, but it would appear much less strange to a philosopher who had grown up in the atmosphere of the Early Pragmatism.

Speaking just after reporting his work on Brownian motion, the autobiography tells us that, just after arriving at MIT, through some older colleagues, especially Henry Bayard Phillips (1915–1947), Wiener "learned the importance to the pure mathematician, of a physical attitude". In particular he became "aware of the great work of Willard Gibbs on statistical mechanics. This was an intellectual landmark in my life" (64g [56g], 34). Just after that Wiener adds: "When I came to M.I.T., I was intellectually prepared to be influenced by the work of Gibbs". (64g [56g], 35).

This statement sounds rather strange. Indeed, in his curriculum up to his entrance at MIT there is no trace of any interest in statistical physics. In the first edition of this book I hypothesized that the implicit reference there is to Royce and his seminar on the scientific method. Now we have further evidence to support this conjecture.

Wiener had always been very reluctant to reconnect his scientific path after 1919 to the previous philosophical one. It was only with regard to Royce that he could not avoid attributing some credit. In particular regarding "Royce's seminar on scientific method, which I attended for 2 years, and *which gave me some of the most valuable training I have ever had*" (64h [53h], pp. 165–166. Italics added). Actually, it is precisely in Royce's seminar that we find the missing link. There is another important aspect of Royce's seminar we should point out, besides what was discussed in the first chapter, in particular regarding the discussion that took place during AY 1913–1914, and which was recorded in a notebook by Harry T. Costello. It relates to the AY following the 2 years attended by Norbert Wiener, but which was attended by his friend Eliot (Cf. Smith 1963 and Skaff 1986).

Reading the notebook we learn that Royce had presented a paper entitled *The Mechanical, the Historical and The statistical*, subsequently published in April 1914. For us the paper is of great importance, because we find in it a methodological philosophy very close to the one that Norbert Wiener applied throughout his entire scientific life. Royce went into the substance of an issue of great relevance at that time, namely, the contrast between vitalism and mechanism. These were terms between which he was looking for a bridge, a medium—as he had done for intuition and scientific conceptualization—finding it again in Peirce, particularly in Peirce's predilection for the statistical method. The paper reviews three methods, described as "historical", "mechanical", and "statistical".

Science, Royce explains, applies the historical method to the subject when dealing with individual events, such as a single solar eclipse or the birth and death of a man; it uses the mechanical method when dealing with laws that are not subject to exceptions; and finally it applies the statistical method to imperfectly uniform behavior. The statistical method appears superior, in Royce's opinion, to the other two, because it is also applicable in their place, both in the historical, social, and biological sciences, and in those concerned with inanimate nature (Cf. Royce 1914, 556). He considers in particular that the turning point occurred in physics with the kinetic theory of gases and the subsequent statistical interpretation provided by Boltzmann's thermodynamics. In particular, the second law of thermodynamics in Boltzmann's interpretation is to be considered as a law concerning mechanical phenomena, but also as an inherently statistical and evolutionary law. Royce writes:

> The second law of energy becomes a principle stated wholly in terms of the theory of probability. It is the law that the physical world tends, in each of its parts, to pass from certain less probable to certain more probable configurations of its moving parties. As thus stated the second principle not only becomes a law of evolution, an historical principle, but also ceases to be viewed as any mechanically demonstrable or fundamentally necessary law of nature (Royce 1914, 561).

Royce points out that a law of this kind is no longer subject to the blind necessity of the old mechanism: if it is true that "energy, according to the kinetic theory, runs down hill as it does for statistical and not for mechanical reasons", going from states with low entropy into those with high entropy, it is also true—Royce argues—as suggested by Maxwell in his image of the demons sorting the atoms of a gas, that such a tendency might make energy run up hill instead of down, without the violation of any mechanical principle. More recently, Boltzmann, in his further development of Maxwell's hypothesis, pointed out that the theory of probability itself requires that, over the course of very long intervals of time, there must occur some occasional concentrations of energy and some sensible un-mixings, i.e., some reversals of the diffusion of gases, if indeed the kinetic theories are themselves true (Cf. Royce 1914, 561).

We are thus faced with a law that is probabilistic in an essential way: processes go in one direction rather than another only statistically, and for intrinsic reasons. This is explicitly reminiscent of Peirce's writings. In particular Royce quoted from:

The architecture of theories (Peirce 1891), *The doctrine of necessity examined* (Peirce 1892a), and *The law of mind* (Peirce 1892b). Royce commented:

> These papers are fragmentary; and yet in their way they are classical statements of the limitations of the mechanical view of nature, and of the significance of the statistical view of nature (Royce 1914, 562).

Royce generalizes Peirce's considerations, stating that statistical methods should not be considered only as a temporary expedient, suitable to remedy the lack of an adequate mechanical description. He vigorously disputes the idea that we should consider it as a description of the canonical form of knowledge (Cf. Royce 1914, 561). On the contrary, statistical methods should be seen as the standard form of scientific knowledge, because

> our mechanical theories are in their essence too exact for precise verification. They are verifiable only approximately. Hence, since they demand precise verification, we never know them to be literally true.

> But statistical theories, just because they are deliberate approximations, are often as verifiable as their own logical structure permits. They often can be known to be literally, although only approximately, true. This assertion is, in its very nature, a logical assertion. It is not any result of any special science, or of any one group of sciences. It solves no one problem about vitalism. It is a general comment on the value of the statistical point of view. But, if the assertion is true, it tends to relieve us from a certain unnecessary reverence for the mechanical form of scientific theory – a reverence whose motives are neither rationally nor empirically well founded (Cf. Royce 1914, 562).

I must confess that the discovery of these pages by Royce filled me with wonder, as ideas about the logical superiority of the statistical method and the conviction of the reality of an inherently probabilistic physical world are the hallmark of all Wiener's scientific research. The physicist Heims, one of Wiener's main biographers, states, without mentioning Royce:

> Wiener's prescription for at least one kind of useful physical-mathematical theory is to incorporate the imprecision of measurements in the mathematical description, for otherwise the theory would be inappropriately definite. Yet such a mathematical description incorporating imprecision could be completely rigorous. (Heims 1984 [1980], 139).

It is interesting here to read certain passages from Wiener's autobiography, where he argues:

> The traditional way of solving the ballistic equation is to assume the initial data as given precisely [...] and we immediately start to revise these with the aid of methods of interpolation or correction, reckoned by a procedure which is entirely distinct from the first.

> In this process we waste a good deal of effort, first in making our data unrealistically accurate, and second in correcting our imperfectly realistic results. (64g [56g], 256).

As an alternative, Wiener proposes to adopt the tools of functional analysis, and in particular, integral equations. He explains:

> In addition to purely computational advantages in the more complicated cases, this method is also essentially superior to the Newtonian method of computation from the logical point

of view. The reason is this: what we put into our problem [...] contains intrinsically the very inaccuracy which hinders our work. We are thus not overcomputing and relieving the effect of this overcomputing by an ad hoc study of its errors, but putting all our cards on the table at the beginning. What we finally get is what we want, neither more nor less. This cuts down a lot of unnecessary effort, but it also increases the real precision of what we are doing (64g [56g], 257–258).

In general, Wiener sees the possibility, and in fact the need, to extend this methodology to science as a whole:

> If this recognition of the statistical nature of all science is already proving to be valuable in the most Newtonian type of mechanical-engineering computation, how much more must it then be the natural method of computation in those fields in which our errors of observation are naturally very large! (64g [56g], 259)

It should be added that this approach considers statistical methods not so much as a useful stopgap given the practical impossibility of achieving a perfect deterministic description of physical phenomena, but rather as the most suitable method to represent a world that is inherently irregular and subject to chance. Wiener recognised this, or at least believed he recognised this, also in quantum mechanics. Indeed, he always strenuously defended the statistical interpretation of quantum mechanics, conceived not as a temporary arrangement of quantum facts, but as their most fundamental nature (Cf. McMillan-Deem 1976).

As we see, the method proposed by Wiener is substantially the same as the one proposed by Royce and derived from Peirce. The difference is that with Royce it was a vague sketch, while with Wiener it become a set of mathematical procedures, ready to be applied, which Wiener obtained in two ways: working on the thinking underlying Gibbs' statistical mechanics; and using functional analysis and in particular integral equations. We will also see that these two aspects were inspired largely by meta-theoretical convictions gained during his time as a philosopher.

Even Wiener's firm belief in the intrinsic irregularity of the universe can be attributed to Peirce and Royce. In *The Doctrine of necessity examined*, Peirce upholds the idea that events can occur "irregularly without definite cause, but just by absolute chance" (Peirce 1892a, 322). He added: "it will be simplest to remark that physicists hold that the particles of gases are moving about irregularly, substantially as if by real chance" (Peirce 1892a, 330). Royce writes: "the average behavior of a very large collection of irregularly moving objects has characters which are decidedly lawful, even though the laws in question are what may be called laws of chance". (Royce 1914, 560). And Wiener would state: "I formed a new respect for the irregular and a new concept of the *essential irregularity of the universe*. (64g [56g], 323. Italics added)

It is very likely that Wiener had read *The Mechanical, the Historical and The statistical* (Royce 1914), written and published by his master Royce. We cannot be sure that he had notice of the rest of the discussions at the seminar of 1913–1914. However, there is a possibility that there could have been other sources of inspiration for his scientific work since 1919.

From the notes we learn that one of the most active students attending the Royce's seminar of 1913–1914 had been a certain Leonard Thompson Troland, who had obtained a BS at MIT in 1912 and an MA at Harvard in 1914. At the seminar Troland gave a paper on statistical mechanics and Peirce's ideas about statistical methods (Cf. Royce 1963, 149). He also spoke about Brownian motion and the way in which Jean Perrin had studied it. The material for his 1914 paper appeared in the book *The nature of matter and electricity* (Comstock and Troland, 1917; cf. Royce 1963, n. 53, 149). In this book we find a reference to *Brownian movement and molecular reality* (Perrin 1910 [1909]), written before *Atoms* (Perrin 1916 [1913]). In his early book Perrin had been very clear about the nature of the Brownian trajectories. Changing the scale of magnification, Brownian motion always maintains its non-differentiable nature. Perrin wrote:

> [Drawings showing consecutive positions of the same granule] only give a very feeble idea of the prodigiously entangled character of the real trajectory. If the positions were indicated from second to second, each of these rectilinear segments [obtained by noting positions every 30 s] would be replaced by a polygonal contour of 30 sides, relatively as complicated as the drawing here reproduced, and so on. One realises from such examples how near the mathematicians are to the truth in refusing, by a logical instinct, to admit the pretended geometrical demonstrations, which are regarded as experimental evidence for the existence of a tangent at each point of a curve (Perrin 1910 [1909], 63–64).

In addition, we have to remember another aspect of Royce's philosophy, about which we have already said something. This is the idea of Royce's *Supplementary essay*, in which he seeks to confute the alleged denial of multiplicity by Bradley, resorting to Cantor's notion of an infinite set that contains a representation of itself. In particular, using the example of a perfect map of England drawn somewhere in England, and therefore representing the whole of England including the map itself, and in the map another image of England containing another map, and so on ad infinitum.

Brownian motion as described by Perrin seems to be a real example of the idea proposed by Royce. And Perrin himself considered the map of Brittany—even if not that of Great Britain—with reference to different degrees of magnification. Perrin stated:

> We must bear in mind that the uncertainty as to the position of the tangent plane at a point on the contour is by no means of the same order as the uncertainty involved, according to the scale of the map used, in fixing a tangent at a point on the coast line of Brittany. The tangent would be different according to the scale, but a tangent could always be found, for a map is a conventional diagram in which, by construction, every line has a tangent. An essential characteristic of our flake and, indeed, of the coast line also when, instead of studying it as a map, we observe the line itself at various distances from it is, on the contrary, that on any scale we suspect, without seeing them clearly, details that absolutely prohibit the fixing of a tangent (Perrin 1916 [1913], IX–X).

In the specific case of Brownian motion described by Perrin and treated mathematically by Wiener to achieve the "Wiener process", the self-similarity is not geometrical as for the Koch or Peano curves, but merely statistical: any degree of enlargement will have the same statistical propriety.

It is clear here that we are dealing with the concept of fractal introduced by Mandelbrot. Actually, just as we found a link between Wiener and Royce, we can find a strict link between Mandelbrot and Wiener. The young Benoît Mandelbrot, 30 years younger than Wiener, had worked between 1947 and 1948 as the editor of the first issue of *Cybernetics*, published by Hermann & Cie. (cf. Segal 2003). Considering the earlier works Mandelbrot published, Wiener's influence on him is clear. Mandelbrot was in some sense a pupil of Wiener. In his last book, *God & Golem, Inc.*, published the year of his death, Wiener quotes enthusiastically one of the first pieces of research made by the young Mandelbrot (1963), concerning time series of the prices of commodity markets and showing a kind of self-similarity akin to the Wiener process. Wiener writes:

> He [Mandelbrot] has shown that the intimate way in which the commodity market is both theoretically and practically subject to random fluctuations *arriving from the very contemplation of its own irregularities* is something much wilder and much deeper than has been supposed, and that the usual continuous approximations to the dynamics of the market must be applied with much more caution than has usually been the case, or not at all. (Wiener 1964a, 92. Italics added).

The passage is not easy to understand. In the phrase in italics, Wiener emphasizes the fact that the randomness discovered by Mandelbrot is made volatile and wayward in this way due to "autocatalysis", in which the curve becomes more irregular and chancelike, "contemplating" its own irregularities.

Four years later Mandelbrot was to discover "fractals". He introduced them through the article: *How Long Is the Coast of Britain? Statistical Self-Similarity and Fractional Dimension* (Mandelbrot 1967), whose central idea is:

> Seacoast shapes are examples of highly involved curves with the property that - in a statistical sense - each portion can be considered a reduced-scale image of the whole. This property will be referred to as 'statistical self-similarity' (Mandelbrot 1967, 636).

Mandelbrot used Perrin's image, but he read it, so to speak, through Wiener's glasses.[4]

3.6 Wiener's Leibnizian Framework

During a recent study on the relationship between Wiener and Leibniz's thought (see Montagnini 2017a), I have shown that one can see in Wiener's mathematics a sort of framework which he elaborated on the basis of the ideas of Leibniz.

Speaking about his second year at Tufts College (1907–1908), Wiener stated: "the two philosophers who influenced me most in my reading were Spinoza and Leibniz" (64h [53h], 109). Wiener's preference for the two rationalist philosophers

---

[4]In the first French edition of his masterpiece on *Fractals* (Mandelbrot 1975), he acknowledged Wiener's work as his main source of inspiration, considering - maybe excessively so—that the "Wiener process" was the "first fractal". In the second French edition Wiener's credit is reduced, and it is eventually confined to a small corner of the big English edition (Mandelbrot 1985, 445).

is easily explained by his youthful enthusiasm for the "discipline of a rigid logic and a mathematical symbolism" (64h [53h], 222–223). On the other hand, this "enthusiasm" for rigor had to reckon with the strong imprinting left on him by pragmatism. And he was pushed to look for a synthesis between the opposing requirements of absolute rigor and pragmatism. Indeed, his entire intellectual itinerary could be described as a search for this synthesis.

The result of the impossible synthesis he sought was an acceptance of the limitations of science, an ontology of contingency, and a moderate fallibilism in epistemology which he carried into his scientific and mathematical research. Wiener took leave early from Spinoza, while Leibniz always remained a close companion. And Leibniz can actually be viewed as one of the best attempts to reconcile logical rigor with a contingent world. In *Back to Leibniz!* (32c) Wiener resumed the Leibnizian stance in this way:

> For Leibniz, all contingent truths, that is, all truths of particular fact, are determined not merely by the principle of contradiction, but by an additional principle which he calls the principle of sufficient reason, which asserts that some particular perfection must be realized by each phenomenon in this world and by this world as a whole to distinguish it from all other possible worlds. This principle is closely connected with Leibnizian optimism (32c, 203).

Wiener was very impressed by this way of thinking. In Leibniz's opinion, on the one hand, God had introduced a set containing al possible worlds, as the result of divine logical rigor, applying the principle of contradiction; on the other hand, God chose only one world, basing himself on consideration of the best compromise, applying the principle of sufficient reason. If we reflect on this argument, it appears to be, in some sense, exactly the best possible synthesis between Rationalism and Pragmatism.

As Wiener explains, the Leibnizian way of thinking had penetrated into physics from Maupertuis. The latter had discovered, in Newtonian celestial mechanics, a quantity called "action", "the integral of the energy with respect to the time is smaller over the actual path of a particle than over any possible alternative path" (32c, 222). Therefore Maupertuis had suggested considering all possible trajectories and choosing the one in which the quantity of "action" is a minimum. In this way he obtained an alternative method for reaching the same results as Newton. Maupertuis thought that this fact was a sign "that God in creating the world has done it with the greatest possible economy of effort". Wiener remarks: "From one standpoint, this is utter anthropomorphic trash, but it leads to sound physics and originates from a very profound principle, which has modified the entire course of science" (32c, 222).

In this way the "Leibnizian framework" worked its way into mathematics and physics, introducing the variational calculus and, more generally, an approach that we also find in functional analysis and Gibbsian statistical mechanics. But how did Leibniz's mental attitude manage to shape the way Wiener did mathematics? Wiener himself explains his path:

> In the form of the principle of least action, the other possible worlds are introduced merely to be rejected, for they do not satisfy the desired principle of minimization. On the other

hand, in a statistical mechanics — the quantum theory reduces the whole of physics to a form of statistical mechanics — these other possible worlds are considered from the standpoint of probability. To put it crudely, the propositions of statistical mechanics assert nothing about any individual possible world, but rather about the overwhelming majority of all of them. This concept, without which modern physics could not have assumed anything like its present form, is a definite part of the philosophy of Leibniz (32c, 224).

When Wiener arrived at MIT, he came into contact with the physics of Willard Gibbs, pervading various departments from engineering to physics. Gibbs, an American physicist, had worked all his life at Yale University, and some of his work in thermodynamics during the 1870s had proved influential to Maxwell himself. However, it was only in the early twentieth century that Gibbs' true value had been fully recognized, in particular regarding his research on the rational foundations of thermodynamics through a statistical procedure. The central idea of Gibbs' statistical mechanics was to represent the thermodynamic system under study by means of a set of all possible systems with given macroscopic characteristics, each of which was associated with a probability, according to a characteristic probability distribution. One could then calculate the average of this particular set of systems to describe the properties of the system in question.

According to the Wiener's point of view, Gibbs' approach was very akin to the one used by Leibniz, even though Gibbs' approach involved chance, which was not considered by Leibniz. As Wiener explains:

> Gibbs' innovation was to consider not one world, but all the worlds which are possible answers to a limited set of questions concerning our environment. His central notion concerned the extent to which answers that we may give to questions about one set of worlds are probable among a larger set of worlds (89, 12).

Leibniz's way of thinking (first considering the set of every possible configuration, and second, making a choice through some rigorous method), when augmented by Peircean and Roycean intuitions (the fundamental role of chance, irregularity, and self-similarity in nature), forms a metascientific background on which Wiener's rigorous mathematics can come to bear. These considerations are useful to explain the astonishment of the mathematician and physicist Mark Kac, who, in 1966, stated:

> In retrospect one can have nothing but admiration for the vision which Wiener had shown when, almost half a century ago, he had chosen Brownian motion as a subject of study from the point of view of the theory of integration. To have foreseen, at that time, that an impressive edifice could be erected in such an esoteric corner of mathematics was a feat of intuition not easily equaled now or ever (Kac 1966, 68).

3.7   Pragmatic Factors in the Fertility of Wiener's Science

Just after his arrival at MIT, Wiener immediately began to publish a series of mathematical discoveries. It is interesting to try and enumerate the pragmatic external success factors, considering that the pre-theoretical background described above could explain the novelty of his approach, but not so much his success.

Even though he shunned overly competitive situations, as shown by the case of the Banach spaces, Wiener was inspired by ambition and desire to excel in his research. It was Wiener himself who drew attention to this factor, as we have seen, speaking of the staff at the University of Maine (Cf. 64h [53h], 10). He devoted himself completely to the new job and study, drawing upon the sense of dedication and discipline in his studies he had learned from his father (64h [53h, 290]). He was at MIT from nine in the morning to five in the afternoon, including Saturdays, and not infrequently even on Sundays; to his 20 h of teaching each week he devoted more time to the study and creation of new mathematics.

In addition, he continued to be a voracious reader of articles and books, assiduously visiting libraries to read up on novelties in his various fields of interest. Wiener was also a great communicator: it was a constant feature of his style of research that he always maintained extensive networks of relationships, starting with simple exchanges of views with colleagues or even students met by chance. At MIT his walks were well known as *Wienerwegs* (Cf. Conway and Siegelman 2005, 82). He tried never to lose contact with the European and also Asian scientific and cultural environments, taking advantage of summer holidays, scholarships, and congresses.

This lifestyle arose from a strong moral principle to do science, and a maxim to define the human being in his/her integrity. In the first edition of *The Human of Human Beings*, he would write:

> For us, to be less than a man is to be less than alive. Those who are not fully alive do not live long even in their world of shadows. [...] for man to be alive is for him to participate in a world-wide scheme of communication. It is to have the liberty to test new opinions and to find which of them point somewhere, and which of them simply confuse us (50j, 217).

Another source of fertility in science was constituted by interdisciplinarity. As with most terms in everyday use, especially when abstract, interdisciplinarity is polysemic and ambiguous. In Montagnini (2010, 2013, and 2017b), I proposed to distinguish *prima facie* two kinds of interdisciplinarity: one individual and another collective. We find both of them in Wiener's scientific method, under specific conditions. Regarding the first kind, Wiener's requirements were for a type of scientist enabled by a wide educational background to interact with scientists of other disciplines, but nevertheless possessing a clear disciplinary identity.

As we shall see, this formula was also typical of many other people who worked with him in several contexts, first and foremost, in cybernetics. Later on, we shall meet many-faceted people like John von Neumann, Warren McCulloch, and Walter Pitts. They were specialized, but had broad bases in other sciences, and believed, as Campbell (2005) states, that interdisciplinary training cannot have a "Leonardesque aspiration". And the same is true of Wiener.

As we have seen, Wiener's zigzagging from philosophy to biology to mathematics "did not result from any particular plan on my part or on the part of my father" (Wiener 1953 [1964, 295]). In fact Leo Wiener wanted only to try and find the field which best fitted his son. Even Norbert did not cherish the ideal of being a universal man. He wanted to become a specialized man, and was very happy to embrace the identity of a mathematician, so much so that the second part of his autobiography was entitled *I am a Mathematician*, beginning with his hiring by MIT. *Nolens volens* from this itinerary emerged a singular many-sided mathematician, about whom Wiener says:

> Indeed, the peculiar advantage of the ex-infant prodigy in science [...] is that he has had a chance to absorb something of the richness of many fields of scientific effort before he has become definitely committed to any one or two of these (64h [53h], 295).

This condition of a man specialized in a field but with a broad cultural base became for Wiener an important prerequisite for doing good science.

With regard to the second, we would need to distinguish between very small working groups, formed by two, three, or four people, to carry out specific projects on experimental and mathematical research, and larger interdisciplinary groups like the one created by Royce that were very useful for discussions and information interchange. For the latter we need to set a limit, which could be of about twenty people. He did not consider it useful to work in a big organization, governed by the parameters of Big Science which would come in with the Second World War, in large laboratories such as the MIT Radiation Laboratory for the development of radar or the Los Alamos Laboratory. On the other hand, Wiener found small groups free from bureaucratic supervision to be more congenial. In 1948, he wrote:

> We had dreamed for years of an institution of independent scientists, working together in one of these backwoods of science, not as subordinates of some great executive officer, but joined by the desire, indeed by the spiritual necessity, to understand the region as a whole, and to lend one another the strength of that understanding (61c [48f1], 3).

His ideal of interdisciplinary collaboration was in fact the result of the merging of the two ideal types, individual and collective. He wrote:

> We need a range of thought that will really unite the different sciences, shared among a group of men who are thoroughly trained, each in his own field, but who also possess a competent knowledge of adjoining fields (50j, 57).

There are also more pragmatic aspects to consider. As reported by Heims:

> While Wiener was in the emotionally strenuous process of establishing himself in the world of mathematicians, he also sought to establish himself in another way by moving out of his parents' house, taking a wife, and starting a family life of his own (Heims 1984 [1980], 24).

Wiener had actually fallen in love with a young European astrophysicist (Cf. Heims 1984 [1980], 24), but his mother arranged a marriage with the more practical German-born Margaret Engelmann, and Norbert accepted his parents' choice, marrying her in 1926. In Heims' opinion "it would be a stable marriage that lasted as long as Norbert lived" (Heims 1984 [1980], 25), from which he had two daughter, Barbara in 1928 and Margaret (Peggy) in 1929.

However, after being taken on at MIT in 1919, Wiener's personal life continued in fact without major shocks at least until the the Second World War, which Heims considers to be a real watershed in what was otherwise a beautiful ascendant career. Presumably, although he was less involved in the war, even the first watershed, that is, the move from Harvard to MIT, from philosophy to mathematics, had more to do with the First World War than mere chronological coincidence.

At the beginning at MIT it was decisive for a young mathematician like him—who could reveal so much brilliance but with an unstable character, always in need of encouragement and comfort—to enter a warm human environment among the mathematicians and physicists, that could help him gain the confidence he needed (Cf. 64g [56g], 32).

Quite unexpectedly the young philosopher soon became well acclimatized to the technical institution he had entered. Despite his sorrow over the closed doors at Harvard, which went on burning for many years, pushing him to cut off all personal relations with philosophers, MIT proved to be the perfect environment for this young mathematician-philosopher, convinced of his pragmatist ideal and of the value, never idolized, of the mathematically rigorous tools he had acquired.

3.8 Researches Alongside Engineers

Galilean physics, Husserl (1970) pointed out, had "discovered" many things through accurate and perfect measurement, but had also "covered" many others, expunging all those phenomena that did not fit into this idealization. Husserl, as we have seen, did not particularly influence Wiener, who did not appreciate this philosopher.[5] Yet even in Wiener we witness the lifelong attempt to put back into science what Galileo had "covered". However, he did not try to create a parallel science of pure essences, Husserl's way, but always kept a firm grip on the science of the day. It is also true, however, that what Galileo had ignored, engineers have systematically reintroduced, because in everyday life one cannot ignore friction, for example.

Wiener gave paradigmatic significance to the research on Brownian motion. It had showed him that a "geometry of disorder" was possible. The contact with the world of engineers was crucial for him, because his aim was to put order in disorder, but somehow without eliminating the disorder that was so deeply rooted in the engineer's practice.

Whether they were aware of it or not, in the everyday operations of the engineers, Wiener would discern something very different from the ideal of classical science, finding in it confirmation of his previous belief in "the essential irregularity of the universe". In addition, the problems raised by his engineering colleagues provided much stimulus for his mathematics. Therefore, neglecting indeed his philosophical past, he could justly state:

> Because I had studied harmonic analysis and had been aware that the problem of continuous spectra drives us back on the consideration of functions and curves too irregular to belong to the classical repertory of analysis, I formed a new respect for the irregular and a

---

[5]Wiener to Russell, June or July 1914, cit. by Russell 1968, 41.

new concept of the essential irregularity of the universe. Because I had worked in the closest possible way with physicists and engineers, I knew that our data can never be precise (64g [56g], 323).

Perhaps his most important work, on "generalized harmonic analysis", which would occupy him for about a decade, arose in the conversations with Dugald C. Jackson, director of the department of electrical engineering at MIT (Cf. 64g [56g], 72). Harmonic analysis had been introduced by Fourier at the beginning of the nineteenth century. It allows one to represent a curve as the sum of a number of sine curves of different frequencies. But Fourier's method applies only in the case of periodic curves. An extension of this idea is the so-called Fourier integral, which applies to non-periodic curves. As Wiener explains, research was carried out mainly by pure mathematicians, whose result "were too new in 1920 to have trickled down to the working electrical engineer". In addition "the standard form of the theory [...] concerns curves which are small in the remote past and are destined to become small in the remote future". The professional mathematician had "completely neglected" the case of main interest to engineers, that is, curves that "keep running indefinitely at about the same scale", "the sort of continuing phenomenon that we find in a noise or a beam of light". (64g [56g], 77)

Wiener worked precisely on the extension of harmonic analysis to the latter case. He did this using the stochastic mathematical tools developed for Brownian motion. In general, Wiener took cues from a practical question, then sought a coherent theory to solve it. (Cf. 64g [56g], 80 ff.).

The gaps he encountered when doing this came back to an earlier question. During his stay in Göttingen, in the academic year 1926–27, the mathematician A. E. Ingham of the University of Leeds, an expert (Cf. 64g [56g], 115) on number theory, pointed out to him that many problems similar to those which still remained open in Wiener's harmonic analysis had been solved by the Tauberian theorems discovered by Hardy and Littlewood. Following this suggestion Wiener not only managed to fill the gaps in the demonstration of his theory, but also to simplify a wide area of number theory (Cf. 64g [56g], 115–116). The final essay on generalized harmonic analysis (30a) was published in 1930 in the important Swedish journal *Acta Mathematica*, and shortly after that the article (32a) on Tauberian theorems appeared in *Annals of Mathematics* (Cf. 64g [56g], 144–145). These publications lifted him up to the empyrean of the most eminent contemporary mathematicians, something that was sealed by his nomination in 1934 as a member of the prestigious US National Academy of Sciences (Cf. 64g [56g], 176).

Another study emerged from engineering requirements, this time manifested by Vannevar Bush, who succeeded Jackson to the direction of the electrical engineering department at MIT. It concerned the formal system of operational calculus, a set of mathematical methods that played an important role in electrical engineering. At that time it remained substantially at the level in which the British engineer Oliver Heaviside had left it 20 years earlier; the result of great intuition, but characterized by a lack of formal rigor. (Cf. 64g [56g], 77–8 and 58f, 4).

Wiener's research resonated with his contemporary work on harmonic analysis, and even here the formalism of Brownian motion proved to be invaluable (Cf. 64g [56g], 78).

3.9 Participation in Projects to Build Technical Equipments

It seems that Wiener usually worked with engineers only with the ultimate goal of building pure mathematics, abstract forms. But in several occasions he did behave less as a mathematician and more as engineer. The main example was the project of automated calculation systems, even if in his autobiography he notes:

> In these my contribution was wholly intellectual, for I am among the clumsiest of men and it is utterly beyond me even to put two wires together so they will make a satisfactory contact (64g [56g], 112).

This design activity was conducted along the lines of analog computing, as pursued at MIT by Vannevar Bush. This approach allows one to obtain solutions to specific differential equation using a physical process that involves a measurable quantity subject to a law which is formally similar to that equation. This line of development was pursued at MIT with great success by Vannevar Bush, who designed and built, among other calculator devices, "Bush's differential analyzer", a mechanical calculator for ordinary differential equations. Such machines responded to the need to supply—within acceptable time delays—numerical solutions to equations that would otherwise have required much longer, if calculated by hand or with arithmetic calculators.

In particular, one idea he was obsessed with consisted in the use of a high speed scanning process, using light rays (Cf. 58f, 5). This is how this idea came to him, according to his own account:

> One time when I was visiting the show at the old Copley Theatre, an idea came into my mind which simply distracted all my attention from the performance. It was the notion of an optical computing machine for harmonic analysis. I had already learned not to disregard these stray ideas, no matter when they came to my attention, and I promptly left the theater to work out some of the details of my new plan. The next day I consulted Bush (64g [56g], 112).

That very Wiener's idea opened up a line of research involving the so called "integraph cinema" which was developed at MIT by two pupils of Bush, namely the engineers Harold Hazen (see Hazen 1934) and Gordon Brown, who worked on "integraph cinema" both for his 1934 MS thesis and his 1938 PhD thesis. In 1940 Hazen and Brown became the most representative members of the newborn MIT Servo Lab.

Still following his interest in harmonic analysis in the late 1920s, Wiener began a collaboration with Yuk Wing Lee, a student of engineering at the time, focusing on the study of electrical circuits using Fourier analysis. The result was the invention of the so called Lee-Wiener network or filter: modular circuits based on Laguerre's Functions. These nets would play a major role in the Wiener's work on predictors during WWII (Masani 1990, 163-5). The collaboration with Yuk Wing Lee continued throughout 1936/1937 in China, where Wiener had been invited to spend a year at the University of Tsing Hua in which Lee was teaching electrical engineering (Cf. 64g [56g], 189–191 and Hongsen 1996). They worked together on the theory of electrical circuits and on an electrical analog computing machine to solve Partial Differential Equations (PDEs) (Cf. 64g [56g], 189). Wiener had begun thinking about how to change Bush's machines to solve PDEs toward the end of the 1920s, while working with Bush on the operational calculus. In the case of PDEs the mechanization was mandatory since the numerical solution of these equations required such a mass of calculations that had never even been solved for many of the simplest equations discovered at the beginning of the nineteenth century. Despite the fact that this project used something similar to Bush's analyzers, the machine presented new operating problems, due to the increased operational speed and the fact that the output data were reinserted into the input (Cf. 64g [56g], 190). It was because of these difficulties that Wiener decided in 1940 to abandon the analog approach and try a digital strategy for calculation.

3.10   Research Alongside Physicists

Another great source of stimuli for Wiener's mathematics was offered by physics, a trend which finds its first example in the research on Brownian motion. Another problem of mathematical physics he devoted himself to in the first half of the 1920s was the theory of the potential, which is the study of how electric charges distribute themselves in differently shaped geometric solids (Cf. 64g [56g], 80 ff.). Once again, in this case, his Brownian motion formalism came into play.

Wiener spent the summers of 1924 and 1925 in Göttingen, where he had the opportunity to give lectures on the results he had achieved so far in his generalized harmonic analysis. Here his work received remarkable interest among mathematicians and physicists, in particular from David Hilbert, Max Born, and Richard Courant, engaged at the time in the interpretation of the contradictions emerging from quantum phenomena. (Cf. 64g [56g], 95 ff.) Harmonic analysis then constituted one of the mathematical tools deemed most useful by the Göttingen physicists for studying quantum problems (Cf. Heisenberg 1982 [1977], 51–68). However, the great interest raised by Wiener's lectures in 1925 is best understood if we consider the actual content of the lectures he gave in Göttingen in 1925. In his autobiography, Wiener explains how he tried to clarify the contradictions emerging in quantum phenomena on the basis of harmonic analysis. Recalling musical examples, Wiener states:

> Ideally, a simple harmonic motion is something that extends unaltered in time from the remote past to the remote future. In a certain sense it exists *sub specie aeternitatis*. To start and to stop a note involves an alteration of its frequency composition which may be small, but which is very real. A note lasting only over a finite time is to be analyzed as a band of simple harmonic motions, no one of which can be taken as the only simple harmonic

motion present. Precision in time means a certain vagueness in pitch, just as precision in pitch involves an indifference to time. [...] It was this paradox of harmonic analysis which formed an important element of my talk at Gottingen in 1925 (64g [56g], 106).

As we can see, Wiener introduced a very similar approach to what, 2 years later, in 1927, would be known as the Heisenberg uncertainty principle. And in Wiener's autobiography there is an awareness of this. Wiener states:

> To see the relevance of my ideas to the actual development of quantum theory, we must step ahead a few years, to the time when Werner Heisenberg formulated his principle of duality or indeterminism (64g [56g], 107).

From the enthusiasm which followed, he got the the idea of a collaboration with Max Born, who was *visiting professor* at MIT in 1925, and with whom he published an article of some importance for a unified description of quantum effects (Wiener and Born 26d, and 26e). In addition, Richard Courant, director of the Institute of Mathematics at the University of Göttingen, helped Wiener to get a Guggenheim scholarship at Göttingen as an annual visiting professor (Cf. 64g [56g], 95–108). Therefore Wiener came back to Europe during the whole academic year 1926–27, with the express purpose of applying to quantum mechanics "the philosophy behind my old paper on Brownian motion" (64g [56g], 109). However, in Göttingen there were disagreements relating to the way his father had acted with the press. Even though the resulting friction was unrelated to theoretical issues, the collaboration became very difficult (Cf. 64g [56g], 114), and Wiener's interests turned away from quantum mechanics. In particular, he took the opportunity of his presence in Europe to deepen the theme of the above-mentioned Tauberian theorems, which he had discovered in that very period. From that point on, he would work sporadically on quantum mechanics problems, although he never ceased to be interested in them. Later on he would work with his dear friend the physicist Dirk Jan Struik on the Schrödinger equation (Cf. 64g [56g], 124–125). He continued to apply his stochastic methods to quantum mechanics, even in the 1950s and 60s.

Among other research on mathematical physics over these 20 years, we should mention his joint work with Eberhard Hopf, a German astronomer visiting Harvard, with whom Wiener developed equations that were useful to describe the balance processes in stellar radiation, known as the Wiener–Hopf equations (Cf. 64g [56g], 142–143). During the Second World War, these equations would play a decisive role in the creation of Wiener's prediction theory.

Wiener's interest in Gibbs, and generally for statistical mechanics, became especially important in the 1930s, following the research on the ergodic hypothesis by John von Neumann and George David Birkhoff. The ergodic hypothesis, as enunciated by both Gibbs and Boltzmann, was supposed to say that in the course of its evolution in time a physical system would go through all of its possible states, according to the probability distribution associated with these (Cf. Masani 1990, 139). The research undertaken in the years 1929–1932 by Von Neumann and Birkhoff showed that the ergodic hypothesis was not sustainable and that only a weak formulation could be accepted; in order to demonstrate their assumption they resorted to functional analysis, that is, the Lebesgue integral and its extensions, the same mathematical tools that Wiener had applied to the study of Brownian motion.

Since the mid-1930s Wiener was very interested in extending these results, strongly believing that—as he says in the introduction to his mathematical work of 1938—statistical mechanics is nothing but "the application of the concepts of Lebesgue integration to mechanics" (38a, 897). This is a very strong statement: it underpins the idea that, through functional analysis applied to statistical mechanics and the results of the research on the ergodic hypothesis, it is possible to rethink the whole of mechanics and indeed the whole of science in entirely statistical and stochastic terms.

Chapter 4
Reflections on Science and Technology

As soon as he entered MIT Wiener stripped himself of his former image of a professional philosopher and also stopped almost all the channels of communication with other professional philosophers. But he nevertheless maintained his reflections on science and technology and their socio-economic dimensions. As he wrote in 1954, when giving the final touches to *Invention*, a work that has only recently been released posthumously:

> The present book is in one sense the result of the reflection of 35 years spent at the Massachusetts Institute of Technology in intimate connection with engineering, scientific, and economic developments (93, 1).

To the period between the two wars belong various writings of unequal value, scattered pieces not easy to fit into a common design without excessive stretching, but which are useful if we wish to examine and understand how the most mature reflection on the years of cybernetics did not suddenly arise out of the blue, but was the climax of the workings of his mind over several decades. It was the impetus given to his research by the Second World War and the urgent need to question the latest innovations and the responsibilities of science in the new age that was opening up with the end of the war, that would set in place the tiles forming the mosaic that Wiener would eventually call "cybernetics".

4.1 Mathematics as a Fine Art

During his first decade at MIT, from 1919 to 1929, Wiener threw himself body and soul into his work as a mathematician. Consequently, during those years, there were only two publications that might be considered philosophical, and both involved mathematical thinking as a subject of reflection, something which Wiener continued to meditate on with great perseverance throughout his life.

The short article *On the nature of mathematical thinking* (23a), written for an Australian psychology and philosophy journal is linked in a non-trivial way to his

© Springer International Publishing AG 2017
L. Montagnini, *Harmonies of Disorder*, Springer Biographies,
DOI 10.1007/978-3-319-50657-9_4

considerations on the absoluteness of mathematical certainty that are to be found in *Relativism* (14d) and *Is mathematical certainty absolute?* (15b).

The author had by now more than a few years of practical mathematical research behind him, and felt the need to establish the relationship and the difference between logic and mathematics. Logic gives validity to the norms of mathematical process, but it is not the mathematician's main heuristic tool for discovering new theorems. One might say with a later terminology: logic is useful in the context of justification, but not for that of mathematical discovery (23a, 268). The logician himself "is a critic, not a creator" (23a, 268). And indeed, the mathematical discovery process can begin in a way that is

> very vague, and of a nature totally repugnant to logical thinking, [...]. There is nothing more surprising than the power of the mind to formulate these vague yet useful hypotheses concerning a subject matter abstract and logical in character (23a, 270).

Sooner or later, though, one must submit one's ideas to logic as if they were on trial, and it is typical of the great mathematician to be able to find a deep theorem to link fragmented and tortuous theories into a "clear, luminous, and simple" whole (23a, 271).

"Mathematics is an experimental science" (23a, 271), Wiener writes using a peremptory hyperbole. Certainly, these experiments are carried out with pencil and paper rather than in test tubes, but the result is the same: what happens is a match between "preconceived ideas" and "hard facts", and this match is checked in the case of mathematics by logic, and in the other sciences by experience. Both experience and logic only respond with a "yes" or a "no", but in the first case the answer is "whispered", in the second "shouted" (Cf. 23a, 272).

Despite the revival of the theme of the vagueness of ideas, all in all here we seem to witness a retreat from the weakening of the notion of mathematical certainty which was typical in his writings from the years 1914–1915. It is true that logic is not regarded as a privileged forum of mathematical discovery, but mathematical propositions must nevertheless be confronted with a logical canon that seems to be considered unique and decisive.

This idea is reiterated in the thicker *Mathematics and Art. Fundamental identities in the emotional aspects of each* (1929h), where it is argued that mathematics is a genuine form of artistic expression, in the same way as architecture or painting, able to produce an emotion indistinguishable from aesthetic contemplation, where the means of expression of a mathematician, equivalent to the marble of a sculptor, is formal rigor. The latter limits it and at the same time allows its expressiveness. The thesis of an analogy between mathematical creativity and art is quite widespread among mathematicians (Cf. e.g. *Il bello della scienza*, Montagnini 2001) . In very similar terms his former teacher and friend Godfrey Harold Hardy (1877–1947) in *A Mathematician's apology* (1941) would assert: "A mathematician, like a painter or a poet, is a maker of patterns. If his patterns are more permanent than theirs, it is because they are made with ideas" (Hardy 1994 [1941], 13).

Wiener tends to depart from Hardy's thought to the extent that he emphasizes the identity with art, and mathematics is framed in its own right in the history of art, with which it shares the evolution of style and canons (Cf. 29h, 129). In his view, considering only modern history, art history can be divided into three periods: Classicism, Romanticism, and Modernism. Classicism went from the beginning of the seventeenth century until the mid-eighteenth century. It was characterized by practical and concrete mathematics, in which the requirement of formal precision certainly existed, as evidenced by the *Letters to an infidel Mathematician*, in which Berkeley criticized the Newton's hypothesis of infinitesimals. However Newton's methods predominated. "Newton's mathematical logic is as dead as the dinosaurs, [...] but Newton's system marches on" (29h, 132). Once again a great practical sense and a disinterest in demonstrations can be found in Euler, whose "idea of a function, of a series, of convergence, is hazy in the extreme, and does not begin to stand up under the criticism of modern rigor. Nevertheless his superb insight has made his collected works a mine of material for the latter day mathematician which is yet unexhausted" (29h, 132).

With the romantic age we witnessing a revival of the need for formal rigor. This age is represented by such figures as Cauchy, Galois, Bolyai, and Abel, whose biographies are themselves Byronic (Cf. 29h, 132). Galois put most of his mathematics in writing on the eve of his death, which took place in a duel over an affair of the heart; János Bolyai, the son of another mathematician Farkas Bolyai, was a cavalry officer who fought seven consecutive duels provided he was allowed to play the violin between the duels. Their mathematics expresses individualism and a sense of revolt: J. Bolyai was the founder of his own non-Euclidean geometry, which he deliberately chose to build in open contrast with tradition. Other authors of this period invented their own disciplines, like Grassmann with his extension theory, and Heaviside with the operational calculus (Cf. 29h, 132 and 160). It is interesting to note in passing that Wiener already compared Heaviside to Prometheus here, as in the later novel *The Tempter* (59e).

That period, Wiener asserted, came through the exhaustion of the romanticism so typical of the late nineteenth century,

> The period of self-satisfaction in mathematics and in physics [...] in which the physicist's ideal was the addition of another decimal place to already established constants in which many a mathematician envisaged the future as an absolutely traditional pursuit of lines of research already mapped out (29h, 160).

With the advent of modernism there was again another revolt, reminiscent of the romantic one, but in this case just involving artistic experimentation, seeking new methods,

> without the act of assent that marks the true romanticist. The romanticist said, 'I believe in this new art, as against the vitiated tradition'. The modernist says, 'This idea looks interesting to me. Let me see where it leads me, even though I may give it no ultimate approval (29h, 160).

Mathematics now, Wiener adds, is the "the science of free logical experimentation with conceptions of order, applied to an indifferent material" (29h, 160). Although he collaborated with physicists and engineers, Wiener always did it in the guise of a pure mathematician, i.e., as a creator of mathematical forms. This remained its fundamental interest, even when dealing with machines and, as we shall see, even when he would work alongside physiologists. Looking for the answers to practical requirements was not his ultimate goal, but always a stimulus to build pure forms. This explains why in *Mathematics and art* we may read the following:

> The applied mathematician is an artisan of mathematics bound by his problem rather than by the free flow of his ideas and receiving this problem from sources beyond his control (29h, 131).

Wilder (1985, 860), the scholar who was asked to comment on the article in Wiener's *Collected Works*, said he was genuinely surprised by such a statement, coming from a mathematician who had for years been engaged in research with engineers and physicists. In the article Wiener seems still to think like Hardy, who wrote:

> But is not the position of an ordinary applied mathematician in some ways a little pathetic? If he wants to be useful, he must work in a humdrum way, and he cannot give full play to his fancy even when he wishes to rise to the heights. 'Imaginary' universes are so much more beautiful than this stupidly constructed 'real' one; and most of the finest products of an applied mathematician's fancy must be rejected, as soon as they have been created, for the brutal but sufficient reason that they do not fit the facts (Hardy 1994 [1941], 41).

At least until 1933, Wiener remained entirely faithful to Hardy, even regarding the ideal of a science not subjected to momentary needs or a petty usefulness, but addressed to "useless" knowledge, or knowledge for knowledge's sake (Hardy 1994 [1941], 41). This is proven by a letter of that year, sent to the writer Paul de Kruif, the author of several biographies of scientists, including the life of Charles F. Kettering (1876–1958), an American inventor of electrical devices for the automotive industry. Wiener wrote to de Kruif to protest vigorously that:

> [the Kettering biography] leaves the impression in the mind of the reader that applied science, followed with an eye to the benefit of humanity, is perhaps the most worthy of all pursuits, but that pure science, the fruit of mere curiosity, is likely to be dilettantish and fruitless.[1]

The letter contains hence a kind of hymn to mathematics as a disinterested search for truth:

> Moreover, a clearly framed question which we cannot answer is an affront to the dignity of the human race, as a race of thinking beings. Curiosity is a good in itself. We are here but for a day; tomorrow the earth will not know us, and we shall be as though we never were. Let us then master infinity and eternity in the one way open to us; through the power of the understanding. Knowledge is good with a good which is above usefulness, and ignorance is an evil, and we have enlisted as good soldiers in the army whose enemy is ignorance and whose watchword is Truth. Of the many varieties of truth, mathematical truth does not stand the lowest.

> Since we have devoted our lives to Mathematics—and she is no easy mistress—let us serve her as effectively as we may. If we work best with an immediate practical problem in view, well and good. If mathematical fact comes to our mind, not as a chain of reasoning, built to

---

[1] Wiener a de Kruif, 3 August 1933 (WAMIT). Cit. by Masani (1990, 340–1). The letter had been published by the *Saturday Evening Post*.

answer a specific question, but as a whole body of learning, first seen as in a glass, darkly, then gaining substance and outline and logic, well and good also. The whole is greater than the parts, and in a lifetime of achievement, no one will care what particular question of practice was in the scholar's mind at such and such a moment.[2]

A small difference of opinion with Hardy that gradually creeps in concerns the idea of equating a mathematics that was born with utilitarian purposes and one conducted in a completely disinterested manner; but the ultimate goal can only be the same: that is, to produce general and deep knowledge.

The issue of the relationship between pure and applied mathematics occupied Wiener considerably thereafter. A direct comparison with the methods of the English school of Hardy and Littlewood must have contributed greatly to his appreciation of this subject. After his period of postdoctoral studies, Wiener never broke off relations with Hardy, his former master. But they became closer when Wiener began to care for the Tauberian theorems, which had been used by Hardy and Littlewood in the context of number theory, and which proved useful to Wiener in filling some formal shortcomings in his generalized harmonic analysis.

During the academic year 1931–32, prior to the Zurich International Congress of Mathematics in the summer of 1932, Wiener almost always resided in Cambridge, where on Hardy's invitation, during the second semester, he lectured on harmonic analysis. During this period, he met R.E.A.C. Paley (1907–1933), a brilliant disciple of Littlewood who obtained a scholarship to collaborate with Wiener at MIT during the academic year 1932–33. Of this collaboration, unfortunately tragically interrupted by Paley's death in a skiing accident, Wiener writes in his autobiography:

> He [Paley] brought me a superb mastery of mathematics as a game and a vast number of tricks that added up to an armament by which almost any problem could be attacked, yet he had almost no sense of the orientation of mathematics among the other sciences. In many problems which we undertook, I saw, as was my habit, a physical and even an engineering application, and my sense of this often determined the images I formed and the tools by which I sought to solve my problems. Paley was eager to learn my ways, as I was eager to learn his, but my applied point of view did not come easily to him, nor I think, did he regard it as fully sportsmanlike. I must have shocked him and my other English friends by my willingness to shoot a mathematical fox if I could not follow it with the hunt (64g [56g], 168).

In this way Wiener began to be more and more aware of how the requirements of applications were not secondary, but essential to his creative work in mathematics. He also gradually came to identify, not only the sociological roots of mathematical purism, but also the very technical roots which allow purism to thrive in some cases. This came through a rethinking of the history of Greek mathematics, which is expressed in the most complete form in *Invention* (93). Here he identifies in the Mycenaean age, the time of Daedalus, a period of close dialogue among inventors, kings, and scholars. Wiener writes:

---

[2]Ibid., 341.

> Daedalus and the artificers of Crete had been able to communicate with kings and learned
> men, but this had become a mere shading memory of the past, and Daedalus is more a
> contemporary of Watt in his ways of thought than he was of Plato (93, 57).

The age of Pericles and Plato followed. The age of classical Greece, centered on
the city-state, where the philosopher belonged to an upper-middle class and was busy
with state affairs and war. "The Greek philosopher was a gentleman, participating in
affairs of state and in affairs of war, interested in reflection, but with soft hands" (93,
56), while manual work was left to slaves and half-breeds. With the advent of the
Hellenistic civilization "the artisan and the philosopher had begun to speak the same
language, and a limited period of scientific and engineering development was initiated". (93, 57). From this was born a favorable period for inventions, conditions that
occurred again between the late Middle Ages and the Renaissance (93, 58). In this
way *Invention* comes to a two-way dialectic between the pure and applied aspects:

> Paraphrasing and modifying Plato's statement that in the ideal state, kings must become
> philosophers and the philosophers, kings. For a great period of invention, the artisans must
> become philosophers or the philosophers, artisans (93, 56).

And more:

> The best use of mathematics as an aid for discovery can be made by the man who does not
> commit himself to either of the two labels, pure and applied, but who is willing to combine
> the mathematical resources of the pure mathematician with the translating ability of the
> applied mathematician (93, 29).

How then was it possible to create a completely pure mathematics? The answer
is found in the very nature of the problems that this type of mathematics can deal
with. In Hardy's obituary, Wiener raises the idea that the latter's purism was related
to his interest in number theory:

> There is nothing better than concrete instances for the morale of the mathematician. Some
> of these concrete cases are to be found in mathematical physics and the closely related
> mathematical engineering, but there is a branch within mathematics which has similar
> merits as a source of actual problems. This field is that of number theory. It is here that such
> concrete cases arise with the greatest frequency and where very precise problems which are
> easy to formulate may demand the mathematician's greatest power and skill to resolve.
> Here Hardy found the central core of his work. It is precisely because Hardy's analytical
> tools are applied to number theory that his work has a freshness and exactness which much,
> although by no means all, of the fashionable work of the present day fails to exhibit. In
> short, Hardy had his feet on the ground (49f, 77).

The same solution is found in the science of the "poleis" as the reason for
"purism" in classical Greece, as he remarks in *Invention*:

> Classical Greek science consists in speculations which can have very little to do with
> manipulation, and much of it is devoted to the logic of number and quantity rather than to
> the measure of quantity and computation by means of numbers (93, 56).

Bringing his Platonism to the extreme, Hardy had come to match the true, the
beautiful, and the good, and therefore to say that, if mathematics is truly pure and
disinterested, there can be no misuse of it, concluding:

> Real mathematics has no effects on war. No one has yet discovered any warlike purpose to be served by the theory of numbers or relativity, and it seems very unlikely that anyone will do so for many years. It is true that there are branches of applied mathematics, such as ballistics and aerodynamics, which have been developed deliberately for war and demand a quite elaborate technique: it is perhaps hard to call them 'trivial', but none of them has any claim to rank as 'real'. They are indeed repulsively ugly and intolerably dull [...]
>
> So a real mathematician has his conscience clear; there is nothing to be set against any value his work may have; mathematics is [...] a 'harmless and innocent' occupation (Hardy 1994 [1941], 44).

No mathematical predictions would ever be more soundly disproved, since there could not have been any atomic bomb without the theory of relativity. The same number theory led to countless applications in war, if only for its use in cryptographic techniques. Wiener, for his part, admits this in his autobiography:

> The very same ideas that may be employed in that Limbo of the Sages known as number theory are potent tools in the study of the telegraph and the telephone and the radio. No matter how innocent he may be in his inner soul and in his motivation, the effective mathematician is likely to be a powerful factor in changing the face of society. Thus he is really dangerous as a potential armorer of the new scientific war of the future. He may hate this, but he does less than his full duty if he does not face these facts (64h [53h], 189–190).

This sense of the moral responsibility of the mathematician does not yet appear, however, in his writings from the period between the wars, where he regards the total freedom of research as essential, and appear to have no qualms about the military use of technology, as illustrated in his paper *Limitations of science. The holiday fallacy and a response to the suggestion that scientists become sociologists* (35b).

In this he distances himself from comments which appeared in *Science* by Dr. Harvey W. Cushing (1869–1939), a renowned neurosurgeon and leading expert on the diagnosis and treatment of intracranial tumors, who, in order to remedy the unemployment caused by technological developments—let's remember that we are in the years of the Depression—had proposed that scientists go on vacation for at least 5 years, devoting themselves to sociology if they felt the need to remain active. On the contrary Wiener believed that it was not right to put a stop to science and technical development:

> For better or for worse, we are destined to live in a world devoted to modern science and engineering. If the road that we are on is slippery, we cannot avoid a catastrophe by putting on the brakes, closing our eyes, and taking our hands off the wheel. Science is a going concern and those who participate in it can only render worthy service by keeping their hand in (35b, 270).

Wiener did not even think advisable to control science through the subjugation of the scientist to a code of ethics of the kind applied to doctors, especially since research is a creative and unpredictable activity.

> Science grows by indirection, and projects to harness its energy and draw it consciously only to the useful may be compared with projects in which an explorer should be sent to an unknown country with the injunction only to explore agriculturally useful land (35b, 270).

As for Cushing's suggestion that scientists dedicate themselves to sociology, Wiener pointed out several critical aspects about the application of mathematical methods in sociology. It will be useful to report his reasons in detail because they will come back several times along his future intellectual path.

First of all, in Wiener's opinion Cushing's suggestion looked like advising patients to start studying medicine in order to heal themselves. In fact, Wiener believed that sociology was for experts, as he explains:

> The difficulties of sociology lie much less in its details than in its ideas and fundamental methods, and will only be resolved by the understanding of people who have devoted their lives to the work (35b, 270).

On the other hand, Wiener was uncertain that social phenomena could find real benefits from being treated by natural scientists and mathematicians. Here he uses an argument that he would never tire of repeating, that of time scale:

> Among the difficulties of sociology is its time scale. The really important phenomena of sociology do not make their meaning clear in less than a generation, and we stand between a past that has not collected the special information which we want and a future which we shall never live to see. Imagine a geneticist whose only biological material consisted in elephants and century plants! You could not expect any very rapid progress from him (35b, 272).

The second argument concerns space scale:

> The space scale of sociology is the entire world. To isolate a sociological phenomenon which runs over a period of years under essentially constant conditions is almost impossible, and the statistical method which is so much in favor at the present day runs against the insurmountable obstacle that its series of data are all too short for any fine analysis (35b, 272).

In both cases Wiener thinks in terms of the statistical mathematical typical of his approach in which social phenomena are not from his point of view handholds. The third argument concerns the scientific immaturity of the social sciences and the people involved in them:

> The fact remains, despite all our wishful thinking, that there are fields in natural science, in medicine, and, above all, in sociology that are not ripe for an attack by the refined tools of modern physical science; that demand the mentality of the general practitioner who is treating the patient rather than that of the specialist who is treating the disease; and where there is no short-cut toward the obliteration of our ignorance. With all respect for sociology, the time has not come for scientists to lead a great trek into its unknown wastes (35b, 272).

4.2 The Relation with the Natural Philosophy of J.B.S. Haldane

During his stay in Cambridge 1931–32, Wiener became acquainted with J.B.S. Haldane, who became a close friend for the rest of life. John Burdon Sanderson Haldane (1892–1964) was an eminent geneticist and physiologist, one of the main creators of population genetics, a statistical theory that merged Darwin's theory with Mendel's genetics. He also studied physiology and biometrics. Just like Wiener, he had been introduced to science from an early age by his father, the physiologist John Scott Haldane. After the First World War he had become an assistant at New College, Oxford, and later he taught at the University of Cambridge (1922–1932), in Berkeley, California (1932), and at the University of London (1933–1957). From the end of the 1920s, he became sympathetic to Marxism; however, he only in fact became a Marxist around 1937, joining the British Communist Party in 1942, distancing himself from it with the rise of Lysenkoism in the USSR. From 1957 he lived in India, where he obtained citizenship, and directed the state laboratory of Genetics and Biometry in Orissa.

Wiener's reflections on the "return to Leibniz" in physics (32c. see above § 3.6) resonated with the thoughts of Haldane, brought together in the article *Quantum mechanics as a basis for philosophy* (1934), in which Haldane speculated about the possible consequences for biology of the quantum mechanics revolution in physics. According to Haldane, by adopting the point of view of wave mechanics put forward by de Broglie and Schrödinger, holistic phenomena and the purpose of life and mind could be made comprehensible. He saw similarities between an animal's ability for repair (even if partial) and that of an atom which regains a lost electron; or between the loss of genetic identity of a cell after reproduction and the loss of identity of elementary particles after collision.

The ability of the teleological mind, according to Haldane, looked similar to the ability of alpha and beta particles to go through the potential barriers of the atom thanks to the wave nature of these particles. This was expressed in the "hypothesis of Gamow" (later called "tunneling" in electronics), which had proved essential to explain the Cockcroft and Walton experiment. According to Haldane, the fact that a wave/particle was able to drill through a potential barrier was analogous to the behavior of an animal which, in a closed place, is able to "see" by thought what lies beyond. The mind could be thought of as a wave phenomenon of the brain, thinking of the latter as being in one of those states that quantum mechanics calls "high degeneracy" or "high internal resonance", in which there is a periodic, summative oscillation making it possible to observe quantum phenomena at the macroscopic level. He believed that matter could be considered as a statistical mixture halfway between two extremes, consisting of a wave system at very high energy, in the case of Newtonian matter, and another at very low energy, in the case of Universals, that is Platonic ideas. Haldane writes:

A universal apprehended by an individual mind, regarded as a wave system of exceedingly low energy, would have a high interchange probability, and a high uncertainty of localization in space-time. [...] Plato's world of timeless unlocalized ideas represents a limiting or asymptotic case which would only be true for objects of zero mass and energy, just as the material world of pre-quantum physics represents the limiting case which would be true for objects of infinite mass. The truth lies between these two limiting cases, but it may well be that just as Newton's theory applies adequately enough for everyday purposes to massive bodies such as a dime or the moon, so does Plato's to its very exiguous objects (Haldane 1934, 90).

Haldane considered that his own materialism could be compatible with most of the assertions of the metaphysical philosophers, that the mind is not a mere epiphenomenon of the body, but that it possesses a reality which is able to interact actively with it; which is in fact able to extend its influence far beyond our bodies, reaching to the farthest galaxies. In this way ideas like divine mind, soul of the world, immortality, and resurrection would certainly not be meaningless (Cf. Haldane 1934, 90).[3]

In a short comment entitled *Quantum mechanics, Haldane and Leibniz* (34c), Wiener showed that there were significant similarities between the thinking of Haldane and Leibniz, since Haldane appeared to him to be substantially Leibnizian. When Haldane considers material particles and universals as nothing more than opposite extremes of the same reality, he is actually following a similar idea expressed in Leibniz' monadology, where the universe appears to consist of monads, some of which, that is, souls, are equipped with apperception, or consciousness, while others with only a blurred perception form material bodies. Even the typical activities of the monads, mirroring the whole universe, are supposed to resemble minds as Haldane understood them, able to extend themselves to the limits of the universe. Certainly, Leibniz' system is generally interpreted as a form of pluralistic spiritualism, but according to Wiener, the difference was very small, and could just as well be understood, as Haldane wished, as a pluralistic materialism (Cf. 34c, 480).

In the essay on *The role of the observer*, Wiener would not completely share Haldane's metaphysical materialistic panpsychism, which interested him mainly for its epistemological consequences. It should also be noted that the encounter with Haldane brought him back to an issue he had neglected for the past 15 years, namely the synthesis between the world of life and mind and the sphere of the inanimate.

4.3   Arturo Rosenblueth

In addition to the relationship with Haldane, Wiener's interest in these issues would be further strengthened by the participation, since the beginning of the 1930s, in the monthly seminars on the scientific method organized at the Harvard Medical School by the doctor and physiologist Arturo Rosenblueth (1900–1970), a close associate of the physiologist Cannon (Cf. Quintanilla 2002).

At these meetings, known as the "neurology dining club", or as the "Philosophy of Science Club" (Cf. Masani 1990, 197), Wiener was introduced by his former student, then colleague at MIT, the Mexican physicist Manuel Sandoval Vallarta. The participants of the "club" were not only physiologists, but also scholars from

---

[3]For a commentary and an updating of Haldane's about quantum phenomena implied by brain, theory alive still today, see Stapp (1997).

other disciplines, at both MIT and Harvard. These meetings involved Wiener deeply, and in *Cybernetics* (1948) he would talk enthusiastically of them (Cf. 61c [48f1], 1–2). He felt immediately in sync with Rosenblueth, and became a friend and collaborator later when working on cybernetics. It should be noted that, despite the commonly made association between Wiener and the concept of feedback, and also with the same physiological concept of homeostasis, an association which spread after the war and which survives even today, it does not seem that these concepts were central in his relations with physiologists in the 1930s. It was only after World War II, in the context of cybernetics, that they took on a significant role, which in any case cannot be overstated. In his autobiography he explains what had previously been the common points of interest to him and Rosenblueth:

> Arturo and I hit it off well together from the very beginning, though to hit it off well with Arturo means not that one has no disagreements with him, but rather that one enjoys these disagreements. One point that we shared in common was an intense interest in scientific methodology; another, that we believed that the divisions between the sciences were convenient administrative lines for the apportionment of money and effort, which each working scientist should be willing to cross whenever his studies should appear to demand it. Science, we both felt, should be a collaborative effort (64g [56g], 171).

These interests were focused therefore on scientific methodology and the organization of research, and it was probably in that context that the fine Oregon metaphor was created for the first time, as it appeared in *Cybernetics*.[4] According to this, excessive specialization made science run the risk of giving different names to the same concepts, as had happened in Oregon when, Wiener explains:

> [It] was being invaded simultaneously by the United States settlers, the British, the Mexicans, and the Russians in an inextricable tangle of exploration, nomenclature, and laws. [...] There are fields of scientific work [...] which have been explored from the different sides of pure mathematics, statistics, electrical engineering, and neurophysiology; in which every single notion receives a separate name from each group, and in which important work has been triplicated or quadruplicated, while still other important work is delayed by the unavailability in one field of results that may have already become classical in the next field (61c [48f1], 2).

A thesis especially dear to Rosenblueth, according to *Cybernetics*, was that of the existence of

---

[4]On the Oregon metaphor, the observations of Antonio Lepschy are interesting: "with regard to Oregon, in the late eighteenth and early nineteenth centuries this name designated a large area, much of the extended US state. [...] During the period Wiener refers to, rights claims existed among the United States, Britain, the Russian Empire, and Mexico, which had taken up those already advanced by Spain. The region was still very little populated, but small groups of settlements from the interested powers were distributed patchily over it" (Lepschy 1998, 187–8). We can therefore imagine a partially colonized territory, where the same areas had different names and were subject to different cultures and jurisdictions. Regarding the history of the term "interdisciplinarity" and the use of geographic methaphors in this context, see Montagnini 2017b.

boundary regions of science which offer the richest opportunities to the qualified investigator. They are at the same time the most refractory to the accepted techniques of mass attack and the division of labor (61c [48f1], 2).

Both Wiener and Rosenblueth were convinced that

a proper exploration of these blank spaces on the map of science [would require a team of interdisciplinary scientists], each a specialist in his own field but each possessing a thoroughly sound and trained acquaintance with the fields of his neighbors (61c [48f1], 3).

In fact, the intellectual relationship between Wiener and Rosenblueth was closely related to methodological issues, focusing on interdisciplinarity, as described so far (Cf. Montagnini 2017b). They would always exclude a holistic-relational approach: the idea that the whole is greater than the sum of its parts, or the tendency to see each item as organically connected with the whole to which it belongs.

Although the holistic-relational approach was supposed to be a necessary part of a physiologist's background, it was little shared by Rosenblueth. For confirmation, see the book *Mind and brain. A philosophy of science* (Rosenblueth 1970). In contrast to a widespread interpretation, the same was true for Wiener. As we have already seen, Wiener remained rather cold towards the ideas of the main champion of this approach in the twentieth century, namely the Harvard physiologist, Lawrence Henderson, whom Wiener had met at Royce's seminar.

It is also important to pay attention to the differences between the approaches of Henderson and Cannon, Rosenblueth's teacher. Since the early twentieth century, both Cannon and Henderson had been interested in the mechanisms by which animal organisms were maintained in equilibrium with the external environment. In the 1930s, to refer to this kind of dynamic rather than static equilibrium, Cannon introduced the term "homeostasis", meaning

the coordinated physiological processes which maintain most of the steady states in the organism [...]—involving, as they may, the brain and nerves, the heart, lungs, kidneys and spleen, all working cooperatively [...]. The word does not imply something set and immobile, a stagnation. It means a condition—a condition which may vary, but which is relatively constant (Cannon 1932, 24).

It was a concept which the two men could share, and which summed up one of the firmer beliefs of physiology at least as far back as the French physiologist Claude Bernard, who considered it essential for the living body to keep its internal environment ["milieu interieur"] constant, while continuously exchanging matter and energy with the outside environment (Cf. Cooper 2008). During World War I, Cannon had studied the shock among troops, a physiological condition in which the body is unable to keep its balance because it cannot manage to control a series of processes. Cannon's thesis was that the mechanism that ensures the balance does not depend only on the action of substances on site, but also on the interaction with the nervous and endocrine systems. Thus, compared to Henderson, who was mainly concerned with adjustments that take place at the biochemical level (as typical in "buffer solutions"; see above § 1.8), Cannon focused on more complex types of adjustments, involving the nervous and hormonal systems, and in particular studied the sympathetic nervous system, which is a part of the autonomic nervous system. He demonstrated the role of that system in the control of various other regulation systems by carrying out a series of experiments on animals in which he removed all parts of the sympathetic system (Cf. Allen 1978).

As a result, the way Cannon and his heir Rosenblueth looked at regulation and body control was connected to the phenomena of communication by nerve impulses, and Rosenblueth moved towards the development of a neurological physiology, supported by a strong focus on experimentation, as clearly shown in the book Rosenblueth 1950. In fact, this approach would become, from 1941 onwards, the basis for his collaboration with Wiener. In Wiener's wartime research, the concept of communication gained increasing importance through his theory of prediction, as we shall see, and so did neurology, through his interest in automatic computing.

4.4 John von Neumann

Wiener had another important and very close partnership, this time with the mathematician John von Neumann (1903–1957), maybe since his stay in Göttingen in the mid-1920s. Actually, in 1955, von Neumann wrote about Wiener in a letter: "I have held him and his work in the highest esteem for almost three decades".[5]

Born in Budapest, von Neumann showed exceptional mathematical and logical skills at a very young age. He had studied in Budapest, Vienna, Zurich, and Berlin and in 1925 he defended a thesis on An axiomatization of set theory [von Neumann 1922]. In this attempt to make a general axiomatic theory for set theory, we already find the paradigm of his scientific style, supported by his powerful personal logical-formal skills, which he would apply in the future to all the issues he would be faced with. Along the way he would develop general axiomatic theories of proof, quantum mechanics, games, digital computers, and automata (Cf. Israel and Ana Gasca 2009).

From 1925, von Neumann began to build relations with Hilbert's school in Göttingen. In 1929 he was asked by Oswald Veblen to teach mathematical physics at the IAS in Princeton. Between 1930 and 1933, he spent time both in Berlin and Princeton, settling in the latter in 1933 when Hitler came to power.

An important chapter in his life, as well as for the history of ideas in general, was a congress on the epistemology of the exact sciences, held in September 1930 in Königsberg. This was organized by the Ernst Mach Society and the Society for the Empirical Sciences. They discussed the foundations of mathematics and compared three positions: logicism, represented by Carnap, a disciple of Frege; intuitionism, represented by Arend Heyting, Brouwer's disciple; and formalism, supported by von Neumann, Hilbert's disciple.

---

[5]Von Neumann to Irving R. Goldstein, 17 August 1955. (VNLC) General Correspondence, Box 7, "Wiener".

At the meeting von Neumann presented Hilbert's theory of meta-mathematics as a way to reconcile with intuitionism. During the conference, Gödel announced the result of his incompleteness theorem. von Neumann realized immediately that the theorem demonstrated the logical impossibility of comprehensively demonstrating the consistency of arithmetic through metamathematics, and that this therefore implied the defeat of his formalist thesis (Cf. Israel and Millan Gasca 2009, 30)

Consequently, von Neumann abandoned active work in logic forever, although he went on to make use of the tools the axiomatic method gave him, as shown by his work on *Mathematical foundations of quantum mechanics* (von Neumann 1932c), in *Theory of games and economic behavior* (von Neumann and Morgenstern 1944), and in the following work on automata.

The theoretical systematization of his work on quantum mechanics was finally completed in 1936 with Birkhoff, with whom he also worked on the ergodic theory.

In the latter, von Neumann came in touch with a theme that was very dear to Wiener through his previous interest in Gibbs's statistical mechanics, and it formed the common ground on which Wiener and von Neumann intensified their relationship, just before and after Wiener's stay in China.[6] As we will see, Wiener would quote von Neumann in his paper on *The role of the observer* (36g), written just before leaving China. Coming back to the USA, on 28 March 1937, on the occasion of a trip to Maryland for a conference at the Johns Hopkins University, von Neumann invited him and his wife "to stay several days"[7] as guests at Princeton. On 9 April 1937 von Neumann adds: "I am looking forward quite particularly to have another mathematical conversation with you".[8] These conversations almost certainly concerned statistical mechanics and the ergodic theorem, as inferred from Wiener's essay of 1938 on "The Homogeneous Chaos", where the name of von Neumann appears repeatedly.[9] In 1937 there was even talk of a possible teaching position for von Neumann as a visiting professor in China, and Wiener was quick to recommend him to his friend Lee, writing:

> Last week I was down at Princeton as the guest of Professor and Mrs. von Neumann [...] they seemed interested in seeing China sometime in the future. Now this is a marvelous opportunity. *Neumann is one of the two or three top mathematicians in the world*, is totally without national or race prejudice, and has an enormously great gift for inspiring younger men and getting them to do research [...]. The Neumanns are quite wealthy. [...] Therefore if he comes to China it will not be to save any money, although he would hope that the salary would meet expenses.[10]

---

[6]The works on von Neumann I mainly followed here are Israel and Millan Gasca (2009), Aspray (1990a) and Heims (1980).

[7]Von Neumann to Wiener, 28 March 1937 (WAMIT). Cit. by Heims (1984 [1980], 176).

[8]Von Neumann to Wiener, 9 April 1937 (WAMIT). Cit. ibid.

[9]In the paper von Neumann's name appears four times, and at note 11 Wiener says that some ideas were raised in discussion with von Neumann.

[10]Wiener to Lee, 4 May 1937 (WAMIT). Cit. by Heims (1984 [1980], 176). Italics added. Two more official letters followed, one addressed to the President of the University of Beijng and another to the Director of the MIT Department of Mathematics. Cf. Ibid., note 42, 465.

Von Neumann was also a guest of the Wiener's, who liked to refer to him as "Gentleman Johnny". (Heims 1984 [1980], 177) This mutual esteem never vanished, even when the war broke out and their scientific collaboration stopped as they found themselves on opposite ethical and political fronts. Such appreciation is also found in the Wiener's autobiography of 1956, which stresses von Neumann's tendency to connect pure mathematical aspects with those applied. Wiener remarks:

> The two mathematicians now or recently active in America who have adopted a similar point of view are—and I believe not by coincidence—*two of the greatest forces in modern mathematics*, namely, Hermann Weyl and John von Neumann (64g [56g], 192) Italics added].

Beyond their controversial collaboration during the Second World War, in Part III we will discuss at length a purely epistemological point regarding their peculiar scientific styles which we need to try to clarify. The book by Heims (1980) which I have widely recognized as being of pioneering importance, presents the two figures in parallel as men and scientists who are clearly in dispute over the dichotomy of "chaos and complexity" versus "logical rigor". Like all simplifications, this interpretation was useful to help understand their different scientific approaches. On the other hand you cannot blame him for the view expressed in Masani (1990), just as important a book about Wiener, which emphasizes the similarities between their scientific itineraries.

Wiener had been a disciple of Russell, between 1913 and 1915, when the *Principia* had just been published. He went to Cambridge (UK) carrying a strong imprint of early American pragmatism. In spite of having acquired Russell and Whitehead's exquisite logical tools, Wiener never shared the logicist project of founding mathematics on logic, reaching a moderate fallibilism more akin to that of Bridgman, which he also transposed to logic and mathematics. Entering MIT in 1919, he abandoned logic in the strict sense. Curiously, in 1914, when Wiener attended Hilbert's lectures in Göttingen, the latter had also temporarily abandoned his interest in logic, after writing his *Grundlagen der Geometrie* in 1899. Not by chance, as Wiener confessed to Russell in 1914 (see above § 2.3), Wiener had not found any real interest for logic among the mathematicians. Hilbert seems to have resumed his interest in logic only in 1917 at the conference "Axiomatic thought" (1996 [1918]), where he began to work on his ideas for a metamathematics (Cf. Bottazzini 2016).

Von Neumann—who was 9 years younger than Wiener—grew up in Hungary, and then completed his studies in Switzerland and Germany. Von Neumann grew up in a time and an environment where there was a great deal of confidence in the foundations of mathematics. And as happened to many of his colleagues, the consequences of Gödel's incompleteness theorems left him stunned forever. It was thus consistent to leave logic, even though his precision of thought and love for logical rigor remained intact.

Considering von Neumann, as Heims does in his book, a follower of logical positivism is incorrect. More than once in his letters, he expressed uncomfortable feelings about some of the leading figures of the movement (Cf. e.g. regarding Carnap, von Neumann 2005, 203–204).

Wiener and von Neumann had grown up and trained differently. They felt mutual esteem and tended to integrate. Von Neumann appreciated, maybe a little envied, Wiener's fertile mind; Wiener had on the other hand held von Neumann's logical abilities in great esteem, and always felt comfortable after submitting his work to him, because von Neumann could reassure him about its consistency.

4.5   "The Role of the Observer"

During his teaching period in China (in the academic year 1935–36), Wiener wrote an essay on *The role of the observer* (36g) which, despite its brevity, is the best expression of his philosophy at that time, not only for its wealth of content, but also for its subtlety of argument. In it Wiener compares the ideas of the scholars he was closer to in those years. The central theme of the essay is the unavoidability of the observer's influence in the cognitive process, in the light of the recent theorem due to Gödel and the Heisenberg uncertainty principle, which was strengthened by the results of the recent axiomatization of quantum mechanics by von Neumann. These outcomes were greeted by Wiener as the final confirmation of his former fallibilism, even in language, his discourse being linked to that of *Relativism* (14d) and *Is mathematical certainty absolute?* (15b).

The beginning of the article deals with the question of the observer in situations drawn from art and the non-physical sciences, especially the life sciences, where there are situations of inevitable dependence on the observer, regardless of quantum mechanics. He discusses the case of theatrical scenography to represent people's social lives in front of an audience of "spectators", mentioning the various tacit agreements, such as rooms with three walls of which two are oblique, tables where the actors only ever sit on three sides, and whispers which are actually shouted. All this is about tacit rules that the spectator is not even aware of, if there are no situations in which they interfere heavily with the message the playwright wants to convey. Something similar happens in literature, where the narrator is nevertheless always a character, although generally invisible, who comes to the foreground in cases such as the psychological novel (Cf. 36g, 307–309).

The observer's intervention—Wiener argues—may be more or less invasive. In psychoanalysis, when one has to bring to light an unconscious motive, one must take into account the fact that "a subconscious motive brought to light is no longer a subconscious motive" (36g, 309). In surgery, in an exploratory laparotomy, surgeons should be aware that what they see is not what they would see without surgery. The anatomy of a corpse does not allow us to know the "living" body. The dyes used in microscopy can modify cell structure (Cf. 36g 309–310).

The Wiener who had long conversations with Haldane and who began regularly to attend Rosenblueth's "neurological dinners" notes that those studying the living world are perhaps the most aware among scientists of these difficulties. But until recently, things had been very different for the physicist, who had always neglected the impulse (force × time = mass × speed) given by light to objects under observation, an impulse which in ordinary mechanical phenomena "is so slight that for long it lay far below the threshold of accuracy of our measurements" (36g, 310). And when the reality of this impulse was understood, it was hoped to reduce the effect of distortion by lowering the light intensity. However, this approach proved to be impossible, because light is made of quanta having not only a well-defined momentum, but also an associated wavelength, which is inversely proportional to it, so that if we want to have a lower momentum to accurately measure the amount of motion, we have to use a quantum of light with a longer wavelength, which makes it difficult to understand where the observed object is. This raises the dilemma of the uncertainty principle, where the accuracy with which the momentum of a particle can be detected is inversely proportional to the accuracy of the position measurement (Cf. 36g, 310–1).

One might still have assumed that what cannot be observed could at least be calculated theoretically, combining the results of two non-simultaneous measurements, one on the momentum and one on the position. But at this point the verdict of quantum mechanics comes into play, according to the final theoretical arrangement which was given by von Neumann:

> von Neumann has shown that this is not the case, and that the indeterminacy of the world is genuine and fundamental. There are no clean-cut laws of motion which enable us to predict the momentum and position of the world at future times in any precise way in terms of any observable data whatever at the present time. In other words, while it is possible to give an account of the world in terms of our observations which themselves disturb our world, this account has only statistical validity, and cannot be brought closer to precision by any chain of observations (36g, 311).

Thus it happens that even physics, the most exact of sciences, "has had to have a thorough logical housecleaning" (36g, 311). This work of logical rearrangement, in Wiener's opinion, would consist in a change of methodology:

> About any proposition of physics, we must ask whether it enables us to predict the result of an actual or possible experiment. If it does, it stands or falls with this experiment; if not, it has no meaning whatever. Physics is merely a coherent way of describing the readings of physical instruments (36g, 311).

Here his discussion involves an extremely delicate step. Wiener seems to put together two by now classic concepts in the epistemological discourse. The first seems to consist in the principle of verification of the logical positivists: in physics, when there is no way to check a proposition by an experimental observation, at least hypothetically, that proposition is to be considered meaningless. The second is typical of operationalism: physics is simply a consistent way of describing the readings of physical instruments. Wiener had had the opportunity to get to know the ideas of the logical positivists, having been in contact with people like the mathematician Karl Menger and the physicist and philosopher Philipp Frank in the early 1930s. But he never went beyond the already typical belief of his pragmatist teachers, and of Bridgman, about the critical need for constant monitoring of the concrete operational content of the propositions of science. In his opinion, this criticism was the normal attitude of a properly aware modern scientist. But logical positivism transformed this sacrosanct need into a dogmatism; it is the criticism that, in a certain sense, Wiener would assert in a 1949 review of a book by Philipp Frank:

> As a general philosophical attitude it represents the accepted point of view of the modern scientist, but it goes beyond his simple attitude in forming a self-conscious body of doctrine and an organized philosophical school (49h).

This movement, in Wiener's opinion, displayed an "excess baggage of organization and propaganda", possibly inherited from "the old frowns of the Habsburg government on politics which tended to make him, like his German colleague, sublimate a desire for political activity in the formation of scientific parties" (49h, 3).

Discussing the case of Euler, *The role of the observer* shows very clearly how the razor of the logical positivists could be of excessive subtlety, with serious consequences for research. Euler—says Wiener—had made extensive use of the notion of mathematical series, with little regard for the kind of proof he adopted. With a clear reference to logical positivism, Wiener remarks that someone could therefore argue that Euler's "work, not being mathematically rigorous, consists of false propositions, or of no propositions at all" (36g, 312). However, this would be "a piece of pedantry" (36g, 313). Euler was a pioneer and "there is too much arduous pioneer work on the frontiers of science to permit all the niceties of civilized intercourse" (36g, 313). Here the principle of verification of the logical positivists is radically questioned: Euler, despite his formal shortcomings, did a huge amount of practical work in mathematics and this cannot be ignored in any way. Wiener recalls that the last two generations of mathematicians had tried to improve the logical rigor of mathematics, and at one point they believed they could assume that they had discovered "the real timeless universals of elementary mathematics" (36g, 314). In reality, the problematic aspects persisted: Zermelo's doubts about the axiom of multiplication, the paradoxes that Russell was trying to avoid with his theory of types, and Brouwer's condemnation of the law of the excluded middle. All these facts eventually testified against that conviction, which collapsed with Kurt Gödel:

> Gödel shows that in any mathematical system there are propositions, the truth or falsity of which can only be determined by a reference outside the system. That is, with the best will in the world to determine a mathematical system categorically, in such a way that its propositions can only have one definite meaning, we can only put off the question of its categoricity to that of some larger, more inclusive system. At some stage, we always run back to that which has no adequate mathematical definition. In every actual mathematical system, there are always details which (as far as we know) may be defined in two or more ways, and an imperfect decision between these ways may precipitate us into new logical paradoxes such as those of Russell. No matter how far we go, our logic will always appear Eulerian to some possible scholar of the future (36g, 313).

This is neither more nor less than a re-reading of Gödel's theorem through the opinions he had expressed 20 years earlier in *Relativism* and *Is mathematical certainty absolute?* Wiener's methodological proposal should not be understood in a logical positivist sense. In reality, he wishes to recall the need for measuring and reporting the experimental or actual demonstration procedures that are subject to the propositions of science. Therefore biology ought to be rewritten as a guide to dissections, biological experiments, observations of animal behavior; psychology as a reasoned history of insights and behavioral observations. The same would happen for mathematics, which

> should be an account of theorems and their recognized criteria of truth or falsity, which will allow us to place new theorems in this respect. Whatever view we have of the "realities" underlying our introspections and experiments and mathematical truths is quite secondary: any proposition which cannot be translated into a statement concerning the observable is nugatory (36g, 311–2).

At the end of this path, Wiener went back to the ideas he had supported 20 years earlier on the impossibility of certain knowledge in mathematics, even rereading Haldane's conjecture: "Thus even in mathematics, precise Platonic Ideas and precise Platonic Truth represent rather asymptotic ideals than anything directly accessible and significant to us" (36g, 314). Now what he is saying has gained strength and he even adds something very specific about the relationship between logic and psychology, through a reconsideration of the concept of universal versus time, which was influenced by the previous reflections on Leibniz and Haldane.

What is logic when all is said and done? Wiener asks himself. It consists of propositions and concepts considered *sub specie aeternitatis*; it possesses a timelessness, or better perhaps, something everlasting. This is the Platonic view, which ignores the roles of both time and the observer:

> The Platonist believes in a world of essence, of cleanly defined Ideas and cleanly defined propositions concerning these Ideas, into which we may enter as spectators, but never as participants. They are out of time, and time is irrelevant to them (36g, 307).

However, Wiener still asks himself, do those "Platonic ideas" exist in the actual reality in which science operates? After all, a logic with any "real interest" will have to deal with this or that universal concept taken from actual science, and what can be discovered in this examination is that, unlike Plato's, the universals found in science always have varying degrees of vagueness and changeability: in concrete science, universals are "evolution", "mathematical function", etc. These are concepts that do not exist *sub specie aeternitatis*, but which have a history: the idea of "evolution" changes over time: it does not mean the same thing for Charles Darwin, for his grandfather Erasmus Darwin, or for us (Cf. 36g, 312).

Logic does not, therefore, allow us to grasp the actual mode of operation of thought. But then neither does psychology, intended as mere induction. We must think of a logic that is midway between logic and psychology, considered as two asymptotic limits. It will be a logic that deals with universals with vague and historicized edges, where the propositions considered are not true with apodictic certainty, but only with a greater or lesser probability:

> Any really useful logic must concern itself with Ideas with a fringe of vagueness and a Truth that is a matter of degree. A logic which ignores the actual history of ideas and the limitations of human faculty is a logic *in vacuo*, and is useless. The distinction between logic, psychology, and epistemology cannot be made absolute (36g, 314).

Venturing into the traditional dispute between rationalism and empiricism, Wiener considers these two philosophies as

> accounts of asymptotic modes of thought toward which we tend in one mood or another, rather than adequate epistemologies of our normal experience (36g, 315). Science is the explanation of process. It is neither possible under a rationalism, which does not recognize the reality of process, nor under an empiricism, which does not recognize the reality of explanation (36g, 316).

After reaching a really interesting climax, Wiener's discussion concludes rapidly with a mention of Kant, regarding certain issues to do with historiography, a science where the concrete gains a very special and important position. He finally arrives at four maxims:

(1) Science is an explanation of process in time.
(2) The act of explanation and of the development of concepts actually consumes a certain minimum of time, and involves a reference to a certain minimum of time.
(3) No concept nor theory is well-defined except in so far as it has been through an actual process of definition.
(4) The time consumed in attaining to a certain degree of definition, as well as the time referred to in a definition of this degree of preciseness, is of an order of magnitude roughly proportional to the degree of precision attained (36g, 318).

The image of a science intended not as an abstract Platonic reality, but as something that exists in time, and of an equally temporalized logic, dealing with concepts which, in order to establish and unfold themselves analytically, require time is evocative indeed. Wiener admits the similarities that exist between his discussion and the ideas of Haldane in *Quantum Mechanics as a Basis for Philosophy*, but in the latter case we have a "conjecture", "a powerful piece of imaginative work, but it is imaginative", that extends quantum mechanics to the physiology of the brain. As such, Haldane's conjecture needs verification. But for his theses, he claims a strictly epistemological nature which does not depend on the fate of Haldane's theory (Cf. 36g, 318–9).

As we will see later on, both Wiener and von Neumann would take up this idea of a temporalized logic in the framework of the theory of cybernetics. Item (4), which is the consideration that the degree of precision of an idea is proportional to the time spent on its elaboration, shows how much Wiener was still tied to a vision of information based on the energy model.

4.6 "The Shattered Nerves of Europe"

From the crisis of 1929 to the advent of Nazism and up until the outbreak of World War II, the 1930s were years full of anxieties, and Wiener increasingly felt the need to raise his eyes from his desk and look at the outside world. In a 1939 book review he says:

> In this epoch of violent change and violent threats we must at all costs read the portents in the heavens and interpret the records of the past. Industrial and scientific development are doing to us something which we do not fully understand, but the understanding of which is vital to our comfort, our prosperity, our happiness, and our continued existence. Thus it is not only natural but most devoutly to be wished that men should devote themselves to the interpretation of the history of science, the history of invention, and the social role these have played in past ages (39f, 115).

During his visit to Europe in 1931–32, Wiener was invited to give lectures in Hamburg by the mathematician Wilhelm Blaschke, in Vienna by the philosopher and mathematician Karl Menger, of whom Gödel had been an assistant, and in the German University in Prague by the physicist and philosopher Philipp Frank (Cf. 64g [56g], 157). Soon afterwards, Blaschke, the author of articles ridiculing American mathematics, would become a Nazi. Menger and Frank had to take refuge in the United States, where Menger eventually found a position in the Illinois Institute of Technology and Frank in Harvard (Cf. 64g [56g], 157). Arriving in Prague, Wiener went to see President Masaryk, an old friend of his father, who received him on a private visit, not failing to express "great worry at the advances of the Nazis" and "very little hope for the future of Europe" (64g [56g], 158-9).

A number of valuable reports by Wiener were brought together in an MIT magazine, where we can find a description of the economic and social crisis of those years. In particular it is worth quoting a passage from the chapter *The shattered nerves of Europe*, written, it should be noted, 2 years before the advent of Nazism. Wiener writes:

> Now as to my impressions: first, as to the general situation in Germany and Austria. [...] The reaction is one of depression, of hopelessness, almost of resignation. Everyone awaits a crash, no one knows from what direction it will come, and no one of my friends has the least expectation that any coup d'état, whether from the right or from the left, will appreciably better matters. I have not been able to discern much enthusiasm for Bolshevism, but what is growing on all sides, with few exceptions, is a consciousness that the present economic order has played itself out. [...] The misery among the German lower classes is intense, and the situation threatens to become as bad in Austria. As to the intellectuals, the strain of a continually uncertain future is telling hard on them. It is impossible to have peace of mind in the Germany of today—perhaps I should say in the world of today, although the older English universities furnish a fair approximation. In Germany there is a continual suppression of assistantships, cutting of salaries, and no man knows one day what is in store for him on the next. What is even worse than personal discomfort and suffering is the uncertainty of the whole order of things, so that no one is sure whether the whole European culture of which he is a part and for which he is working may not be crossed out with a few strokes of a dictator's pen, or effaced by the bloodshed of a revolution. [...] If by some miracle it should become possible to restore economic and

social prosperity to the world, it would still be a matter of 20 or 30 years to restore
confidence and peace to the shattered nerves of Europe (31c3, 218).

The passage can help us to understand the alarming tones Wiener would use in
the period immediately after the war, faced with the evidence in his mind of the
possibility of a generalized replacement of the traditional factory by automated
factories, utilizing "docile robots" instead of workers: these were the fears of a man
who had lived in full awareness of the anxieties of the 1930s.

4.7   Thinking Technology: Matter and Energy

During those years Wiener came to have a keen interest in the history of technology
and its social impact. These reflections are somehow in tune with the English
scientific literature, typical of professional scientists, led generally in Cambridge,
engaged in a discussion about the social consequences of science and technology,
without disdaining futurology and narrative. He had a particular sympathy for the
left wing of the movement, represented by J. B. S. Haldane, John D. Bernal, Hyman
Levy, Lancelot Hogben, the author of *Science for the citizen* (cf. 38h), William H.
George, the author of *The Scientist in action*, "those unknighted—as Wiener writes
—hard-brained sons of toil who attempt to convey to their nonscientific comrades
the ethics and the methods and the purpose of the working scientist" (39g). These
men are representatives of a sort of "leftist science", and somehow, of a peculiar
"scientific socialism" (Cf. Nacci 1991; Werskey 1988). Several of them, such as
Haldane and Bernal, were Marxist. But Wiener would never be Marxist, remaining
faithful to the spirit of American radicalism.

Anyone who knows Wiener's work after World War II may be surprised, but in
Wiener's reflections of the 1930s there is still no trace of any intuition about the
"cybernetic revolution" (Cf. Montagnini 2001–2002b). The idea, that is, of the
advent of a new society that focuses on information. In this period, in contrast, we
still find the notion of energy at the center of his preoccupations, and the technologies which most interest him are those that do not concern communication, but
electrical and thermal machines, and in particular aviation. One of his most significant writings in this sense is the article *Putting matter to work. The search for
cheaper power* (33f), mingling the typically Wienerian issues, elaborated partly
through his contacts with MIT engineers, and partly through insights coming from
Haldane.

Wiener believed that in order to understand technology, a grasp of its evolution
is indispensable, and as he wrote in 1938, one should examine "the way in which
engineering development has been limited by the materials at its disposal, [...] the
way in which it has invented its own materials, [...] the way in which tools are the
products of tools" (38h, 68). *Putting matter to work* considers technological design
and materials as two developments of engineering that are both essential and
complementary to each other. In fact, invention is not just the result of the scheming

genius, but has to deal with the substantive conditions that hinder the realization of the project. Mechanical engineers in the nineteenth and twentieth centuries, he explains, were nothing but the heirs of the makers of precision instruments and watches in the eighteenth century:

> The desire for accurate timepieces and instruments of navigation perfected the art of machining metals, and in particular cast metals, to the point where the accurate cylinders, pistons, shafts, and gears required by Watt's steam engine could for the first time be produced commercially (33f, 47).

He returns to the subject in a 1939 review of a book on the history of technology, (cf. 39f) where, according to Wiener, the author wrongly ridiculed the awkwardness of the attempts of artisans in the preliminary stages of planning. For example, he was surprised about Watt's machine, as Wiener comments:

> Burlingame [the author] is much struck by the inability of the ironmasters of Watt's time to bore a true steam cylinder. Here he has hit upon the right clue, in that he emphasizes that the machine-tool technique of the late 18th and 19th centuries owes its genesis to the craft of the watchmakers and, although he does not mention this explicitly, of the instrument makers of the preceding generation. But he is not awake to the time and hard work of experiment necessary before these small-scale methods could be applied to the large-scale needs of heavy industry. In this connection he curiously misses a point vital to the history of invention in the Industrial Revolution: the tremendous effect on the clockmakers and instrument makers of the navigator's need for a method of determining longitude at sea (Cf. 39f, 115–6).

Compared to *Putting matter to work*, in this review we can add that the really critical time for the invention of the steam engine was represented, and it should by no means be taken for granted, by the transition from the small-scale processing methods of precision mechanics to the large-scale ones required by heavy industries. In *Cybernetics*, Wiener would retain some elements of those considerations. We find in particular the link between the techniques used for clocks and the requirements of sea trade, through the importance of determining longitudes. However, in that case, the transition from mechanical precision to heavy industry would be seen as an epochal leap, from the "age of clocks" to the "age of the steam engine", according to a net periodization of which there is no trace in the writings of the 1930s, where on the contrary the opposite need prevails, namely to mark the continuity between pre-industrial craftsmanship and the engineering of the industrial age; an idea that ultimately reflects his desire to show how theory and practice are inextricably linked.

Putting matter to work was influenced by the reading of *Daedalus, or, Science and the Future* by Haldane (1924), a paper that discussed the problems that the science of the future would probably have to deal with. Among the aspects listed by Haldane, what strikes Wiener the most is the issue of exhaustion of coal and oil, and hence, the practice of energy conservation. According to Haldane, before civilization could collapse, coal and oil would be replaced by hydropower, which, however, is subject to seasonal fluctuations, but also solar energy and wind power, which are even more erratic energy sources. The main problem, therefore, was the storage of surplus production, which according to Haldane could be stored by using the excess electricity to generate oxygen and hydrogen from water through electrolysis, then stocking these products in sturdy containers. Haldane also hints at the fact that, in the case of air transport, the use of hydrogen would constitute an advantage because of its low weight (Cf. Haldane 1924).

In *Putting matter to work*, interweaving his own ideas on the evolution of technology with Haldane's own considerations, Wiener asks himself about the search for cheap energy and the relationships between energy and mass. He suggests an exhaustive classification of the possible ways to store energy: gravitational, e.g., weights in clocks, dams; kinetic, e.g., flywheels; elastic, e.g., springs, rubber bands, compressed air; magnetic (inductors); dielectric, e.g., capacitors; thermal (for storage boilers); chemical with electrical usage, e.g., batteries; chemical with thermal usage, e.g., fuel and explosives; atomic, not used in practice (Cf. 33f, 48). After a careful examination of the various solutions, he concludes:

> There thus seems to be comparatively little hope for the improvement of the storage of energy by what may be called classical methods, at least until materials have been discovered of strength hitherto undreamed of. In the absence of such materials, our one hope probably lies in atomic energy (33f, 49).

Wiener did not disdain the chemical methods, but he did not dwell on them. Instead he focused on electrical power and thermal mechanical techniques. At that time, as today, electricity storage represents one of the main problems of power electrical engineering. Batteries have a limited capacity to store electrical power, so the only solution lies in distributing it immediately for use. It followed, according to Wiener, that energy sources such as wind and sea waves were inadequate, failing to ensure stable delivery systems, at least until the problem of electricity storage could be solved (Cf. 33f, 47).

As regards the techniques developed for heat engines, Wiener was particularly attracted by aeronautics. Here, too, he tracked down the relationship between design and practical possibilities that determine the success of an invention: how the steam engine was in fact made possible by the availability of a cheap technology to carefully melt and shape mechanical parts, so that planes were made possible by the availability of light engines (Cf. also Buchanan 1996). Many flying machines invented in the nineteenth century would have flown if light engines had been available. And it was precisely the introduction of a light internal-combustion engine, spread by the automotive industry, that satisfied the conditions for the advent of aviation.

And here, Wiener predicted a bright future. He saw it as being connected to the generalization of transcontinental air transport. However, a new and urgent question thus emerged, concerning the relationship between energy and mass transported, according to which a plane could not carry more than its pilot and the fuel. This was clear in the first transatlantic flights (Charles A. Lindbergh's enterprise was made in 1927).[11] In order to allow the transport of passengers and goods, Wiener proposed

---

[11]The first project for an aeroplane goes back to 1903 and the brothers Wilbur and Orville Wright, and it was implemented in 1905. The flight across the Atlantic from New York to Paris on board the legendary "Spirit of Saint Louis" achieved by Charles A. Lindbergh dates to 1927.

as an interim solution the creation of an aircraft carrier, anchored in the middle of the oceans, to allow stopovers.

Putting matter to work ends with a very precocious reflection on nuclear energy. During his stay in Cambridge, the experimental physics laboratory of the university, the Cavendish Laboratory, was the location of an experiment carried out in April 1932 by Cockcroft and Walton, which was celebrated as the first example of atomic fission. Wiener was able to observe their experimental setup, and was impressed by the scarce resources and the great ingenuity poured in the experiment (Cf. 64g [56g], 154–5). Cockcroft and Walton were able to obtain the fission of lithium atoms by means of a stream of ionized hydrogen atoms that were positively accelerated by a high voltage field; from this lithium fission, helium atoms resulted. It was found, furthermore, that the sum of the mass of the starting atoms (Li + hydrogen) was slightly less than the sum of the masses of the helium atoms resulting from the fission, and this difference was explained by the transformation of mass into energy according to the well-known relativistic equation $E = m \times c^2$. The experiment also showed, therefore, that it was possible to obtain energy from matter. The result caused a great clamour in the press regarding the possibility of using nuclear power for civilian purposes, although Rutherford, the director of the Cavendish Laboratory, had expressed doubts about it. Wiener commented in *Putting matter to work*:

> A pound of matter has 11,300,000,000 kilowatt hours of energy in it, while a pound of gasoline as used in an internal combustion engine has at most about 1 kWh. The ratio is so enormous that if even a modest part, an almost microscopic part, of the energy in matter can he made available for mechanical use, all other forms of storage will be put far in the shade. It is true that Lord Rutherford has thrown some cold water on our hopes in this direction, but the matter is still *sub judice*. What the hope of achieving this in practice may be cannot be said in advance. Mr. Robert J. Van de Graaff, of the Institute's Department of Physics, with whom I have talked on this subject, is quite reasonably optimistic as to the ultimate possibility of realizing, mechanically at least, a large part of the energy which appears as a change of mass of the nucleus when one element is synthesized into others, as in the work of Cockcroft and Walton. This energy is only about one per cent of the total mass in the formation of helium from hydrogen, but is still far in excess of the energy available in the existing fuels. Mr. Van de Graaff has a fairly definite idea of what type of apparatus offers most promise of the mechanical utilization of this energy (33f, 49–50).

It was of course, Wiener added, only "pious hope", but it was necessary to consider whether sooner or later the goal of obtaining sources of cheap energy and a high amount of energy per unit of transported substance could ever be reached. Atomic batteries purchased in a grocery might perhaps come out of it, or heavy systems able to release the enough energy to transport themselves and much more. In any case a generalization of air transport over any distance would spring from it. Moreover, there would no longer be any need for power grids and electrified railways.

In *Daedalus*, Haldane had talked about the possibility of obtaining aluminum from clay by electrolytic methods and food from the synthesis of organic molecules; Wiener ends up by saying that cheap energy would also subvert mining activities as much as the food industry.

What aroused the most concern in him, however, was the fact that the tendency to replace workers with machines as sources of energy, especially in heavy industry, would be amplified. The danger for employment was seen as coming from the replacement of workers as sources of energy, according to a point of view that *The human use of human beings* (1950) would stigmatize, noting that the "industrial revolution up to the present has displaced man and the beast as a source of power, without making any great impression on other human functions" (50j, 180).

It is useful to stress that in *Putting matter to work* the danger does not come from technological research itself, but from the imbalances that can follow too rapid innovations. We read in the article:

Energy would become a drug on the market, and the present tendency of replacing the laborer by the machine as a source of energy, as distinguished from the laborer as a source of judgment, would be intensified a thousand fold. Particularly would this be the case in the heavy industries. With an ample supply of power, the metallurgical industries would introduce electrolytic methods to a far greater degree than at the present time and there would be a great possibility of the use of low-grade ores and other raw materials which are now unused because of the great energy expense in handling so large a bulk. I refer, for example, to the possibility of obtaining aluminum from common clay.

Any such great cheapening of energy, therefore, could immediately result in a corresponding cheapening of materials. As far as engineering and industry are concerned, the disturbances introduced would be analogous to that overproduction which has so much to do with the present era of depression, but they would be on an incomparably greater scale (33f, 70).

This brings to mind a passage of the report from Europe in crisis in which Wiener cited Haldane's opinion in the face of the imbalances due to depression:

[Haldane] said that if anyone could persuade him that the equations governing the capitalistic system had a stable solution, containing no exponentials in the time with exponents with positive real part, he would vote conservative from then on, but that since he was convinced that there is nothing to prevent economic cycles from becoming progressively more extreme, he was forced to be a socialist. In any case, there is a widespread feeling that only in a world-wide, planned, economic system can there be any security, and that the era of free competition is dead, and only awaits a decent funeral (31c3, 218).

The real danger for the capitalist society would come, therefore, from the excessive instability of the economic system and its cycles. Haldane had begun to look with increasing benevolence to the solution offered by Soviet planning. Wiener, agreeing with the diagnosis, seems more inclined to Roosevelt's cure according to the principles put forward by Keynes:

Our present government is convinced, and I believe rightly, that the economic disturbances of the day can only be ameliorated by organized industrial planning. This is because the equilibrium on which the old-time economics was postulated, which is due to the statistical interplay of the businesses of a large number of small competitors—so small that no single competitor dominates the situation—has been destroyed by modern invention. Cheap energy would destroy the possibility of such an equilibrium even more completely and would precipitate the need of a central control. With this control, cheap energy might easily lead to a period of prosperity such as has never yet been seen. Without this control, cheap energy must be completely disastrous (33f, 70, 72).

It is a thesis that probably did not change after the war, when faced with the development of cybernetics; Wiener would then pin all his hopes on democratic control on the part of well-informed citizens. However, during the 1930s, Wiener believed that the main factor to watch, as potentially a harbinger of unemployment from technology, was energy.

His special interest in aviation is reiterated in *The Decline of cookbook engineering*, a brief account of the congress on applied mechanics in MIT and Harvard in September 1938 (Den Hartog and Peters 1939). Wiener writes:

> The airplane is a structure which very nearly fails to work, and only by supreme intelligence in design can it be made to work even in an approximately satisfactory way. The result is that good old-fashioned cookbook engineering, which went by rule of thumb and covered up all incompetencies of design by a bang-up big factor of safety, has gone into the discard. However, the new investigations arising out of aeronautics are quite as useful in old fields of work. Economically speaking, aeronautical research has paid the overhead for a general improvement of engineering research all along the line (38g, 23).

Aviation then was seen as a leading sector for the optimization requirements it imposed and the interests that it attracted into its orbit. Giving us the opportunity to measure the extent to which his reflection on the relation between pure and applied mathematics had become more complex as compared to 1929, Wiener noted that the union of pure and applied mathematics in mechanics dated back to the time of Archimedes, and observed that over the centuries there had been a change in the focus of interest: in the eighteenth century it was about clocks, geodesy and navigation techniques, and in the nineteenth century it was about the study of the body roll and pitch of ships, the design of bridges, and the development of devices for heat engines. But finally, these centers of interest had become something else again:

> These subjects have not died out as matters of interest to a congress of the present day, but they have been pushed into the background by topics more than half of which arise in connection with aerial navigation. Turbulence in wind tunnels, the examination of materials by photo-elastic methods, refined theoretical studies in elasticity, the determination of the lift and drag coefficients of airfoil sections (38g, 23).

The conference was attended by aircraft designers, fluid dynamics experts like Ludwig Prandtl (1875–1953) and his former co-worker Theodor von Kármán (1881–1963), mathematicians involved in research on turbulence like Geoffrey Ingram Taylor, and of course Wiener himself, along with meteorologists and civil and mechanical engineers.[12]

---

[12]According to Gianni Battimelli, at this congress of applied mathematics in 1938, for the first time, an entire session was devoted to discussion of the mathematical theory of turbulence developed by Taylor, in which the generalized harmonic analysis played a decisive role, and so in general did the use of stochastic mathematics introduced with the Wiener process. See Battimelli (2002). As Battimelli (1986) has shown, Taylor felt indebted to Wiener, while Wiener felt indebted to Taylor.

Looking more closely, Wiener's interest in aviation led him toward the war projects in which he would be involved soon afterwards: on the one hand, the design of high speed digital electronic computers for the solution of partial differential equations, to be applied in particular to the numerical solution of problems of fluid dynamics; (Cf. 85b) on the other, the work on automatic anti-aircraft systems able to predict the trajectories of aircraft to be shot down. It was through these studies that he came to his idea of cybernetics and to his conception of the advent of a new technological age whose main interest lies in information.

Part III
(1940–1945)

> *We have contributed to the initiation of a new science which, as I have said, embraces technical develop-ments with great possibilities for good and for evil. We can only hand it over into the world that exists about us, and this is the world of Belsen and Hiroshima.*
> Norbert Wiener, *Cybernetics* (61c [48f1], 28)

Officially the United States entered World War II only at the end of 1941, with the declaration of war on Japan on 8 December of that year, the day after the attack on Pearl Harbor. However, US science had been down in the trenches at least since 1940, when by a decree of June 27, President Roosevelt had established the National Defense Research Committee (NDRC), to integrate and coordinate the research of the Navy and the Army with those conducted for military purposes by civilians. At the head of the NDRC had been placed Vannevar Bush, the engineer with whom Wiener had worked closely for years, and MIT was one of the major institutions where the NDRC research would take place.

The NDRC was structured in 4 main Divisions. In particular, Wiener worked for Division D (Detection, Control and Instruments), under the Section D-2 (Controls), headed by the physicist Warren Weaver (1894–1978), who since 1932 was the Director of the Division of Natural Sciences of Rockefeller Foundation. Wiener worked also under Section D-1 (Detection), which carried out research on Radar, mainly at the Radiation Laboratory, which was established at MIT, and directed by the physicist and businessman Alfred Loomis. On 28 June 1941 the NDRC was incorporated into the Office of Scientific Research and Development (OSRD), of which Vannevar Bush became president. 19 Divisions of NDRC were established. Section D-1 became Division 14 (Radar), coinciding with the MIT Radiation Lab and headed by Loomis. Section D-2 became Division 7, headed by Harold Hazen and continuing to deal with fire control. On 9 December 1942, the Applied Mathematics Panel (AMP) was created to coordinate mathematicians serving the various divisions. It was chaired by Warren Weaver, who retained a position in Division 7. Hence Wiener was assigned to AMP.

In 1940 Wiener was a 46-year-old mathematician. He was at the height of his career, backed by a number of very significant successes. The war would see him involved in strategically important military projects at MIT, in which he would engage with passion, spending whole nights doing calculations, and not disdaining the use of amphetamines to combat fatigue (Cf. 64g [56g], 249). He probably continued in this way for the duration of the conflict, even though the period between 1943 and 1944 is surrounded by a rather dense fog. His research activity catalyzed most of his previous intellectual and personal career: his peculiar scientific style, his previous mathematical results, his research on automatic computation, his relationships with engineers, physicists, and physiologists, his reflections on science and technology and their impact on society, his whole past, in short, would pass through the bottleneck of the war, conspiring to compose the pieces of a mosaic that would gradually assume a definite form and, in 1947, give the title to a book that would be released the following year—and also a name: "cybernetics", the "science of control and communication in the animal and the machine". We shall deal with "cybernetics" in this and the following chapters, and also discuss a definition that remains highly problematic. The subject that Wiener would ultimately decide to call "cybernetics" began to take shape in his mind in 1941 or 1942 and underwent rapid metamorphosis, but already stabilized in 1944–45. Although he would never arrive at a precise definition of this nebulous concept, either at that time or later, he become precociously aware that behind it lay a scientific and technological revolution of almighty proportions, involving deep consequences for the future of society. Very early on he began to predict the advent of a new era based on control and communication in the animal and the machine, of a cybernetics era par excellence. The act which brought the war to an end, the nuclear bombing of Japan, served to complete the panorama of the cybernetic age which was being drawn up in Wiener's mind. This event shattered his belief that the freedom of research should not be subject to constraints and above all connoted all his postwar reflection with tragic shades.

Chapter 5
Wiener and the Computer. Act One

In the 1950s and 60s, the word "cybernetics" was not only associated with curious automated turtles able to search by themselves for a socket to stock up the power they needed. It also evoked the early giant "electronic brains". Indeed even the "cyber" prefix, resurrected at the end of the twentieth century with the complicity of science fiction and journalism, remains closely related to computer science. Yet, when one thinks about Wiener and his cybernetics, the relationship with the computer becomes evanescent, whence histories of computing did not recognize him as having any role in the gestation of the digital electronic computer. There is no word about that, e.g., in good accounts of the history of computer science such as Akera and Nebeker (2002), Agar (2001), Ceruzzi (1998), or Williams (1997). A very rare exception is Breton (1987). I am not interested in opening a trial over priority that even Wiener would never have wished to see. However, in order to get a better understanding of his thinking, one cannot avoid trying to establish as clearly as possible, within the limits of the available documentary evidence, the relationship he had with computer science in the making.

5.1 Wiener's Principles for a High Speed Computing Machine

Wiener tells us in *Cybernetics* (1948) that, in the summer of 1940, "I turned a large part of my attention to the development of computing machines for the solution of partial differential equations" (61c [48f1], 3), and after a short time, "sent to Vannevar Bush for their possible use in a war" a *Memorandum* with some "recommendations, together with tentative suggestions for the means of realizing them" (61c [48f1], 4).

Let us see how these ideas came together in Wiener's mind. His project starts with a concrete problem: looking for computing machines able to solve partial

© Springer International Publishing AG 2017   103
L. Montagnini, *Harmonies of Disorder*, Springer Biographies,
DOI 10.1007/978-3-319-50657-9_5

differential equations (PDE) with sufficient accuracy. In the summer of 1940, Vannevar Bush had sent, on behalf of the NDRC, a questionnaire to the most eminent scientists of the country, requesting suggestions about research projects for military purposes (Cf. 64g [56g], 231). Wiener answered on 21 September 1940, sending a *Memorandum on the mechanical solution of partial differential equations* (85b), attached to a long and detailed "covering letter" (85a).

Since the nineteenth century, many mathematical problems of fluid dynamics, in particular those related to phenomena subject to turbulence and non-linearity, but also other more simple ones, had been formulated in terms of PDEs, but in most cases it was not possible to achieve an acceptable numerical solution, except through long chains of calculations, using for example difference equations that reach a solution with the necessary accuracy asymptotically. In the case of the problems Wiener had to cope with, to obtain an accuracy of 1 on 1000, the computational process implied 1 and a half million arithmetical operations. Consequently, the only workable way to succeed in this task consisted in mechanizing the calculations.

Wiener thought that these machines could be easily reconfigured in such a way as to treat the three classic types of problems involving PDEs: (1) Hyperbolic problems. Otherwise known as wave problems. (2) Parabolic problems. Otherwise known as heat flow problems. (3) Elliptic problems (Cf. 85b). In addition, there were "many non-linear problems such as those of hydrodynamics, and many problems of higher order, such as those of elasticity" (85b).

The cited mathematical problems do not necessarily raised for military purposes, but in Wiener's opinion they did indeed entail several military aspects. He argued:

> If machines of this sort can be devised, they will be of particular use in many domains in which the present theory is computationally so complex as to be nearly useless. This is true of all but a few of the simplest problems in hydrodynamics. Turbulence theory, the study of waves of shock, the theory of explosions, internal ballistics (the study of the motion of a projectile above the speed of sound, etc.) suffer greatly for the lack of computational tools. There are many cases where our computational control is so incomplete that we have no way of telling whether our theory agrees with our practice (85b, 134).

This speech clearly belongs to a mathematician who knew the problems of fluid dynamics, the importance of which had been emphasized at the congress on applied mechanics of 1938 (see above p. 99).

5.2 Wiener's Numerical Method

The proposed machines would be special purpose machines, that is, machines whose structure was designed to fix only certain types of problem. This explains why they did not entail programs, as would have been necessary if the aim had been to create "general purpose" computers (Cf. Ferry and Saeks 1985; Randell 1985).

The first step was the adoption of a numerical method. In Randell's opinion, it coincided with the relaxation method introduced by Richard Southwell in an article published in 1938 with his assistant Derman Christopherson (Cf. Randell 1985, 135). On the contrary, Wiener states that he inspired himself to write his own article on *Nets and the Dirichlet problem* (23b) in 1923, in collaboration with the MIT mathematician Henry Bayard Phillips (Cf. 58f, 112, 64g [56g], 138–139; Masani 1990, 171). It is also possible that Wiener knew Southwell's method in detail by talking with the author himself, because they were both present at the 5th International Congress of Applied Mechanics in 1938 Cf. Den Hartog and Peters 1939).[1] However, Southwell's relaxation method in its turn is inscribed in a tradition of numerical iterative methods initiated precisely by certain pioneering articles which include the one by Wiener and Phillips (23b). In addition, according to the testimony of David M. Young, Southwell was convinced that "any attempt to mechanize relaxation methods would be a waste of time" (Young 1987, 119).[2]

The most natural interpretation is, in my opinion, that—Southwell or not—Wiener had his own method in mind since 1930 when Bush had first raised the issue. And he had finally found an effective practical expedient (see above § 3.9).

5.3   The Choice of a Digital and High Speed Architecture

Actually, during the 1930s, Wiener had tried several times to extend the Bush analyzer method—analog systems designed for the solution of ordinary differential equations—to solve PDEs. He had met with his last failure in China when working with Lee, and this had convinced him that the problem was inherent to the analog method. As explained by Wiener himself, analog devices do not "count" but "measure", therefore they are subject to "measurement errors". Since the numerical method Wiener had chosen required him to reintegrate the output data in the input time and time again, the system produced an excessive accumulation of errors.

In order to cut the problem with errors at its root, he suggested adopting a numerical method based on the binary number system. In fact, if one represents a magnitude through a succession of "all or nothing" electrical states, i.e., "on" or "off", the possibility of a measurement error becomes very unlikely. The *Memorandum* therefore adopted a central apparatus for addition and multiplication of numerical (or digital) type, and a tape to be written in a binary way, with the presence or absence of just one kind of sign, arranged to form a combination of 10 bits (hence to symbolize 1024 different configurations).

---

[1]Southwell had been a member of the international organizing committee of the conference and gave a paper as mentioned in the "Report of the Secretaries" (Hunsaker and von Karman 1939, XXI), but it does not appear in the official proceedings, nor even does the title. Wiener gave the paper *The use of statistical theory in the study of turbulence* (39b) and published an enthusiastic account of the congress for the journal of MIT: *The decline of cookbook engineering* (38g).

[2]Curiously, in this paper, where Young reconstructs the history of iterative methods, he forgets to mention Phillips and Wiener (23b), although he did mention it in the bibliography of his well known doctoral dissertation (Young 1950).

In his autobiography, Wiener explains that the idea of adopting the digital approach emerged in 1940, at the summer meeting of the American Mathematical Society held in Dartmouth College in Hanover, New Hampshire, between 9 and 12 September 1940 (Cf. *Meetings* 1940). Here he was introduced to a machine for calculating complex numbers (numbers, that is, consisting of a real part and an imaginary one), created by the mathematician George R. Stibitz of the Bell Telephone Laboratories (BTL), and publicly presented at the congress (Cf. 64g [56g], 229–232). It was a digital binary calculator (i.e., based on the binary number system), using electromechanical relays. The BTL had also prepared a teletype connected by telephone to Stibitz's computer, which remained physically at the BTL's headquarters in New York (Cf. Stibitz 1980, 481. Cf. also among others Masani 1990, 172; Pratt 1987; Williams 1997) .

As described by Tropp (1980, 120), the future ENIAC designer John Mauchly, then only a young physicist, inexperienced in electrical engineering and computer design, used to recount that, while wandering through the lobby at the congress, he had seen Wiener thrashing around and struggling with the input and output devices of Stibitz's teletype, and looking "very angry"; and that when he was asked what was happening, Wiener had replied that there was no problem with the machine, but that he was just trying to divide by zero.

Wiener's apparently eccentric behavior is very easy to explain. The congress had created a committee divided into three subcommittees, one on mathematical research for military purposes, and the other two for support and educational functions. For the subcommittee on research, several chief consultants had been appointed; John von Neumann for ballistics, Harry Bateman for aeronautics, T.C. Fry for industry, S.S. Wilks for probability and statistics, and H.T. Engstrom for cryptanalysis. Among them, Norbert Wiener had been appointed "Chief Consultant for Computation (numerical, mechanical, electrical)" (Morse 1940, 500–2). Since the discussion about the committee and the appointments had taken place on 9 and 10 September, Wiener's interest in Stibitz's machine, presented at the congress on 11 September (Cf. Williams 1997, 223–4), must very likely that the newly appointed Chief Consultants for Computation had wanted to evaluate its performance. And he did that by subjecting the machine to an impossible problem: division by zero. The printer of the teletype would not have answered with the classic "error" message, an expedient introduced only later, but would have produced endless pages filled with nonsense.

To design his project, Wiener must have talked not only with his former student and now colleague Norman Levinson, but also with several other people, including MIT engineers and maybe outsiders, too.

After many years of hindsight, in a time when digitization has invaded most of the human technological universe, it is not easy to realize how much the choice of a digital technology might then have seemed revolutionary: in fact the expedient of representing numbers by exploiting only two states of a physical quantity, instead of the endless possibilities offered by the continuous (think for example of a lamp actionated with a variable switch, as compared with a lamp that has a classic on–off switch), appeared highly uneconomical. In the case of the BTL, the use of electromechanical relays was consistent with the fact that Bell already used relays for its own automatic telephone exchange units.

The problems that Wiener's machines had to cope with required a very high number of operations, and he was confident in the fact that the price of the accuracy obtained using the binary systems would be paid by the much increased speed in every aspect of the computation. So speed became the crucial point in the design philosophy of these machines, and soon afterwards people would begin to call them *high-speed electronic arithmetical machines*.

To increase the speed of computation he suggested excluding human intervention from the computation; to use vacuum tubes (or "electronic tubes", from which came the name of "electronic computers") instead of electromechanical relays; to use a large and fast storage system, able to record and delete results very quickly, most of these being only temporary during the computation.

It is little known that, in 1938, Vannevar Bush had already started a project for a Rapid Arithmetical Machine at MIT. It would have been a numerical electronic computer, provided with a decimal arithmetic calculator and decimal devices for number storage made by special electronic tubes, together with three paper tapes for automatic control of the computation. In a *Memorandum* of 2 March 1940, Bush describes the status of the design and explains about the paper tapes that "each of these will be a punched tape, although magnetic tapes or tapes with pencil marking to be picked up by feelers may be substituted" (Bush 1982 [1940], 337).

On 21 September 1940, in the *Memorandum on EDPs*, Wiener recommended to Bush not to adopt paper tapes. He explains in the covering letter:

> I have here assumed a magnetic printing and scanning. It is perfectly possible to replace this by a photoelectric scanning combined with a printing by some such device as an electric spark. The difficulty in this latter case is that erasure is impossible and excessive quantities of paper tape must go to waste. However, you will see that the variations in method keeping the same idea are enormous (85a).

In his *Memorandum* Wiener adds:

> (A) A quick mechanism for imprinting numerical values on a running tape. [...] The signal might be magnetic – either DC mark or an AC hum; mechanical – a puncture in paper made by a spark; phosphorescent – stimulated by light, cathode rays, or X-rays; a state of ionization in the tape – stimulated by cathode rays, light, or an electrostatic field; or it might be none of these.
> (B) A mechanism for reading [...]. Again, the reading may be magnetic, photoelectric, dielectric, or something still different (85b).

Wiener's preference was for a system based on cathode rays, used in the television iconoscope, which had also been the apparatus that had inspired him to pursue the proposed numerical method.

Regarding the originality of his proposal, Wiener never laid any claim to priority about the aspects that characterized his proposed machines. He says in *Cybernetics*:

> These notions were all very much in the spirit of the thought of the time, and I do not for a moment wish to claim anything like the sole responsibility for their introduction (61c [48f1], 4).

In fact he had collected technological elements available at that time, but at the highest level of technological advancement and often confidential. We need only consider the MIT Rapid Arithmetical Machine, decimal but electronic, and Stibitz's computer, binary but electromechanical.

However, nobody had seen the need for a fast machine, something which Wiener's *Memorandum* describes as "imperative". For the MIT Rapid Arithmetical Machine, Bush considered sufficient a speed of one multiplication every 1/5 of a second (Bush 1982 [1940], 341). On the other hand, for the tape Wiener had required "a rate of scanning of 10,000 impulses per second" (85a), and for the arithmetic device said: "I am told that vacuum tube-capacitance mechanisms of this type exist, with an overall speed of 1/50,000 s. for operation" (85a).

These speeds could be attained using the counters for nuclear detection. In late 1940, the National Cash Register Company had developed a decimal counter of 150,000 counts per second, which had been followed shortly thereafter by a binary counter of one and a half million counts per second (Desch 1942a, 3–5). Transforming a counter in an arithmetic calculator was not difficult. And that would be the way followed later, in both the USA and the UK.

As one can see, therefore, the originality of Wiener's proposal lies primarily in bringing together several elements under a unified technological philosophy, in which the speed offered by electronics becomes a strategic feature presiding over the entire machine design.

5.4 The PDE Project Packed Away in Camphor Balls

In 1940, Wiener's proposals were rejected by the NDRC. "At that stage of the preparations for war", he explains in *Cybernetics*, "they did not seem to have sufficiently high priority to make immediate work on them worthwhile" (61c [48f1], 4). Further evidence arises from the dense correspondence between Bush and Wiener between September and December 1940. Despite the fact that Bush was otherwise interested in Wiener's proposal, its practical implementation would have taken too long and committed researchers needed to meet more urgent needs.[3]

The rejection of Wiener's project should not surprise us, when we consider that it was only the first in a long series of rejections by the NDRC, particularly on the part of Section D-2, which then became Division 7, against projects for electronic computers. The members of the NDRC (we can imagine also Caldwell and Weaver, in addition to Bush) were not in principle opposed to the digital approach in computing. During the war years, the NDRC would subscribe several contracts with BTL for the construction of relay calculators. In principle, they were not even

---

[3]See the letters: Wiener to Bush, 20 and 23 September 1940; Bush to Wiener, 24 and 25 September; 7, 19 and 31 December 1940 (WAMIT, Box 4, ff. 58). Cit. by Piccinini (2003, 38 and notes 30, 31, 32 on page 38).

opposed to the use of electronics in general (see the applications of electronic digital systems in the fields of radar, cryptology, and nuclear detection). They were specifically very much contrary only to the use of electronics for the purpose of digital computing. During the war years they never saw any military urgency (or maybe even industrial interest) to justify financing projects relating to the application of electronics to digital computing. It is emblematic the case of the Bush's Rapid Arithmetical Machine which had been under the care of the engineer Samuel H. Caldwell since 1939, director of the newly established "MIT Center for Analysis". In February 1942, immediately after the US entered the war, it was decided to postpone the project until after the war. In 1943 Caldwell would write that they had it "at present packed away in camphor balls".[4] The MIT Center for Analysis was also responsible for realizing a new Bush integrator, the Rockefeller Differential Analyser,[5] which had to be used as soon as possible for the main military task of the center, namely computing ballistic and bombing tables. This task had the highest priority even in 1940, so we may assume that the same fate had befallen Wiener's own project: it was not completely rejected, "but packed away in camphor balls" to be resumed after the war.

The story we have reconstructed here was still only the first act of Wiener's involvement in research that would lead to the invention of the modern computer. There would in fact be times when the quick calculation of PDE would become highly strategic for the US and the UK, in the Manhattan Project. But before talking about this second act of the story, we must deal with the other war project in which Wiener was engaged from 1940 until the very beginning of 1943: the automation of anti-aircraft artillery.

---

[4]Memorandum by S.H. Caldwell to Harold L. Hazen, 23 October 1943, NDRC. Cit. by Goldstine 1980 [1973], 151; I complete using the quote in Wildes and Lindgren (1985, 231).

[5]So named because it had funding to the tune of $100,000 from the Rockefeller Foundation.

Chapter 6
The Research Project on "Anti-aircraft Directors"

6.1 Anti-aircraft Directors: A Strategic Project

Regarding his PDE computing project, Wiener tells us:

> Bush recognized that there were possibilities in my idea, but he considered them too far in the future to have any relevance to World War II. He encouraged me to think of these ideas after the war, *and meanwhile to devote my attention to things of more immediate practical use.* (64g [56g], 239). Italics added].

Following Bush's advice Wiener looked for other research fields. He tried cryptology. However, he did not feel at ease in this area, considering in retrospect that "the design of ciphers is not a matter into which one can go cold, without knowledge of the existing tradition and of the practical demands for each particular use" (64g [56g], 240). This statement confirms an idea we have already encountered: Wiener preferred to work on problems whose history he knew very well; or better, whose history was part of his own intellectual itinerary.

He finally found his rightful place in a project on predictors for artillery, that is, devices able to predict the future position of a moving target. Actually, on the one hand, the project gave Wiener the chance to use many of his previous mathematical discoveries and insights, as noted by Norman Levinson: "The mathematical problem of prediction as he [Wiener] formulated it was solvable by a synthesis of his own previous work. He could handle it readily any time after 1931, had he conceived of the problem" (Levinson 1966, 26).

On the other hand the theme fitted perfectly with the strategic views of the NDRC. As shown by the *Report of the NDRC for the First Year of Operation* which Bush sent to President Roosevelt on 16 July 1941, the NDRC had been established to find quick technical solutions for the creation of military equipment that could be put into mass production as soon as possible. A large part of the effort was focused on the detection and interception of targets, especially aircraft, using radar, on the basis of a basic consideration that saw in this recent invention the main reason for

© Springer International Publishing AG 2017
L. Montagnini, *Harmonies of Disorder*, Springer Biographies,
DOI 10.1007/978-3-319-50657-9_6

the success of the UK in winning the "Battle of Britain", namely that the systematic strategic bombing of Britain which took place in autumn and winter 1940–1941, undertaken by Hitler to prepare the invasion, was stopped because Britain was covered by radar, which gave advance warning allowing the Royal Air Force to send fighters against the bombers of the Luftwaffe. This experience, Bush states, "convinced the Committee that it could contribute to defense in an almost immediate manner by concentrating on this problem" (Bush 1941, 7).

It was no accident that the two "more important programs of research and development" by the NDRC, according to Bush, were first "aircraft detection", which meant creating an increasingly long-range, precise, and light kind of radar, all aspects of which involved very high frequency radio emitters, and second, "gun control", that is, systems helping gunners to adjust their aim towards moving enemy targets (in particular, but not exclusively, aircraft), called directors or predictors (Bush 1941, 7), and servomechanisms to move the guns automatically. Up to then these systems had all been mechanical, but now the NDRC was beginning to explore the fastest and cheapest electrical technology which could in addition work together with radar detection, based on "ultra-high frequency [...] for this purpose which, when directed at a plane target, will automatically continue to point at that target with sufficient precision to control anti-aircraft fire" (Bush 1941, 16–17; cf. also 23).

6.2   An Eccentric Idea

Wiener's research project "originated in a suggestion" which he made "at a conference on servo-mechanisms in November 1940",[1] and which had been prompted by "certain questions that were put to [him] [...] concerning servomechanism design" [*Wiener's Final report*].

The method proposed by Wiener exploited the theory of operators, which was born in the field of communications and was still considered as exclusively a matter for the communication engineer. Speaking in modern terms, he decided to deal with prediction as a problem of signal theory. In particular, he saw the predictor as a black box. The curve drawn by the past path of a target was considered as an input signal, while the future path would be the output signal. Note that in this approach, input and output are expressed as a function, not of time, but of frequency, requiring a complex notation.

The same is true—still in modern language—for the transfer function, i.e., the ratio between the output signal and the input signal, which represents the behavior of the black box. Knowing the transfer function, one had only to multiply the input signal by the transfer function to obtain the output signal, that is, the future path of the target.

---

[1]*Summary Report for Demonstration*, 1. Cit. by Hellman 1981, 146.

That is easier said than done, because it clearly violates the principle of causality. Consequently, the method is in theory not realizable. But Wiener thought that it would be possible to approximate the transfer function by means of realizable filters. More precisely, he proposed to use the Wiener–Lee networks.[2] These were passive modular circuits which, according to the particular situation, one could increase or reduce in number to obtain the required transfer function. Alternatively, a similar transfer function could be obtained by an equivalent assembly of integrators of a Bush analyzer. According to Wiener:

> I suggested these notions to Professor Caldwell [...]. After the custom of those times, Caldwell immediately put a classification on my ideas, so that thereafter I could no longer speak freely of them to anyone with whom I wished to talk.

> For a trial setup for my problem, Caldwell and I were tempted to make use of Bush's differential analyzer because of the ease of assembling its parts for the simulation of a large range of different problems. [...] At this stage, prediction theory was made a government project (64g [56g], 241–242).

The results obtained with the Bush analyzer were very encouraging; therefore Wiener was invited to write a report on *Principles governing the construction of prediction and compensating apparatus* to send to the NDRC, together with a note of approval by Caldwell.[3] The proposal, immediately classified, was accepted in December 1940 by the NDRC with the title of *Anti-aircraft Directors* (*A.-A. Directors*) D.I.C. project 5980 (Cf. Masani and Phillips 1985, 143).

The idea of applying this in the field of control appeared eccentric. On 9 November 1940, Edward Poitras, member of Section D-2, wrote in his official diary about the idea put forward by Wiener that he "wants to tackle the problem of solving for the controller of servos in terms of the input as the frequency spectrum, [...] he believes that considerable of the present network theory could be applied to the servo problem".[4]

In the *Final report* on this research, Wiener describes the idea from which it had take form in the following terms:

> At that time the author [Wiener] was very much interested in what has become a routine matter since, namely, the use of the methods of communication engineering in servo design (*Wiener's Final report*, 1).

What most puzzled the audience about the frequency was probably the fact that the Wiener approach had been proposed for automatic controls implemented not only with electrical technologies, but also with mechanical ones. As we read in the first memorandum sent to the NDRC for approval:

---

[2]Based on the theory of Laguerre functions.

[3]Cf. S. H. Caldwell, *Proposal to section D2*, NDRC (3 pages), 22 November 1940. This is the covering letter of the report: Wiener, N., *Principles governing the construction of prediction and compensating apparatus* (8 pages). Cit. by Masani 1990, 181.

[4]Edward J. Poitras, "Diary of Poitras", 9 November 1940, OSRD 7 GP, box 70, Collected Diaries, vol. 1.; cit. by Mindell 2002, 277 and note 2 p. 383.

The proposed project is the design of a lead or prediction apparatus [...]. The principles of design are those of electric networks in general, although the realization may be by mechanical equivalents to electrical networks.[5]

In January 1941, Julian Bigelow, a young engineer with a good mathematical background, was assigned to Wiener's project. Even after acceptance by the NDRC, the proposal remained somewhat "eccentric", considering that, when inviting Bigelow to take on the new appointment, his chief at IBM had said:

Nobody at MIT knew whether what Wiener was saying was *sensible or not or feasible or not*, and they were looking for a man with an engineering and mathematical background who would serve as an interpreter and serve as a colleague for Norbert Wiener (Bigelow 1971, Italics added).

6.3   Carrying out Research

Wiener's research lasted just a little over a year, between the end of 1940 and 1 February 1942, that is, until the classified publication of his book nicknamed *Yellow peril.*[6] It contained the essential of the deep mathematical theory he went on to develop during his research, and also the pivotal ideas of what he would eventually call "cybernetics", in the summer of 1947. In 1942 the small research group, in a context of mysterious obstacles created by the NDRC, was dedicated mainly to building the prototype predictor and testing it.

The rapidity with which Wiener moved on can easily be explained by the fact that he was a "volcanic" thinker, who had devised the Wiener–Hopf equations in one night and who would write his book on *Cybernetics* over a single summer, perhaps spending even less time for the *Yellow peril*. On the other hand, as remarked by Norman Levinson, in his research, Wiener managed to bring together most of his previous studies. In *I am a Mathematician*, he writes:

My war work on filtering and prediction of time series had represented an extension of my earlier work on generalized harmonic analysis and on the Brownian motion as tools for the study of irregular phenomena distributed in time. For years I had the intention of using these tools in every region in which they seemed apt (64g [56g], 288).

At its inception the aim of the research was merely to explore the possibilities that the proposed method offered for a curvilinear prediction, thereby improving Sperry's mechanical predictor, and the electrical one then being developed at the BTL, both based on a method producing a linear extrapolation.[7]

---

[5]Caldwell, *Proposal to Section D2*, p. 2–3, cit. by Masani–Phillips 1985, 143.

[6]*Summary Report for Demonstration*, p. 2. Cit. by Hellman 1981, 157.

[7]Cf. Bigelow, J. *To predict non-uniform curvilinear performance of the target. Report of conference*. Cit. by Bennett 1994, 60.

Very early on, the research was extended to statistical issues, so that the method proposed by Wiener would soon become known as the "statistical method" of prediction, rather than the "curvilinear method" of prediction. Probably already in the test phase with Caldwell, Wiener became aware of the fact that, if one chooses a high order of approximation (corresponding to higher sensitivity), a good prediction was only made in the case of smooth curves, such as a line segment or a portion of a sine curve. On the contrary, in the case of a coarse curve, the same setting made the apparatus extremely unstable, and the prediction was improved by decreasing the order of approximation (corresponding to a lower sensitivity) (Cf. *Final Report*, 1–2; 61c [48f1], 9; 64g [56g], 243–244).

Wiener came to think that the phenomenon reflected an intrinsic difficulty of the problem at hand. Based on the type of account which we have seen was typical of Wiener even before Heisenberg had formulated the uncertainty principle, rather than insist on finding a perfect universal predictor, Wiener considered that one should only look for the "best" possible, i.e., the one able to offer the best compromise between errors due to inaccuracy and errors due to excessive sensitivity in the case of trajectories with abrupt changes (64g [56g], 244).

At this point Wiener was dealing with a statistical problem and decided to interpret the term "best" in the sense of the minimum quadratic error in the prediction, that is, the value minimizing the average of the sum of the squares of the differences between the predicted value and the true value (64g [56g], 244–245). This "problem of minimization" could be handled using the calculus of variations, and in this particular case by an integral equation of the Wiener–Hopf type that was not difficult to solve analytically (Cf. 64g [56g], 245).

The next problem Wiener faced was prediction based on inaccurate data, i.e., associated with errors, or noise, considering the frequency approach. This was treated once again by seeking to minimize the error function, and thus getting another equation of Wiener–Hopf type (Cf. 64g [56g], 246). The sources of noise in Wiener's considerations included human factors. In general, Wiener understood that:

> The concept of predicting the future of a message with a disturbing noise on the basis of the simultaneous statistics of the noise and message turned out to contain in itself the whole idea of a new method for separating noises and messages in what would be in some sense the best possible way (64g [56g], 246).

Wiener's theory of prediction and filtering was born.[8] On 1 October 1941 Warren Weaver made the following remark about Wiener's research:

---

[8]Still in use. See, e.g., Gomez 2016, Chap. "Wiener–Kolmogorov Filtering and Smoothing" 449-519. In addition to the model introduced by Wiener and Kolmogorov, of "input-output" type, the theory based on "state variables" is then included in the model (Ibid. 213 ff.).

[It] probably represents about the ultimate that could be accomplished in designing a predicting system which will take into account all ordinary geometric and dynamic factors, will do the best possible job in filtering out errors, and will take proper account of any statistical trends which may exist in aerial tactics and/or in the habits of aviators.[9]

This is an important testimony, which shows not only the appreciation for the work done, but which allows us to be quite certain that all developments mentioned so far had emerged by then. On the other hand, at least the mathematical aspects of both prediction and filtering would be presented in a systematic form in *Yellow peril* (49g).

Very likely in the late autumn of 1941, the mathematical statistician William Feller warned Wiener that his results converged with those obtained by the Russian mathematician Andrey Kolmogorov (Cf. Kolmogorov 1941; but also Kosulajeff 1941). Wiener was not immediately aware of the similarity of Kolmogorov's work with his own. In the *Final Report*, he wrote that "the theory arising from these conclusions [the *Yellow Peril*] [...] is essentially similar to a method carried out somewhat earlier by Kolmogorov in Russia but not known to the author until the pamphlet in question was well under way" (*Wiener's Final Report*, 3).

Wiener always recognized the chronological priority of Kolmogorov's discovery, although defending the independence of his work (Cf. 49g, V; 61c [48f1], 11; 64g [56g], 261). Nowadays, it is usual to speak of the Kolmogorov–Wiener theory of prediction. Engineers also speak of Kolmogorov–Wiener filters, although Kolmogorov never dealt with the filtering matter.

However, it seems to have been precisely Wiener's intellectual honesty which, by leading him to grasp the substantial identity between the two mathematical approaches, helped him to understand the core feature of what would later be called cybernetics, namely that the subject of communication theory is not the electrical signal, but the message to be understood as a statistical time series, measured on an *ensemble,* that is, information.

Wiener was asked to write a book on the whole matter and he completed it on 1 February 1942. It was entitled *Extrapolation, interpolation, and smoothing of stationary time series*. It was reproduced in 300 copies, classified as "restricted", and bound by a yellow cover that earned it the nickname of *Yellow peril*. It was diffused according to a strictly defined distribution list (Cf. 64g [56g], 255).[10] In 1949, Wiener managed to get it taken off the military secrets list and it was published (49g).

---

[9]Weaver, Warren, Summary of Project #6, Section D-2 NDRC, 1 October 1941, OSRD, E-151, AMP, General Records, box 24. Cit. by Mindell 1996, 436.

[10]As pointed out to me by Antonio Lepschy when I was writing *Le Armonie del Disordine* (Montagnini 2005), the expression 'Yellow Peril', suggested in that case by the color of the cover and the great complexity of the mathematical treatment, had become idiomatic during the 1930s, and almost certainly even earlier, in the United States and in other places, with racist nuances, to refer both to the low-cost production (and the consequent dumping effects) of many artefacts, especially Japanese, while far eastern immigration was growing in importance, mainly from Japan and China, and especially to California, with disruptive consequences in the labor market and phenomena related to the penetration of mafia-type organizations (such as the so-called Chinese "triads").

We have to consider that up to this moment Wiener had dealt only with the linear approach (that is, black boxes with mathematically linear behavior). However, I have found a "restricted" report from April 1942 (see 42c) approaching the non-linear problem with Wiener's then classic method of using white noise as input signals. This testifies just how precocious Wiener's research was for the non-linear problem, which would lead conclusively to his *Nonlinear Problems in Random Theory* [58i].

6.4   The First Synthesis: Inclusion of Control in Communication Engineering

On closer examination, the first clear exposition of what in the summer of 1947 Wiener would call "cybernetics" is contained in the long introduction to the *Yellow peril*. In fact "cybernetics" was in his opinion a "generalized theory of communication" that was born through two successive syntheses between theories. Firstly, communication engineering had incorporated control engineering. Secondly, the statistical theory of time series had incorporated communication engineering, discovering that the subject matter of the latter was not the electrical signal, but information.

Let us give a brief introduction to control theory for the layperson. One of the most common control systems in use since the end of the eighteenth century is the "centrifugal speed regulator" or "governor".

Usually a Watt steam engine provided the power for several looms, distributed by the rotation of shafts, bound to a crankshaft. When one or more looms were detached, the rotation speed of the crankshaft tended to grow; conversely when looms were added it tended to decrease. The Watt "centrifugal speed regulator" was designed so that, when the crankshaft speed increased, it would open a valve decreasing the steam pressure and hence decreasing the speed. On the contrary, when the speed tended to decrease, the device would shut the valve, increasing the pressure and therefore the speed.

This kind of phenomenon, in which the growth of a quantity becomes the cause of its decrease and vice versa, is now called "negative feedback". When the growth of a quantity is the cause of further growth, and vice versa, we now speak of "positive feedback". However, these two terms only appeared during the Second World War. Previously, the important concepts underlying them had remained for centuries cocooned in the practices of control engineers.

Following Watt's device, various types of governor became more common for a range of different uses. In his article *On Governors* in 1867, James Clerk Maxwell discovered the linear (or linearizable) ordinary differential equation, in the time domain, which represents the behavior of governors and which serves to determine whether the system is stable, that is, whether or not it is able to return to its normal operating state after being subjected to a disturbance. The stability condition that Maxwell found was as follows:

That all the possible roots [*i.e.*, the real solutions], and all the possible parts of the
impossible roots [*i.e.*, the real parts of the complex solutions], of a certain equation shall be
negative (Maxwell 1867–1868, 271).

Shortly afterwards, in 1877, his former student Edward John Routh (1831–1907)
introduced an algorithm to determine when Maxwell's stability condition was
fulfilled, without having to go through the analytical solution of differential equations involved (Cfr. Routh 1877). Independently of Routh, Adolf Hurwitz reached
similar results, and we now speak of the Routh-Hurwitz criterion. The approach had
become a classic in control engineering by WWII (see, e.g., Tolle 1905).

As shown, for example, in a paper on *Thermionic vacuum tubes and their
application* by the engineer Robert King, which appeared in the "Bell System
Technical Journal" in 1923, the concept of feedback had arisen in communication
engineering, but outside the control context, as an expedient to obtain several types
of desired behavior in electric circuits provided with thermionic tubes (now called
electronic circuits) through specific assembly.[11]

Since the nineteenth century, especially through research by Heaviside, communication engineering had developed a rich network theory, part of which was
based on what we may call the black box approach, working in the frequency
domain. Among other methods, it developed the "frequency response", which for
linear systems offers the possibility of determining the transfer function, giving a
sinusoidal signal as input. But the most beautiful methods were those for studying
stability, in particular the one introduced by Harry Nyquist (1932) at BTL, and
similar methods that came after, due to Bode, etc. Unlike the methods in the time
domain, such as the Routh-Hurwitz method, used up until WWII by control
engineers, these provided not only a procedure for determining whether or not a
system was stable, but also a way to calculate the degree of stability, and a guide for
designing stable systems.[12]

However, until WWII, communication engineers were unaware that everything
developed about feedback had significance for control, and the same was true for
the control engineers.

---

[11]When the grid of a triode has a positive voltage curve, an amplified voltage appears on the
anode: this is the amplifier operation of a simple triode. If we connect the anode and grid, the two
signals are subtracted. This assembly is called negative feedback, and we have obtained a negative
feedback amplifier. If, however, we put a second triode as an amplification stage in cascade with
respect to the first, and couple the anode of the second triode with the grid of the first, the two
signals are added, and we obtain positive feedback. Feedback could also occur due to undesirable
parasitic coupling phenomena. Around 1912 it was discovered that positive feedback could be
used to improve the gain of triode amplifiers and also to create oscillators (Cf. Rubin 1968, 14). In
the *Bell System Technical Journal* we find an extensive article discussing all vacuum tube
applications (King 1923).

[12]It was later proved by MacColl 1945 that this result can be easily obtained by the argument
principle (also known as Cauchy's argument principle).

What role did Wiener play in this context? In his autobiography Wiener tells us:

> Servomechanisms for the control of gun turrets and other pieces of heavy apparatus were naturally assumed to belong to power technique rather than to communication technique. The whole tradition of power technique was to consider electric currents and voltages as varying in time, while the whole tradition of communication technique, particularly under the influence of Heaviside, had led to consideration of a message as a sum of a large or infinite number of different frequencies (64g [56g], 265).

The MIT servomechanism engineer Harold L. Hazen (1901–1980) confirms that, at MIT before WWII, the situation was indeed as described by Wiener, if not worse. Hazen stated:

> At this time I was not aware of Routh's and Hurwitz's work. I knew vaguely of Nyquist and Bode's frequency domain work at B.T.L. which I mentally associated only with communications network theory.[13]

Wiener's autobiography goes on:

> It was not easy to see that the frequency treatment, rather than the time treatment, was just as appropriate for the servomechanism as for the telephone, the telegraph, and television.

> I think that I can claim credit for pointing this fact out and for transferring the whole theory of the servomechanism bodily to communication engineering (64g [56g], 265).

Wiener's statement could easily be misunderstood if one did not understand his highly advanced mathematical way of thinking. We have seen that, at the inception of Wiener's research on predictors, there was his insight that one should deal with the "servo design" using the "methods of communication engineering", a practice— he added—that has "become a routine matter since" (*Wiener's Final report*, 1).

Wiener had developed a broad insight. That opened the way for further developments, leading during the five years of war to the understanding that feedback in communication corresponds to a "trick" adopted for centuries by craftsmen and engineers dealing with control systems. It was clear that the field of control and servos could be treated in a very elegant way with concepts and methods from the field of communications. In April 1943, Warren Weaver established an interdisciplinary committee on servo and commissioned the mathematician Leroy A. MacColl of BTL to write a report about the entire matter (Cf. Segal 2003, 101). The result was the book *Fundamental theory of servomechanisms* (MacColl 1945), in which the synthesis glimpsed by Wiener in 1940 appeared in a complete and rigorous form.

In that moment was also possible to describe a servo system by means of "*block diagrams*" until then without a control meaning by communication engineers.

This new way of looking at things introduced by Wiener led not only to prediction theory, but more generally to application in the field of control of all the techniques that had been introduced to study and design electrical circuits

---

[13]Hazen interviewed by Bennett, 22 October 1975. Cit. by Mindell 1996, 253.

Fig. 6.1 Feedback block
diagram

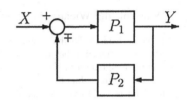

incorporating thermionic tubes (today called "electronic circuits"), including those using feedback, developed for communication purposes.

Block diagrams were already used by communication engineers, with and without feedback. Figure 6.1 shows a simplified block diagram representing the two possibilities for a system with negative or positive feedback.

X is an input signal and Y an output signal. The block P_1 could be an electronic amplifier. Feedback is obtained by bringing back the output signal and adding it to or subtracting it from the input signal. If it is subtracted, we have negative feedback, if added, we have positive feedback. It is important to stress again that, before the Second World War, feedback block diagrams served only to represent electronic circuits, used as amplifiers or oscillators. Then they could also represent systems such as servomechanisms or regulators.

During the Second World War a second way to represent feedback systems was introduced, called *flow diagrams* or *flow graphs*,[14] with a more obvious significance for control.

Block and flow diagrams are formally equivalent, but they allow one to see the same phenomenon from different perspectives. However, having understood this equivalence, all the tools discovered by communication engineers to deal with feedback in their circuits also became available to the control engineer.

6.5   The Second Synthesis: Inclusion of Communication in Statistics

The *Yellow peril* presents the theory of prediction and filtering as the result of a second synthesis between two branches of science which had remained isolated from one another until then and which were thereby brought into mutual synergy: on the one hand, the statistics of time series, that is, continuous or discrete series expressing the evolution of a statistical variable in time, like, for example, the time course of the closing prices of a commodity traded on the Chicago stock exchange or the time course of the temperatures in a town; and on the other hand,

---

[14]*Flow diagrams* were introduce by the British servo-engineer Arnold Tustin (Bennett, S. 1993, 136). The American Samuel J. Mason (1953) called them *"signal flow graphs"* (see also Percival 1953 and 1955). Polemizing with Mason, Tustin interviewed by Bennett (1993) strongly reaffirmed that "they are not signals and they are not flows, they are cause and effect diagrams".

communication engineering (Cf. 49g, 1), whose purpose, Wiener explains, "is the study of messages and their transmission", where the message is no more than an "array of measurable quantities distributed in time. In other words, [...] the message [...] is developed into a time series" (49g, 2).

So the various operations which a message undergoes in the devices of communication engineering "are in no way essentially different from the operations computationally carried out by the time-series statistician with slide rule and computing machine" (49g, 2–3).

In *Cybernetics* the concept would be expressed in similar terms:

> What is not generally realized is that the rapidly changing sequences of voltages in a telephone line or a television circuit or a piece of radar apparatus belong just as truly to the field of statistics and time series [...]. These pieces of apparatus — telephone receivers, wave filters, automatic sound-coding devices like the Vocoder [...] — are all in essence quick-acting arithmetical devices, corresponding to the whole apparatus of computing machines and schedules, and the staff of computers, of the statistical laboratory (61c [48f1], 60).

The "Vocoder" was a radio-telephone device, also known as the X-system or SIGSALY (project X-61573), developed by BTL on behalf of the US Army Signal Security Agency (SSA). It consisted of an analyzer and a synthesizer of the human voice based on analysis of harmonic frequencies, and included a system to make the message safe, called a "scrambler" (Gladwin 2004). It was used for telephone communications between the highest authorities of the United States and Britain during the war. In 1948, these devices were still classified, and Wiener's reference to them is therefore strange.

However, in this way Wiener understood that the essence of communication engineering consists in dealing, not with electrical quantities, but with time series, which in their turn are nothing more than messages; for the *Yellow peril* these were essentially time series associated with probability distributions (Cf. 49g, 9). It should be noted that these messages/time series were framed by Wiener in a statistical theory which he called the "statistical mechanics of Gibbs", although he referred to a highly evolved form of it, which assimilated the results of George D. Birkhoff's ergodic theorem (1931), and the further development of ergodic theory by von Neumann, Wiener, and others. In this way Wiener arrived at the statistical notion of information, the same notion that would later be taken up by Claude Shannon. In the *Yellow peril*, we read:

> No apparatus for conveying information is useful unless it is designed to operate, not on a particular message, but on a set of messages, and its effectiveness is to be judged by the way in which it performs on the average on messages of this set (Cf. 49g, 4).

The same idea appears on the first page of Shannon: "the significant aspect is that the actual message is one selected from a set of possible messages". (Shannon 1948, 379). Shannon does not cite Wiener here. An explicit acknowledgment of Wiener's precedent appeared only in one note in the third part of Shannon's paper, where we read:

Communication theory is heavily indebted to Wiener for much of its basic philosophy and theory. His classic NDRC report 'The Interpolation, Extrapolation, and Smoothing of Stationary Time Series,' to appear soon in book form, contains the first clear-cut formulation of communication theory as a statistical problem, the study of operations on time series. This work, although chiefly concerned with the linear prediction and filtering problem, is an important collateral reference in connection with the present paper. We may also refer here to Wiener's forthcoming book 'Cybernetics' dealing with the general problems of communication and control (Shannon 1948, note 4, 625–6).

And he added in the "Acknowledgments"

Credit should also be given to Professor N. Wiener, whose elegant solution of the problems of filtering and prediction of stationary ensembles has considerably influenced the writer's thinking in this field (Shannon 1948, p. 652).

Actually, after the *Yellow peril* (1 February 1942) Wiener never carried out a real systematization of information theory. He did work on it, and made an aborted attempt at publication on the topic with Joseph Doob, between 1945 and 1946 (Snell 1997, 306). Finally, the closest he came was in the third chapter of *Cybernetics* entitled "Time series, information, and communication" (61c [48f1], 60–94).

In some ways the relation between Wiener's ideas on information and Shannon's article is similar to what happened in the case of classical control theory: Wiener introduced a new point of view, after which his ideas were systematized by others, especially in the last three years of the war.

In the *Yellow peril* there appears for the first time a reflection on the typical German engineering distinction, which we also find in *Cybernetics,* between *Starkstromtechnik* [engineering of strong currents] and *Schwachstromtechnik* [engineering of weak currents], corresponding to what are known in English as "power engineering" and "communication engineering". The first type of engineering deals with the production of electrical energy by means of big systems such as alternators, its distribution by means of high-voltage transmission lines, and its use in electric motors and similar devices: in all these cases the electric currents are treated at very high energy levels. The situation is completely different in the case of the current flowing in a telephone wire, or indeed a television antenna or a radar, which involve very low energy levels. Basing ones considerations on the relevant energy levels can be useful with regard to, e.g., the choice of materials, but according to Wiener, it masks the essential features of the problems treated, consisting in the first case in technologies to produce, process, transmit, and use energy, and in the second, technologies aiming to produce, process, transmit, and use messages, namely information, regardless of the energy level.

This new way of looking at things is the result of the two successive theoretical syntheses carried out during the research. Many aspects of power engineering can be treated with the elaborate techniques of telephony and radar. For example, if noise is interpreted as measurement error, it becomes possible even to apply a noise filter to mechanical devices using knobs and gears (Cf. 49g, 9–10). The message does not even need to "be the result of a conscious human effort for the transmission of ideas". (49g, 2) It may well be constituted by the measurement of a current or a voltage sent from an automatic control unit to another control device, such as a servomechanism when it meets a flaw and sends a message stopping a machine. In general, according to this approach, all the science of control and servomechanisms could be viewed as part of a new generalized form of communication engineering, considered as a theory of message processing. It also became possible to extend this new theory to living organisms. And we may say that this generalized theory was actually what Wiener had in mind when thinking about cybernetics. As he writes in *Cybernetics*:

> On the communication engineering plane, it had already become clear to Mr. Bigelow and myself that the problems of control engineering and of communication engineering were inseparable, and that they centered not around the technique of electrical engineering but around the much more fundamental notion of the message, whether this should be transmitted by electrical, mechanical, or nervous means. The message is a discrete or continuous sequence of measurable events distributed in time—precisely what is called a time series by the statisticians (61c [48f1], 7–8).

6.6 The Role of Living Organisms in this Research

In addition to what were classically considered as communication systems, and also servomechanisms, Wiener's generalized theory of communication now included living organisms. In *Muscular Clonus* (85c), probably written in 1946, Wiener would explain:

> During this war it also become apparent that in a large variety of pieces of apparatus, human messages and automatically produced messages, human effectors, and mechanical effectors, were to be used in the same piece of apparatus. The gun chaser is more or less interchangeable with a radar or photoelectric pickup for the purpose of holding a gun on the target. The crank worked by the gun-pointer may be replaced by an electric motor actuated by an amplified message coming in through a radar pickup. Thus, no thoroughly complete theory of communication and control engineering is possible which does not contain the theory of the human being as an element in the communication and control chain (85c, 489).

This "theory of communication and control engineering" would be called "cybernetics". And it is no accident that the subtitle of the book *Cybernetics* was "*Control and Communication in the animal or the machine*".

Actually, from the very beginning of his research on predictors, Wiener had taken human factors into account. One consideration involved from the start was the pilot of the enemy plane. If he had had an endless range of possibilities for diversionary maneuvers, prediction would have been almost impossible. It became possible due to a series of constraints. In fact, an enemy pilot cannot take evasive action at will because of the limitations on the accelerations he can withstand, panic, and resistance to change which the plane opposes to such diversionary maneuvers. Just these limitations suggested studying statistically the real behavior of various kinds of planes, to improve the accuracy of prediction.

And this was what Wiener explained during a meeting of 4 June 1941 with the mathematicians and engineers of the BTL working on another project, anti-aircraft directors: H. W. Bode, R. B. Blackman, C. A. Lovell, E.C. Wente, and Claude E. Shannon (Cf. Segal 2003, 98; Mindell 1996, 467; 2002, 318–320; Hagemeyer 1979, 356). Wiener stated:

> To predict non-uniform curvilinear performance of the target [...] involved a knowledge of the probable performance of the target during the time of shell flight and that his method proposed to evaluate this probable performance from a statistical correlation of the past performance of the plane. This involves a statistical analysis of the correlation between the past performance of a function of time, and its present and future performance.[15]

In addition, it was necessary to take into account the gunners employed to drive telescopes which were originally used in directors to chase aircraft manually, with rather jerky maneuvering knobs. This was a source of error which, according to Wiener, could be filtered by treating it as noise. Even in this case it was useful to know their performance characteristics (Cf. 61c [48f1], 6).

As evidenced by a letter to Haldane, Wiener went on to consider the human operator like any other electrical or mechanical component of the entire anti-aircraft system, adopting a frequency black box approach,[16] which effectively produced a sophisticated behavioristic method,[17] as was perfectly well understood by Stibitz, the project supervisor on behalf of the NDRC. In a note in his diary dated 1 July 1942, Stibitz writes:

> W[iener] points out that their equipment is probably one of the closest mechanical approaches ever made to physiological behavior. Parenthetically, the Wiener predictor is based on good behavioristic ideas, since it tries to predict the future actions of an organism not by studying the structure of the organism but by studying the past behavior of the organism".[18]

The idea that one could have a real transfer function for the human operator considered as a black box began to catch on. As claimed by Hazen after listening to Wiener at another, less technical conference held at the beginning of May 1941 in Fort Monroe (Cf. 64g [56g], 249):

> The idea struck me more and more forcefully that we should know as much as possible of the dynamic characteristics of the human being as a servo and therefore his effect on the dynamic performance of the entire control system. [He envisages a study of the] fundamental mechanical parameters of the human operator.[19]

---

[15]Wiener in Bigelow, *Report of conference*. Bennett 1994, 60.
[16]Wiener to J.B.S. Haldane, 22 June 1942. WAMIT, Box 4 (or 2?), folder 62. Cit. by Piccinini (2003), 38–39.
[17]Ibid.
[18]Stibitz's "Diary of Chairman," 1 July 1942. Cit. by Galison 1994, 243.
[19]Hazen, H. *The Human Being*, report cit by Mindell 1996, 441; Cf. also Mindell 2002, 276 and note 1, on p. 383.

Between 1944 and 1945, in the Applied Psychology Unit at Cambridge in the UK, K. J. W. Craik[20] dealt with human-machine interactions and collaborated with the servo engineer Arnold Tustin to determine the characteristics of the frequency response of human operators assigned to various forms of instrumentation.[21] This took place in close connection with the research in the United States. The *Yellow peril* and other reports by Wiener were also crucial landmarks in the UK.

6.7 "Behavior, Purpose, and Teleology"

In July 1942 Wiener asked Weaver's permission to discuss feedback mechanisms in the human being with the Harvard neurophysiologist Arturo Rosenblueth (Cf. Bennett 1994, 61). The collaboration between Wiener's team and Rosenblueth had already begun some time before, as evidenced by various documents, including *Cybernetics* itself, from which we learn that the results of the talks with Rosenblueth had already been "disseminated by Dr. Rosenblueth at a meeting" (61c [48f1], 12) on "cerebral inhibition" (Cf. Heims 1991, 14–15 and note 1, page 289). The meeting was held at the Beekman Hotel in New York, from 13 to 15 May 1942, under the patronage of the Josiah Macy Jr. Foundation. As Bateson would declare: " 'Cerebral inhibition' was a respectable word for hypnosis", adding that "most of what was said about 'feedback' was said over lunch" (Bateson and Mead, 1976). On 22 June 1942 Wiener had also informed his English friend J. B. S. Haldane about: "Some biological work which I am carrying out together with Arturo Rosenblueth."[22] Then these results would flow back into the joint paper by Bigelow, Rosenblueth, and Wiener on *Behavior, Purpose and Teleology* (43b).

Wiener, Bigelow, and Rosenblueth discussed in particular the analogy between intentional behavior, such as moving an arm to grab a glass of water and drink it, and the behavior of an anti-aircraft system. Wiener and Bigelow's hypothesis was that in this kind of behavior our actions are constantly controlled by the return to the brain of information about the incompleteness of these actions, information which could be visual, but which was on the whole proprioceptive (i.e., based on feelings we have about the relative position of our limbs).

---

[20]Cf. The war report "The Psychological Laboratory. University of Cambridge [UK]. Unit in Applied Psychology," 1946. Website of the Medical Research Council Cognition and Brain Sciences Unit: http://www.mrc-cbu.cam.ac.uk/history/electronicarchive.

[21]On the frequency response of the human operator see Porter 1965, 332; Bennett 1993, 167. Tustin, A. 1947a and 1947b; Uttley 1944; Craik and Vince 1945a and 1945b; Craik 1944; Bates, 1947.

[22]Wiener to J.B.S. Haldane, 22 June 1942 (WAMIT), Box 4, folder 62. Cit. by Piccinini 2003, 38–39; and Galison (1994, 242).

To evaluate their hypothesis they considered "pathological" aspects of the two types of system, following a classic approach in the school of Cannon that had been inherited by his assistant Rosenblueth. In other words, it was considered useful to look at dysfunctional behaviour in order to understand normal behavior. Now, Wiener and Bigelow's anti-aircraft system was constantly subjected to correction by negative feedback coming from a radar. In certain cases it missed the target and began to oscillate. The question to Rosenblueth was formulated more or less like this:

> Are there any known nervous disorders in which the patient shows no tremor at rest, but in which the attempt to perform such an act as picking up a glass of water makes him swing wider and wider until the performance is frustrated, and (for example) the water is spilled? (64g [56g], 253; cf. also 61c [48f1], 8)

Rosenblueth's response had the flavor of the success of a crucial experiment. The neurophysiologist replied that such a pathological condition did actually exist and was known as "purpose tremor", associated with a dysfunction of the cerebellum, the organ that controls both the organization and the intensity of muscle activity (Cf. 61c [48f1], 96) and 64g [56g], 253–4). In *Behavior, Purpose and Teleology*, the authors write:

> The analogy with the behavior of a machine with un-damped feed-back is so vivid that we venture to suggest that the main function of the cerebellum is the control of the feed-back nervous mechanisms involved in purposeful motor activity (43b, 20).

To understand this passage properly, we cannot ignore control theory. The phenomenon constituted by oscillations in unstable systems, in devices such as governors, was known as "*hunting*" (Bennett 1979). In *Cybernetics* Wiener showed that he knew this phenomenon well, and quoted it just in this context, thus confirming my interpretation (61c [48f1], 7).

To give stability to systems undergoing the *hunting* effect, one usually introduces dampers, i.e., plungers immersed in a cylinder filled with liquid, and/or, if necessary, other devices such as springs and masses. So when they used the locution "undamped feedback", Wiener and Bigelow were supposing that the cerebellum performed a similar stabilizing function.

Wiener and Bigelow probably did not know it yet, but "*hunting*" corresponds in electronic feedback amplifiers to an analog phenomenon called "*singing*" (Bode 1960). In this case, the amplifiers are stabilized by adding passive circuits built with resistors, capacitors, and inductors, which are the electrical analogs of dampers, springs, and masses, respectively, in the mechanical context. This correspondence was discovered by studying the phonograph, and Wiener would show for sure that he knew about this by 1945–47.[23]

---

[23]Wiener writes: "the improvement of the phonograph, has shown that there is nothing exclusively electrical about the ideas used. Where the telephone circuit uses the concepts of resistance, induction, and capacity, the phonograph circuit employs the analogous concepts of friction, inertia, and stiffness of a spring" (*Muscular Clonus*. 85c, 489).

It is good to keep in mind the technical notes just offered, in order to grasp the meaning of an article which has been easily misunderstood by people who knew nothing about control theory, because it is a text describing technical discoveries, then highly advanced and also under military secret, but still in need of further improvements. In the paper, the word feedback is applied, not only to amplifiers, but also to servo systems; and it was probably the first time that this meaning appeared in a paper not subject to military secrecy. This perhaps explains why many laypersons have attributed the discovery of feedback to Wiener himself.

What I wish to clarify here is that the paper did not aim to "mechanize teleology", to use the words of McCulloch (1965, 16), but to introduce a new method for investigating living organisms through a comparison with what one discovers in parallel by building artificial systems, and then carrying out the necessary experimental check. In the article the authors propose a classification (43b, 21) of organisms and machines with growing "purposive" complexity, getting to the following scheme (Cf. Ibid.):

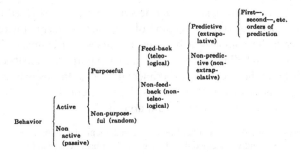

Let's examine only some of the most significant cases. In particular, two kinds of purposeful behavior: (1) "feedback behavior", also called "teleological behavior", and (2) "non-feedback behavior", also called "non-teleological behavior".

The first case is precisely the one considered in research on prediction. The authors take as an example the action of a man bringing a glass of water from the table to his mouth. In this case, they argue, there must be continuous feedback coming from the goal to modify the action and guide the man.

Now consider the second case. That is a purposive behavior without feedback (or non-teleological purposive behavior). The authors propose the example of a snake striking "at a frog, or a frog at a fly" (43b, 20). Here, they claim, the very high speed of the movement involved makes it difficult to suppose that the behavior is guided in a continuous way by feedback from the goal. Persuasively, they explain that

> it is not likely that nerve impulses would have time to arise at the retina, travel to the central nervous system and set up further impulses which would reach the muscles in time to modify the movement effectively (43b, 20).

Even in this case, several interesting ideas arise, and in particular a picture that enables one to guess when feedback-controlled behavior is likely to be present and when it is not. In fact, one would expect it for not very fast, accurate, purposive movements, while for faster movements, one should expect some preordained action plan.

Now, both the first and the second type of behavior are characterized by purposiveness. And on this the authors are clear: they discuss two types of animal behavior that are blatantly purposive. The novelty is not in explaining unintentional phenomena through feedback (although without the word)—as physiologists had done for centuries—such as the fact that an individual begins to sweat to keep its body temperature constant when it's hot. This is the case which Kant called "natural teleology" in his *Critique of Judgment* (2007 [1790], 210), and which he admits as scientific, in contrast to unscientific teleological statements such as saying that the purpose of the sun is to illuminate the day and the moon the night, assuming some higher willful entity. The novelty of the article lies, paradoxically, in supposing the intervention of feedback in behavior that is incontrovertibly teleological, that is, aiming at some specific purpose, because intrinsically willed, such as the act of a man bringing a glass of water to his mouth.

Wiener, Rosenblueth and Bigelow make a scientific study. As I have stressed in the past (Cf. Montagnini 2010), the typical method of cybernetics emerged here: creating hypotheses through the comparison between machines and animals—in this case about the role of the cerebellum as an organ stabilizing purposive feedback behavior—then going on to do the necessary experimental checks. In this way the paper gave the author a glimpse of a model for possible theoretical and experimental collaboration in the future. As Wiener would argued:

> Dr. Rosenblueth and I foresaw that this paper could be only a statement of program for a large body of experimental work, and we decided that if we could ever bring our plan for an interscientific institute to fruition, this topic would furnish an almost ideal center for our activity (61c [48f1], 8).

The author did not seek to explain how individual decisions were taken. They directly assumed behavior connoted by "awareness of 'voluntary activity' ", in which, as they explain, one chooses the purpose, but cannot know how the body pursues this purpose. The study concerns just this aspect (43b, 19). However, there is another task about which they say nothing for the moment, namely, how the subject deliberates in order to choose a purpose. A task, actually, that would have been implied by research on the central nervous system, compared with computing machines, which Wiener, Rosenblueth, and others would undertake later.

6.8 The Epilogue of the Research on A. A. Directors

After the publication of the *Yellow Peril*, it seems that Wiener had found himself marching upstream. The initial contract had included not only exploring "the purely mathematical possibilities of prediction by any apparatus whatever", but also developing "a physical structure whose impedance characteristic is that of this rational function", that is, a trial prototype, and finally, constructing "the apparatus".[24] Wiener had tried therefore to respect this roadmap.

Since February 1942, Wiener, Bigelow, and their assistant Mooney had been working on a "rough" prototype predictor to be tested. The demonstration took place on 1 July 1942 at MIT in the presence of all the members of the NDRC Section D-2: Weaver, Poitras, Fry, and Stibitz. The efficiency demonstrated by the system amazed them. Readjusted to real life scale, the device could predict the target position 20 s later to a good approximation from a record of the aircraft's route during the previous 10 s.[25] Stibitz wrote in his diary that "taking into account the character of the input data, their statistical predictor accomplishes miracles".[26]

At that point Wiener's team should have moved on to "the design of a complete apparatus for the anti-aircraft control and prediction" (64g [56g], 254). Given the statistical nature of the approach, this last phase of development would require an adequate knowledge of the real characteristics of the behavior of the components of a real system. But, rather inexplicably, Wiener was left without further instructions. Notwithstanding the good results of the demonstration, it seemed that those in charge wanted him to stop, although Wiener was never clearly told so. Weaver repeatedly exerted pressure in this direction on Bigelow, as happened on 10 November 1942. Bigelow and Weaver had a private meeting, in which Weaver curiously insisted that Wiener's prediction method was not suitable for practical applications at that time.[27] On 19 July 1942, Wiener wrote to Weaver to say that he was waiting, "vegetating here, chopping wood, walking n miles a day, and haunting the RFD box in the hope of further orders from D-2".[28]

On 22 July 1942 Weaver answered in a rather cryptic way. He believed that the theoretical work had been successfully completed and that he would find wide applications, but he did not know if the theory could be applied to the specific problem of anti-aircraft prediction. Wiener went on his way. He presented a proposal dated 31 July 1942 to the D-2 section, asking for permission to collect realistic data on detection errors and other things. After that, he began a broad round of consultations with Bigelow.

It seems that at least at first he received no approvals from Weaver. On 1 September 1942, using somewhat disrespectful language, Weaver wrote in his diary:

> [Wiener and Bigelow] have gaily started out on a series of visits to military establishments, without itinerary, without any authorizations, and without any knowledge as to whether the people they want to see (in case they know whom they want to see) are or are not available.

---

[24]Wiener, N., *Principles governing the construction of prediction and compensating apparatus*, 22 November 1940, Report for the Section D-2 of NDRC, 2 and 3, cit. by Masani 1990, 182.

[25]Cf. Stibitz, G. R., *Diary of Stibitz*, 1 July 1942, NARS, cit. by Bennett 1994, 60-1 and note 14.

[26]Cf. Stibitz, G. R. *Diary of Stibitz*, 1 July 1942, NARS, cit. by Galison 1994, 243.

[27]Cf. Weaver, Warren, Diary of Weaver, 10 November 1942. Cit by Bennett 1994, 61 and note 19.

[28]Wiener to Weaver, 19 July 1942, cit. Bennett (1993), note 28 p. 183.

W[arren] W[eaver] is highly skeptical about this whole business. [...] Inside of 24 h my office begins to receive telegrams wanting to know where these two infants are. This item should be filed under 'innocents abroad'.[29]

In one way or another, several centers opened their doors to Wiener and Bigelow. At Princeton and Tufts College, they obtained information about the errors in survey procedures; at Langley Field, data on the regularity or otherwise of aircraft motion; more information was gleaned from several naval and army establishments. Among the latter, the most important source was the Army Anti-Aircraft Board at Camp Davis, North Carolina, which provided them with radar tracking data about two test flights, "flight 303" and "flight 304", with data at one second intervals (Cf. Galison 1994, 244). It seems that the purpose of the data collection changed during the work. Wiener writes:

> It was decided by Section D-2 that before any definitive conclusion should be arrived at concerning carrying our theory into practice we should inform ourselves more specifically as to the actual statistical character of airplane flight and of the tracking of such flight (*Wiener's Final Report*, 4).

Wiener had been seconded, but the purpose of collecting realistic data was transformed from a step toward the building of the real A. A. Director to an evaluation of the efficiency of different methods based on real data. The collected data actually allowed Wiener and Bigelow to make a statistical study of aircraft trajectories, comparing the prediction methods for high altitude flight paths that they were familiar of with the two methods of the BTL group.

Using real data from flights 303 and 304, Wiener discussed two versions of a new linear prediction method developed by BTL, comparing them with his own method. He did that in the *Final report* of 1 December 1942, where he notes that his system would have been very efficient for flight times of less than 20 s, that is, for lower altitudes, where bigger guns were coming into competition with smaller heavy machine-guns, for which simpler predictors were available (Cf. *Wiener's Final Report*, 7). For the high altitude problem, however, Wiener's statistical method was not better than the BTL memory-point method, although it would have been possible in any case to improve the former by including data about the various elements in it. He closes by saying that "there is less scope for further work in this field than we had believed to be the case" (Cf. *Wiener's Final Report*, 8).

In a subsequent and definitive letter to Weaver on 15 January 1943, Wiener urged even more emphatically the adoption of BTL's "memory-point method", saying that his own method would not improve significantly, while the latter was by far the simplest and most practical. He reasserted that one should nevertheless take into account the actual data.[30]

---

[29]Weaver, Warren, *Diary of Weaver*, 1 September 1942, GP, Office Files of Warren Weaver, Collected Diaries, Box 71. Cit. by Mindell 2002, 281, and note 17, p. 384.

[30]Wiener to Weaver, 15 January 1943. In *Report to the Services n. 59*.

6.9   The Paternity of Discoveries and Inventions

Of all the events we have mentioned, we do not know anything of Wiener himself. In his autobiography of 1956, a story that unfortunately suffers from too strong a self-censorship, perhaps also because of editorial pressures, he only tells us:

> When I began to emerge from my sheltered life into the scientific confusion of wartime, I found that among those I was trusting were some who could not be held to any trust. I was badly disillusioned more than once, and it hurt (64g [56g], 272).

Personally, in the past, I interpreted this passage in terms of the crisis of conscience of 1945, which Wiener went through after the news of Hiroshima and Nagasaki, as we will see later. However, that "more than once" entitles us to think that it referred to more than one episode, and most likely also to the difficulties that emerged in 1942. Wiener would often complain after the war that his findings had been only partially recognized, if not at all. On the other hand, as Kuhn explains, new paradigms are fixed through manuals, and these are written by "designers" and "developers", not by the brilliant pioneers. Furthermore, this research, which Wiener would call cybernetics, had a Janus-faced nature, like a sort of platypus, being half pure science and half engineering. And cybernetics was destined to become the land of encounter, but also of confrontation, between the different practices of scientists and engineers. For example, in the career of the mathematician and the pure scientist, patents are rarely a matter for concern, while they certainly are for the engineer. Wiener had jumped headlong into research; in the most intense period he had spent whole nights alone making calculations, even taking Benzedrine, an amphetamine (Cf. 64g [56g], 249). He worked with the same passion that had always animated him, and with the knowledge that he was fighting the Nazi enemy from his trench at MIT. He never thought at all about patent issues, at least until March 1942, when a physicist informed Warren Weaver that:

> [Wiener] seems in an unusually bad nervous state the last few days, and I have been trying to get him to take a few days' rest. He had an unfortunate clash with the cleared patent attorney whom M.I.T. had asked to study some of his ideas on circuit theory […].[31]

We do not know in detail the matter of the litigation. However, we must note that the first patent applications for recognition of predictors by BTL date back to 1 May 1941, and those of the Servo Lab to 20 September 1941, while there are none in the name of Wiener and/or Bigelow. On the one hand, there was a lack of interest in the subject on Wiener's part, at least until March 1942, and on the other, engineers who dealt with similar research considered pure science to be unpatentable. This was made clear during the negotiations over the contract on predictors with Gordon Brown, who wished to use Wiener's ideas.

---

[31]Boyce to Weaver, 23 March 1942, cit. Galison 1994, 241.

Brown had written to Nathaniel Sage, head of the D.I.C. (MIT's Division of Industrial Cooperation)[32] about the need to separate Wiener's "fundamental work" from his own "practical work",[33] which was to be ascribed to Sperry as leader of Lab Servant, "since mathematical analyzes are not viewed as patentable material".[34]

When the engineer prioritizes invention, it does not matter who had the first idea. Vannevar Bush, for example, argues that despite the fact that Peter Tait had expressed "the fundamental idea" underlying the differential analyzer, he was not the inventor of it, as inventors must produce functioning results (Cf. Bush 1970, 181–195).

On the contrary, as Wiener recognized, for a pure mathematician it is essentially the first to have the idea that gets the priority for a discovery. Regarding the discovery of "prediction theory" by Kolmogorov, Wiener observes that:

> Kolmogorov confined himself to discrete prediction, while I worked on prediction in a continuous time; Kolmogorov did not discuss filters, or indeed anything concerning electrical engineering technique; and he had not given any way of realizing his predictors in the metal, or of applying them to anti-aircraft fire control.
>
> *Nevertheless, all my really deep ideas were in Kolmogorov's work before they were in my own although it took me some time to become aware of this* (64g [56g], 261. Italics added).

---

[32]D.I.C. was the department through which the MIT regulated contractual relationships with external subjects, such as individuals, private companies, or the government, as in the case of the NDRC.

[33]Brown to Sage, 18 December 1940, OSRD7 GP, box 4, project file 6. Mindell 2002, note 2, 383.

[34]Brown to Sage, and p.c. to Hazen and Caldwell, 9 December 1940, cit. by Bennett 1993, 173 and note 32 p. 183.

Chapter 7
Wiener and Computers. Act 2

7.1 From the NDRC to the OSRD: The Time of Designers and Developers

In 1940, Bush had circulated a questionnaire among scientists, looking at the NDRC's future activities. Wiener suggested:

> [...] the organization of small mobile teams of scientists from different fields, which would make joint attacks on their problems. When they had accomplished something, I planned that they should pass their work over to *a development group* and go on in a body to the next problem on the basis of the scientific experience and the experience in collaboration which they had already acquired (64g [56g], 231–2. Italics added).

This *modus operandi* was typical of Wiener. And at least until the *Final Report*, he actually managed more or less to work according to this ideal. From the beginning of 1943, no similar project would take place during the war, at least not overtly.

On the other hand, compared to 1940, the organization of American science had in the meantime been changing radically. An executive order by Roosevelt of 28 June 1941 had reformed the NDRC, and along with two other committees, it was incorporated into the larger structure of the OSRD (Office of Scientific Research and Development). Bush became president of the OSRD, while management of the NDRC passed to James B. Conant, dean of the University of Harvard.[1] Nineteen divisions were introduced. Section D-1 became Division 14 (radar), still under the direction of Loomis, and still coinciding with the MIT Radiation Lab. D-2 became Division 7, continuing to deal with gunfire control. It was placed under the leadership of Harold Hazen, director of the MIT Department of Electrical Engineering, who had covered an interim as head of the Servo Lab.[2]

---

[1]Established by Franklin D. Roosevelt with Executive Order 8807 of 28 June 1941.
[2]Cf. *Summary Technical Report of the Division 7. Part 1: Gunfire Control.*

© Springer International Publishing AG 2017

L. Montagnini, *Harmonies of Disorder*, Springer Biographies,
DOI 10.1007/978-3-319-50657-9_7

In general, in this new phase, decision-making and state funding were given to a broader range of industrial companies and universities, breaking the monopoly of AT&T, MIT, and Harvard which had characterized the NDRC in 1940–41. This coincided with US entry into the war, on 8 December 1941, which brought a tremendous increase in government funding to the OSRD: during the first year, the NDRC had spent just over 6 million dollars, while its expenses arose to 150 million dollars per year, over the three years 1943–45 (Stewart 1948, 322).

The passage from the NDRC to the OSRD marked a transition from a freer and more fluid phase to one in which research became even more applicative, parceled out, and sectored, although less tied to the rigid anti-aircraft philosophy originally prevalent at the NDRC.

As Wiener would state in his autobiography: "As the war went on, the work left the hands of the *pure scientists* like myself for those of the *designers*, and I was at loose ends" (64g [56g], 276. Italics added). We have already discussed the specific work on control theory, especially by Leroy MacColl, and on the "mathematical theory of communication" or information theory by Claude Shannon. In addition, during the period 1943–45, many mathematicians worked on the *Yellow peril* to translate it into a language understandable to engineers and applied mathematicians, or to fix some missing mathematical details. John Tukey, the mathematician credited with the introduction of the word "BIT", says that during wartime there circulated at least a dozen "simplifications" or "explanations" of the book (Cf. Tukey 1952, 319; see, e.g., Bergman, 1942; Levinson 1942). Several people also worked to extract practical applications from the theory of the *Yellow peril*. On 3 October 1944, there was even a "Conference on the methods of N. Wiener".[3]

On 9 December 1942 the mathematicians' activities were coordinated by the Applied Mathematics Panel (AMP), headed by Warren Weaver, who had, however, retained a place in Division 7. The AMP's mission was to satisfy the mathematical needs arising from the various NDRC divisions (Cf. Wildes and Lindgren 1985, 185; Stewart 1948, 84–97; Owens 1988).

Wiener wanted a job that would accord with his style of research and, as reported by an annoyed Weaver, his "business consisted of 'holding myself in readiness in case other jobs turned up'". On the contrary, Weaver remarked that he "had not been doing any of the things we particularly wanted him to do".[4] The kind of collaborator Weaver desired was accommodating with the military, up-to-date on the development of armaments, tolerant, cooperative, and altruistic, in short, a person who could play in the team and was not convinced that his ideas were transmitted directly from "God the Almighty".[5] However, Weaver was forced to admit that:

[3]Cf. "Report of conference on the methods of N. Wiener," 3 October 1944. Cit. by Bennett 1994, 60–61 and note 15.

[4]Weaver to Boyce, 24 Mar. 1942, Record Group 227, OSRD, Division 7, General Project Files, 1940–46, General Mathematical Theory of Prediction and Application, MIT, Wiener, NDCrc-83, NA-LC. Cit. by Galison 1994, 241.

[5]Weaver to Stone, 6 December 1943, cit. by Owens 1988, 291.

It is unfortunately true that these conditions exclude a good many mathematicians, the dreamy moonchildren, the prima donnas, the a-social geniuses. Many of them are ornaments of a peaceful civilization; some of them are very good or even great mathematicians, but they are certainly a severe pain in the neck in this kind of situation.[6]

What did Wiener do during 1943? We know that between 1942 and 1943 he took part in a "Statistics Seminar" organized at MIT. Among others, there was the statistical mathematician William Feller, the mathematician, engineer, and expert in fluid dynamics Richard von Mises, statistical mathematicians interested in economics such as Harold Hotelling and Abraham Wald, and economists in the strict sense like Harold Freeman and Paul Samuelson.[7] Wiener gave a talk on *Ergodic Theory* (91). He expressed a rather optimistic opinion about the applicability of the ergodic theorems to "meteorological time sequences because of the stability of the underlying forces" (91). On the other hand, regarding their application to economics, he was rather doubtful, explaining that "for economic time series, however, this condition is often not fulfilled and caution must be taken in applying the theorem". As we see, Wiener continued here to have the same hesitation regarding the mathematization of the social sciences as he had expressed in the 1930s. These questions would have important consequences in the relationship with social scientists in the postwar discussion of cybernetics.

In this period Wiener began to investigate more and more aspects of living organisms which had emerged during his research on prediction, intensifying his relationships with neurophysiologists. We have already spoken of *Behavior, Purpose and Teleology* (43b); it represented a model of collaboration between mathematics and neurophysiology, to be pursued further with Rosenblueth after the war, in which they applied knowledge of machines to the understanding of living organisms.

The second milestone in the development of this comparative approach involving animals and machines came from work by the psychiatrist Warren McCulloch and the young mathematician Walter Pitts, who were working on a logical neural network model to explain human thought. The notions that emerged from this study would prove extremely useful in the development of the von Neumann computer architecture.

---

[6]Ibid.

[7]Among the other papers presented at the seminar, we should at least mention the following: William Feller, *Stochastic Processes*; Harold Hotelling, *Some Unsolved Problems of Statistical Theory*; Abraham Wald, *A Problem in Multivariate Analysis*; Paul Samuelson, *A. Gram-Charlier Series*; Richard von Mises, *The Probability of Occupancy*; Dirk J. Struik, *The Foundations of the Theory of Probabilities*.

7.2 McCulloch and Pitts

Warren S. McCulloch (1898–1969), born in New Jersey, was a psychiatrist with neurophysiological and philosophical interests, as we may deduce from some of his publications, e.g., *What Is a Number, that a Man May Know It, and a Man, that He May Know a Number?* (McCulloch, 1941). He preferred the construction of simplified and elegant models to the meticulous experimental research that typified Rosenblueth's approach.[8] At the end of 1920s McCulloch introduced the notion of "psychon" to refer to an irreducible atomic psychic event caused by previous psychons and, in turn, producing subsequent psychons (Cf. McCulloch, 1941 (1965b, 8)).

During the 1930s this theory evolved when McCulloch discovered the propositional calculus of *Principia Mathematica* by Russell and Whitehead. This happened at Yale University where he was influenced by the psychiatrist-philosopher Eilhard von Domarus, a pupil of the philosopher Filmer S.C. Northrop (1893–1992) (Cf. McCulloch 1965b [1941], 2), and also by Joseph H. Woodger (1894–1981), then visiting Yale.[9] Woodger had tried to axiomatizing genetics and other theories of biology on the basis of the logic of the *Principia* (Cf. Woodger 1937).

McCulloch began to identify the psychon as an atomic proposition of the propositional calculus (which could have only two values: true or false) (Cf. McCulloch 1965b [1941], 8). He thought that a psychon interpreted in such a way could be matched neurophysiologically with the yes/no behavior of neurons. Basically, he conceived of the human brain as consisting of a relay network, and thinking as being implemented by a succession of stimuli switching these relays on or off (Cf. McCulloch 1974).

McCulloch met Wiener in the spring of 1941, through their common friend Rosenblueth.[10] As McCulloch tells us:

> I first laid eyes on him [Wiener] at dinner with Rosenblueth when they, with Bigelow, were mechanizing teleology. He told me promptly what I could expect of my own theories of the working of the brain. Time proved him right (McCulloch 1965a [1941], 16).

---

[8]This is the theme of some letters: Cf. Rosenblueth to McCulloch, 21 June 1941; 5 September 1941, 3 December 1941; McCulloch to Rosenblueth, 1 May 1941 (MCAPS), ff. Rosenblueth. Cit. by Piccinini 2003, p. 28.

[9]Cf. McCulloch to Ralph Lillie, ca. February 1943 (MCAPS, ff. Lillie). Cit. by Piccinini 2003, 28–9.

[10]The date of spring 1941 can be considered pretty safe. In fact, McCulloch says he does not remember whether the meeting took place in the spring of 1940 or 1941. He remembers, however, that it had definitely happened before Pearl Harbor (December 1941). We know, in addition, that in the spring of 1940, Wiener had not yet started working on "predictive filters", a theory which, in McCulloch's historical reconstruction, seemed to be in a fairly advanced stage. Finally, McCulloch added that Wiener was already working with Bigelow, assigned to the project in January 1941. [Cf. McCulloch 1974] McCulloch had known Roseblueth at least since 1938 (Cf. McCulloch to J. F. Tönnies, 11 April 1938 (in MCAPS, ff. Tönnies); McCulloch to Rosenblueth, 22 December 1939 (in MCAPS, ff. Rosenblueth). Cit. by Piccinini 2003, 27 and note 38.

McCulloch remained very surprised by the accuracy and clarity of Wiener's neurophysiological ideas. Wiener "was happy with" McCulloch's "notion of brains as, to a first guess, digital computers, with the possibility that it was the temporal succession of impulses that might constitute the signal proper" (McCulloch 1974).

To put his ideas into practice, McCulloch needed a first class logician, and this he discovered in the young Walter Pitts (1923–1969). Pitts was born in Detroit. He was a very brilliant young man, in particular in logic and mathematics, but was also provided of a vast culture that he had built up by teaching himself. Due to family problems, he had become more or less homeless, wandering the streets of Chicago. Pitts knew the *Principia Mathematica* as well. He had actually discovered the book in a quite thrilling way. To escape a gang of thugs who had come after him, he had hidden among the shelves of a public library. At night he picked up the first volume of the *Principia*. After three days of deep study, the boy sent a letter to Russell in which he pointed out the mistakes he had found in the book. Russell immediately replied, inviting him to study with him in Britain, but Pitts went on with his wandering life. However, he then began to attend several kinds of lessons at the University of Chicago, without regular enrolment. In 1938, when he was fifteen years old, Pitts read the latest book by Rudolf Carnap (1891–1970), a leading exponent of logical positivism, and a refugee in the USA since 1935, for racial reasons.

As was his habit, Pitts underlined what he considered mistaken in Carnap's book, very likely *The logical syntax of language* (Carnap 1937), quoted also in McCulloch-Pitts 1943. Pitts entered Carnap's room at the university and gave him the book without introducing himself. The professor then spent two months trying to find the boy, in which he had discovered such extraordinary talent, and got him a menial job at the university Cf. Lettvin 2000, 2–3). From 1938 until the summer of 1943, Pitts was a student of Carnap, also beginning to work for Rashevsky's biophysics school, and publishing some very important papers.

Nicolas Rashevsky was a physicist of Russian origin who had arrived at the University of Chicago in 1934, where he worked in the departments of psychology and physiology. He wanted to create a mathematical biology, inspired by mathematical physics [cfr. Cull 2007; Rapoport and Landau 1951, Abraham (2000, 2002, 2004)], following in the wake of research by Alfred J. Lotka (1925) and Vito Volterra (1926a, b, 1931). His approach is well represented by the various attempts to find mathematical models of a single neuron, by means of the theory of diffusion of substances and electrochemical gradients. His research (Rashevsky 1933, 1934 and 1936) was published in the volume *Mathematical Biophysics* (1938). In 1939 he founded the "Bulletin of Mathematical Biophysics".

McCulloch made Pitts' acquaintance in autumn 1941, after moving from Yale to the Illinois Neuropsychiatric Institute, in the University of Illinois, via Pitts' close friend Jerome Y. Lettvin. From 1942 until 1943, McCulloch and Pitts worked on the idea of the neural net. They investigated both the functioning of neurons as relays, and the notion of feedback known from Wiener, Bigelow, and Rosenblueth. McCulloch tells us:

> Since there obviously were negative feedbacks within the brain, why not regenerative ones?
> For two years Walter and I worked on these problems whose solution depended upon
> modular mathematics of which I knew nothing, but Walter did (McCulloch 1974).

McCulloch was thinking here of positive or "regenerative" feedback. Interest in
neural loops had emerged simultaneously among other scholars of the brain. They
had played an important role in the epileptic activity of a surgically isolated brain
and in causalgia following amputation. In the future logical model that emerged
from his collaboration with Pitts, they would take into account positive feedback
using modular arithmetic (or modular mathematics), in one of the most difficult
parts of their paper *A logical calculus of the ideas immanent in nervous activity*
(McCulloch and Pitts 1943).

7.3   "A Logical Calculus" and Turing Machines

It will be useful to outline the fundamental thesis of the paper by McCulloch and
Pitts. To do so, I will mainly follow *Theory of self-reproducing automata* (von
Neumann 1966), a very clear and rigorous exposition.

First of all, one must mention the content of another pivotal article, besides the
one by McCulloch and Pitts 1943, namely, Turing's *On Computable Numbers,
with an Application to the 'Entscheidungsproblem'* (Turing 1936). Turing shows
that the concept of effective procedure, e.g., the array of rules we follow when
carrying out a multiplication with pen and paper, can be represented in a rigorous
manner by introducing a set of machines *A* which will be called "Turing machines".
Without going into the details of how a specific Turing machine is made, it will
suffice to think of it as a specific program for a computer, although Turing thought
rather in terms of abstract machines. We obtain many specific Turing machines for
each possible program, each distinguished by a unique number.

Turing then introduces the concept of a machine *U*, called Universal Turing
Machine, able to perform the behavior of any specific Turing machine *A*, i.e., any
program. The Universal Turing Machine does this simply by calling the number
associated with the specific machines *A*. This way one gets to the concept of Turing
computability and to the Church-Turing thesis, according to which the Universal
Turing Machine can compute each computable problem, and if a problem is
unsolvable by the Universal Turing Machine (i.e., it is not Turing computable), then
there is no other way to solve it.

What did McCulloch and Pitts (1943) prove? They avoided embroiling themselves in complex biochemical issues, as Rashevsky had done, and conceived of the
neuron as a simple element, a "formal neuron", that can only have two states, "on"
or "off". It can be turned either on or off by stimuli coming from other neurons, and
in turn, it sends a stimulus that turns other neurons on or off, until the stimulus
reaches something that is not a neuron, e.g., a muscle or a gland.

Now, formal neurons can be combined to give rise to very complex neural nets. And McCulloch and Pitts prove that each specific Turing machine A can match one of these neural nets. Consequently, if it is assumed to work by means of McCulloch-Pitts neural nets, a brain can compute any computable problem, that is, a brain so conceived is a Universal Turing Machine (Cf. von Neumann 1966, 44–5).

By the way, *A logical calculus* does not prove that a brain, in particular the human brain, is made up of McCulloch-Pitts neural networks. It only proves this *if* it is conceived in such a way, which is enough for it to operate as a Universal Turing Machine, that is, as a system able to compute any computable problem.[11]

7.4 The Collaboration Between Pitts and Wiener

It is important to note that, when *A logical Calculus* came out in December 1943, Walter Pitts had already been the assistant to Norbert Wiener for several months. So Wiener, who had been familiar with McCulloch's ideas since 1941 and had probably had similar ideas independently before that, witnessed the gestation of the article.

Wiener had met Pitts in August 1943. He always needed to work with young mathematicians with a fresh mind, to help him to fix his complex and daring mathematical discoveries, always handcrafted and in need of being made more rigorous. After a brief discussion of the ergodic theorems, Wiener became convinced that Pitts would be his man. And Pitts become his assistant that autumn (61c [48f1], 13–14; Lettvin 2000, 3–4).

There are some interesting letters which tell us how Pitts' separation from the Rashevsky circle had occurred.[12] The brilliant mind of Pitts was disputed by several scholars of high intellectual stature, such as Rashevsky, McCulloch, and Wiener. The latter managed to secure his services because the others, although reluctant, agreed that it was for the young man's own good. Indeed, Wiener could procure Pitts a scholarship and let him get his *Ph.D.* in mathematics at MIT in a couple of years, so settling his lack of a formal curriculum.

---

[11] Actually, after the discovery by McCulloch and Pitts, in *Representation of Events in Nerve Nets and Finite Automata,* Kleene (1951) proved that the neural networks of McCulloch and Pitts can compute only a sub-class of Turing-computable problems, i.e., the so-called "regular events". However von Neumann corrected Kleene's conclusions. Although Kleene's proof was correct, the McCulloch-Pitts neural nets do constitute a Universal Turing Machine under certain broader conditions, in particular, if they have an unlimited memory (von Neumann 1956, 56). Then another condition was that the neurons should be able to self-refer.

[12] Rashevsky to McCulloch, 9 August 1943; McCulloch to Wiener, 1 September 1943. Warren S. McCulloch Papers. p. 67 MCAPS, ff. Rashevsky; McCulloch to Wiener, dated 27 August 1943. Cit. by Piccinini 2003, 66–67.

Wiener was also very interested in the project on McCulloch and Pitts' neural nets, which had been born to provide neurophysiological explanations for thinking processes. But in *Cybernetics*, Wiener tells us:

At that time Mr. Pitts was already thoroughly acquainted with mathematical logic and neurophysiology, but had not had the chance to make very many engineering contacts. In particular, he was not acquainted with Dr. Shannon's work, and he had not had much experience of the possibilities of electronics. He was very much interested when I showed him examples of modern vacuum tubes and explained to him that these were ideal means for realizing in the metal the equivalents of his neuronic circuits and systems (61c [48f1], 14).

McCulloch and Pitts had most likely worked without being aware of the work by Claude E. Shannon (1916–2002), a student at MIT since 1936, and author of an article on *A symbolic analysis of relay and switching circuits* (Shannon 1938). Shannon had proved that the whole Boolean algebra could be implemented through switching circuits, and in particular relay nets (Note that the Boolean algebra is the equivalent in algebraic form of the propositional logic adopted by McCulloch and Pitts, following the ideas of the *Principia* and then of the logical positivists.)

Wiener here thought to implement McCulloch and Pitts' neural nets, not by means of elecromechanical relays, but using thermionic tubes (also called electronic tubes), which operate as fast relays, to obtain an "electronic brain", a term that has no journalistic origins, as is commonly believed, but was coined by the early cyberneticists.

So with the project of an electronic brain, Pitts was becoming part of the Wiener dream of a cross-fertilizing collaboration between experts on machines and experts on living organisms. Pitts began to attend "a seminar on scientific mathematics" held privately at Rosenblueth's home.[13] Pitts' collaboration with Wiener went on for several years in a continuous manner, although it followed twisting roads.

In fact, in early 1944, equipped with a "top-security clearance", Pitts went to work for the Manhattan Project at the Kellex Corporation in New York.[14] At the same time Rosenblueth went to work in Mexico. Cannon and Rosenblueth had been accused of communist sympathies and the approaching retirement of the first, in 1942, made it impossible for Rosenblueth to remain at the Harvard Medical School.

Having excluded the possibility of moving to the Medical School at the University of Illinois via McCulloch,[15] which would have required the acquisition of American citizenship, Rosenblueth preferred to return to Mexico City. Here at the very beginning of 1944, he entered the Instituto Nacional de Cardiologia (INC), a recently established institute, where he was appointed director of the department of physiology and pharmacology. He also installed a laboratory with equipment purchased from the Harvard Medical School, obtaining help from the military

[13]Wiener to McCulloch, 30 August 1943 (WAMIT) Box 4, ff. 65, cit. by Piccinini 2003, 67.

[14]Actually, Pitts had been visited by doctors and judged "pre-psychotic", (on the basis of interviews with Pitts' friends, Smalheiser 2000, 221).

[15]Cf. Rosenblueth to McCulloch, 26 October 1943 (MCAPS) cit. by Piccinini 2003, 67.

physician Juan García Ramos. Meanwhile, the time had come in which the contribution of the "moonchild" Wiener—to use Weaver's definition—had once again become pivotal, in particular regarding the computational requirements at Los Alamos.

7.5 The Manhattan Project and the Los Alamos Laboratory

It is very important to understand whether Wiener had worked for the Manhattan project alongside von Neumann, and if so in what way, for at least two reasons: on the one hand, to establish the extent of Wiener's contributions to the creation of the von Neumann computer architecture, and on the other, to explain the crisis of conscience he experienced in 1945, after the nuclear bombing of Japan. We possess a number of documents that would tend to confirm that Wiener actually played an important role in the effort led by von Neumann to provide the computations needed in Los Alamos, although they do constitute rather fragmentary evidence from which we can only draw a plausible scenario.

First of all it is useful to have clear picture at least of the main lines and the timing of the research carried out in the Manhattan Project. After a period of broad-spectrum research under the supervision of the NDRC Uranium Committee, activities working toward the construction of nuclear bombs entered the operational phase at the end of 1942. On 17 September 1942, the task of making the bombs was assigned to the US Army, under the direction of General Leslie Groves, who led the Manhattan Engineer District from an office in New York, hence the name of the entire project.[16]

The Manhattan Project was divided into four sub-projects. Two of these concerned methods for separating uranium 235, the isotope useful for nuclear fission, from the other isotopes of uranium, namely, 234 and 238, with which it occurs naturally: the method of gaseous diffusion, studied at Columbia University, under the direction of Harold Urey, and the electromagnetic one, developed at Berkeley under Ernest O. Lawrence (Frank, Cochran and Norris 1996). A third sub-project, assigned to the University of Chicago under the direction of Arthur H. Compton, concerned the production of plutonium 239, another fissile isotope, obtained through nuclear reactors. After the positive result obtained by Fermi in that University on 2 December 1942 (Greco and Picardi 2005, 52), the so called "Chicago Pile-1" or "Fermi's Pile", were created at middle-sized reactors in Oak Ridge, and finally, from January 1943, industrial-sized reactors at Hanford, in Washington State.

---

[16]Comments here on the history of the Manhattan Project are derived from Frank, Cochran and Norris 1996, Hawkins 1946, Rhodes 1986, Fitzpatrick 1999, Maurizi 2004, Greco and Picardi 2005.

A fourth sub-project had the task of designing and constructing nuclear bombs. This research was conducted in the purpose-built laboratory at Los Alamos, New Mexico, near Santa Fé. On 1 November 1942 Robert Oppenheimer was appointed director there. The effective operation started in the late spring of 1943 (Frank, Cochran and Norris 1996) .

In particular, Los Alamos had to build a uranium 235 bomb, nicknamed "Little Boy", that would be dropped on Hiroshima, and a plutonium 239 bomb, nicknamed "Fat Man", which would destroy Nagasaki. Edward Teller was the head of a small team in charge of studying the hydrogen fusion bomb, a third type of bomb, which would release thousands of times the energy of the other two bombs. This last project received low priority.

Paradoxically, obtaining a nuclear chain reaction of an explosive kind is much harder than obtaining one in a nuclear reactor, because the critical mass, i.e., the sufficient amount of fissile material to start the reaction must be reached instantly. The method for triggering the explosion of the uranium 235 bomb, the so-called "gun method", did not pose particularly hard theoretical problems or require overly complex computation. In contrast, the plutonium 239 bomb entailed highly involved theoretical questions and extremely complex computation. The method seemed unsafe and a "predetonation" very easy, with consequent waste of fissile material. As stated by Nicholas Metropolis "initial concern at the laboratory was for quick assembly because of predetonation, but then von Neumann stressed the added advantage of an increase in material density resulting from the implosion." (Aspray 1990a, note 16, 259). Actually, the very high pressures obtained by implosion reduce the critical mass needed to trigger the explosion.

In spite of that, the managers of the Manhattan Project appeared to be particularly interested in realizing it, if we consider their commitment and the high priority they gave to all that concerned it (Frank, Cochran and Norris 1996).

From late 1943, studies of the trigger method for the plutonium bomb, called "implosion", had top priority at Los Alamos. From spring 1944, most of the personnel of the Armaments Division would work on it. In August 1944 almost all the scientists in Los Alamos, about a thousand, except Teller and those who had to refine and mould the uranium for Little Boy, were included in the work program for implosion.

The implosion method had been proposed in April 1943 by Seth Neddermeyer. In autumn 1943, John von Neumann came to Los Alamos and immediately set to work on implosion.

Nelson and Metropolis (1982) , obviously within the limits still prescribed today by the security classification (Ibid. 348), explained that:

> The implosion simulation required solution of hydrodynamic equations with shock-boundary conditions. The Euler and Lagrangian form, a second-order hyperbolic partial differential equation, was selected as most appropriate for numerical solution (Nelson and Metropolis, 1982, 354).

That was just one of the mathematical problems considered by Wiener in his 1940 *Memorandum on PDEs*. The numerical procedure that Los Alamos adopted to get a sufficiently accurate solution was also very similar to the one proposed in Wiener's *Memorandum*. In fact, Nelson and Metropolis (1982) added:

> The numerical procedure evaluated the differential equation for a sequence of points covering one-space dimension and then integrated ahead one step in the time dimension (Nelson and Metropolis 1982, 350).

Unfortunately, Los Alamos had not planned to tackle such computationally difficult problems. In the scientific community, only astronomers had been able to achieve such accurate calculations by that time. For physicists up to then, highly inaccurate slide rules had always been enough. In fact, in Berkeley in 1942, during the preparatory phase at Los Alamos, physicists had realized the need to perform lengthy calculations. So it was decided to entrust the young physicists Stanley Frankel and Eldred C. Nelson with setting up a small team to calculate using Marchants and Fridens desktop calculators (Cf. Metropolis and Nelson 1982, 348). For security reasons, the team was made up of scientists' wives, integrated with the army staff at the end of the summer of 1943.[17] In the fall of 1943, when he was still struggling with the calculations for the "gun trigger", the physicist Dana Mitchell suggested ordering IBM punch card machines, inspired by the "Thomas J. Watson Astronomical Computing Bureau" at Columbia University (Cf. Metropolis and Nelson 1982, 350). However, the IBM machines did not represent an adequate response to the new problems posed by implosion and arrived only in the spring of 1944.[18]

When it finally arrived, the IBM system was set aside for computations involved in the implosion simulation. The other calculations were undertaken by the team equipped with desktop calculators (Cf. Metropolis and Nelson 1982, p. 350 and p. 354).

The computations were carried out more or less in this way. Each punch card represented a point on a line, so to represent the state of the implosion at each instant, several decks of cards were necessary. The multiplier performed an addition and a multiplication and integration of each step involved reading the deck of cards, the data going to a multiplier and adding machine that was drilling a new deck of cards, while the card steps were repeated a dozen times for each time step. During the procedure it was also necessary to reprogram the computers using a new wiring. Note that, to set up an IBM machine at that time, one had to use plugs and cords as in a manual phone switchboard (Cf. Metropolis and Nelson 1982, 350).

The procedure impressively matched the iterative work Wiener had imagined that his computing machines could cope with. In 1940, he had already suggested not using paper tape. In fact, the IBM machines required mountains of punched cards which had to be thrown away after use, containing as Wiener had pointed out, only provisional data.

---

[17]The manual computer service was referred to as T5. Cf. Metropolis and Nelson 1982, 349; Cf. also Hawkins 1946, 84.

[18]Cf. Telegram from J. Robert Oppenheimer to S.L. Stewart, 28 January 1944, B-9 Files, Folder 413.51, Drawer 96, in LANLA, It was still "Secret-RD" in 1999 Cit. by Fitzpatrick 1999, 64.

This desperate situation was already clear to von Neumann and a few other physicists in January 1944. Ulam tells us:

> I participated, in January 1944, in some of these discussions with von Neumann and [il mathematician Jack] Calkin and I well remember the schematized and over-simplified way by which one first tried to compute the course and results of implosions. From the beginning the schematizations seemed to me too crude and unreliable. I remember discussions with von Neumann in which I would suggest proposals and plans to calculate by brute force very laboriously—step by step—involving an enormous amount of computational work, taking much more time but with more reliable results. *It was at that time that von Neumann decided to utilize the new computing machines which were "on the horizon".*[19]

And as Ulam testifies: "we received administrative support for getting all possible means to enable one to calculate implosions more exactly" (Ulam 1980, 96).

7.6 Von Neumann's Search for Computing Resources

According to an uncommon practice, von Neumann was permitted to reside only part-time in Los Alamos. Indeed, he was allowed to move with some freedom around the country to take part in different projects, which still all seemed somehow related to the problems faced in Los Alamos, a situation that lasted until the end of the war. As Nelson and Metropolis note, von Neumann "consulted for several government projects at such a pace that he seemed to be in many places at the same time" (Nelson and Metropolis 1982, 352).

On 28 January 1944 Robert Oppenheimer sent a telegram to urge the IBM supply request (Cf. Fitzpatrick 1999, 64). A few days earlier, on 14 January, von Neumann had written a letter to Warren Weaver in his capacity as head of the Applied Mathematics Panel,[20] asking what computing machines were available at that time.[21]

Weaver suggested contacting Aiken at Harvard, Stibitz at BTL, and the astronomer Schilt at Columbia University.[22] This was a reasonable reply because he mentioned all the machines already working or close to operational. The astronomer Jan Schilt (1894–1982) was in charge of the "Thomas J. Watson Astronomical Computing Bureau", at Columbia University, after its founder Wallace J. Eckert moved to Washington. This computing center had been equipped with an IBM 601 multiplier, among other machines, of the same type already required from IBM for

---

[19]Ulam 1969, cit. by Aspray 1990a, note 21, p. 260. Italics added.

[20]Cf. Warren Weaver to Marshall Stone, 19 January 1944, and Bigelow to Weaver, 22 April 1944, (NARA) AMP. Cit. by Owens 1988, 291.

[21]Cf. von Neumann to Weaver, 14 January 1944 (UPA). Cit. by Stern 1981, 71.

[22]Cf. Weaver to Emory L. Chafee of Harvard University, 22 March 1944; Weaver to J. von Neumann, 22 March 1944 (VNLC); von Neumann to Weaver 27 March 1944 and 28 June 1944 (HSRC). Cit. by Aspray 1990a, 30, and notes 22 and 24 on page 260.

Los Alamos. In April 1944 von Neumann met Schilt, with whom he agreed to do a computation for the Los Alamos project (Aspray 1990a, 32).

The creation of this computing center had drawn inspiration from a similar one created by Leslie J. Comrie (1893–1950) at the British Nautical Almanac Office using commercial computing machines, among which some from IBM, for creating astronomical tables (Cf. Comrie, L.J., 1943 and 1944).

Even Harvard University had a plan for establishing a computing center for scientific purposes in 1939. The idea had come to the engineer and physicist Howard Aiken (1900–1973), who asked IBM to build a single multi-purpose machine, able to perform various operations. Known as Harvard Mark I, it was delivered in 1944. The machine was actually an electromechanical version of the calculator suggested by Leibniz in the seventeenth century. The numbers were stored in counters consisting of wheels with ten positions. The most significant innovations were: (1) substitution of the classic IBM cards with a punched paper tape (the idea may have come from telegraphy); (2) the adoption of 330 relays, not for binary counting, but to control the rotation of the wheels [Goldstine 1973 (1980, 111); Randell 1982a, 191–194].

In the spring, von Neumann told Nelson and Metropolis about the Aiken's Harvard Mark I and asked them to set, in unclassified form, a problem concerning the implosion already planned for the IBM machine at Los Alamos. This meant setting the Los Alamos problem in such a form that it was impossible to work out its physical nature. Talking about a problem that would be run on the ENIAC (Electronic Numerical Integrator and Computer Analyser), Herman Goldstine explained:

> The Los Alamos problem was classified as far as the underlying physical situation was concerned but not as regards the numerical or mathematical form of the equations to be solved. This policy of keeping the numerical equations unclassified was a wise one that was long maintained. It made possible doing calculations for Los Alamos without obtaining clearances for any of the personnel involved and without having to maintain elaborate security measures in the ENIAC room itself (Goldstine 1980 [1973], 226).

This practice reveals how, outside Los Alamos, a number of people may well have made contributions to building the bomb without being informed about the real purpose. Research on implosion demanded by the end of the war a total of more than twelve computations,[23] most of which were run outside Los Alamos in this way. Early in April, von Neumann also met Stibitz. He considered the conversation with him a real revelation. On 10 April 1944, he wrote to Weaver:

---

[23]We read at the official Los Alamos website: "The first implosion calculation showed that the fissile material would be strongly compressed and that a high yield would result from assembling a relatively small amount of fissile material if a spherically symmetrical implosion was produced. Although much work on explosives, lenses, detonators and other components of the device was required to accomplish this, the Trinity test July 16, 1945, showed that the calculation was correct. About a dozen other calculations of implosion were done to refine it before the end of the war." [50th Anniversary].

I spent the better part of a day with Stibitz, who explained to me in detail the principles and the working of his relay counting mechanisms, and showed me the interpolator as well as the almost-finished anti-aircraft fire-control calculator. We had a very extensive discussion of the possibilities of these machines compared with the IBM type, and their adaptability to the problems in which I am interested. Dr. Stibitz even suggested, what went far beyond my expectations, that he may seek permission for an experimental computation of the kind I suggested on the big machine in the process of its breaking in. [...] At any rate I am staying in contact with Stibitz, and I am extremely glad that you made this possible.[24]

The Stibitz approach, although still based on electromechanical relays, was much more evolved, compared with the IBM computers. Stibitz machines, for example, were provided with three tapes used to enter data and commands into code in a very similar way to what was provided by the MIT Rapid Arithmetical Machine project. Stibitz had also introduced techniques to reduce read and write errors.

On 1 August 1944 von Neumann sent Oppenheimer a detailed report "on the calculating machines" (von Neumann 1944) to bring him up to date on the results of his reconnaissance trips.

7.7   Was Wiener Involved in Work for the Manhattan Project?

In his autobiography, Wiener wishes to make it clear that he was not involved in the Manhattan Project. He tells us that one day, while he was still working on predictors, probably at the end of 1942 at the latest, he was called to Washington by Bush, who suggested a meeting with Harold Urey, director of the sub-project on gaseous diffusion of uranium. After that, Wiener went to New York to talk with Urey, but everything stopped there, maybe because Wiener had shown little enthusiasm for the proposal (Cf. 64g [56g], 293–294).

Nevertheless, Wiener may have been involved indirectly, considering the practice of resorting to external aid, using cryptic methods that concealed the true purposes. Aspray notes, e.g., that introducing von Neumann to the circles of the most advanced research in automatic calculation, Weaver had never mentioned von Neumann's work at Los Alamos (Cf. Aspray 1990a, note 24, page 260).

Concerning the possibility of some theoretical help given by Wiener without his being aware of its true aims, we can only conjecture on the basis of certain clues. One passage in Wiener's autobiography states:

Later on, various young people associated with me were put on the Manhattan Project. They talked to me and to everyone else with a rather disconcerting freedom. At any rate, I gathered that it was their job to solve long chains of differential equations and thereby handle the problem of repeated diffusions. The problem of separating uranium isotopes was reduced to a long chain of diffusions of liquids containing uranium, each stage of which did

---

[24]Von Neumann to Weaver, 10 April 1944, VNLC. Cit. by Aspray 1990a, 32.

a minute amount of separation of the two isotopes, ultimately leading in the sum to a fairly complete separation. Such repeated diffusions were necessary to separate two substances as similar in their physical and chemical properties as the uranium isotopes. I then had a suspicion (which I have, though I know nothing of the detail of the work) that the greater part of this computation was an expensive waste of money. It was explained to me that the effects on which one was working were so vanishingly small that without the greatest possible precision in computation they might have been missed altogether.

This however did not look reasonable to me, because it is exactly under these circumstances of the cumulative use of processes which accomplish very little each time that the standard approximation to a system of differential equations by a single partial differential equation works best (64g [56g], 294).

In this passage Wiener seems to interpret the physical problems underlying the mathematical problem being tackled by "young people associated with" him in the light of his knowledge about uranium separation by diffusion as mentioned by Urey in 1942. But the theoretical and practical problem concerning the separation of uranium had been solved in the early stages of the research and, as far as we know, it had not involved enormous mathematical difficulties, unlike the issues implied by the plutonium project. Actually, a serious theoretical question had arisen at Los Alamos concerning diffusion, but it did not relate to isotope separation by gaseous diffusion. As James Gleick tells us, "the central issues of Los Alamos, too, were problems of diffusion in a new guise. The calculation of critical mass quickly became nothing more or less than a calculation of diffusion." And it is precisely when speaking about this theme that the name of Wiener appears. Gleick adds:

The difference in brightness between the sun's center and its edge gave an indirect means of calculating the nature of the internal diffusion. Or should have - but the mechanics proved difficult until a brilliant young mathematician at MIT, Norbert Wiener, devised a useful method (Gleick 2011 [1992], 172).

Without expecting to reach a definitive conclusion about this episode, which seems to confirm the "extortion" of a contribution from Wiener, it will be useful to relate other details from Metropolis and Nelson. They tell us that, in order to calculate the critical mass, the physicists at Los Alamos had first used "the differential-diffusion approximation". However, although this was "adequate for heatset transfer problems, [it] was grossly inadequate for nuclear reactions," and it overestimated the true value by about eight times. They then adopted "the integral-diffusion equation, in the form of the Wiener-Hopf equation", an equation which, unlike the typical Wiener-Hopf equations, such as those that Wiener had encountered in research on prediction theory, could not be easily "solved analytically". In fact in this case an analytical solution was available only for "a half-infinite plane", while the "solution for finite boundary conditions required numerical methods." (Metropolis and Nelson 1982, 354) .

This account may be accurate, but I would not suggest that it should be considered at a more than vaguely speculative level. What does seem certain, on the contrary, is that Wiener had offered von Neumann an important contribution with regard to the problem of automatic calculation.

Stanislaw Ulam (1909–1984), a mathematician who played an important role in US nuclear research during and after the war, and who had close relations with both von Neumann and Wiener, told of conversations between the two men, most likely in 1938, on turbulence problems in fluid dynamics and their numerical treatment (Cf. Ulam 1980, 94). Note that, in 1938, von Neumann could be considered a layman compared to Wiener, on both fluid dynamics and its numerical treatment. The name of von Neumann does not appear among the participants in the Congress of Applied Mechanics in 1938, while Wiener was among the speakers at the "Symposium Turbulence". In general, as Ulam testified on another occasion:

> For many years von Neumann was very much a pure mathematician. It was only, to my knowledge, just before World War II that he became interested not only in mathematical physics but also in more concrete physical problems (Ulam 1980, 95).

In addition, Wiener's ideas about numerical methods for the solutions of PDEs were very advanced. As already stressed, Richard Southwell himself, unlike Wiener, considered that "any attempt to mechanize relaxation methods would be a waste of time" (Young 1987, 119).

In 1940 Wiener had been appointed Chief Consultant of the War Preparedness Committee for Computation (numerical, mechanical, electrical); and at the same time, von Neumann had become Chief Consultant in Ballistics. In 1941, the report on the activities of the War Preparedness Committee after one year quotes him on the relations subsisting between ballistics and aerodynamics, due to the high speed of bullets fired from planes. Immediately after that, the report added that "Professor Wiener is working on the problem of using this machine or similar machines to solve partial differential equations" (Morse-Hart 1941, 296 and *War Preparedness Committee* 1940).

The friendship, or at least the relationship based on mutual esteem, between these two outstanding mathematicians was very remote, and in early 1944 the discussion on numerical methods to solve the mathematical problems involved in implosion might have been just one stage in a long dialogue, without von Neumann having ever clearly spoken of implosion.

7.8   A First Phase of Cooperation: January–July 1944

In particular Wiener may have helped von Neumann to identify the correct numerical procedure to run on the Los Alamos IBM machines and the Harvard Mark I, to calculate the solution of the PDE describing implosion. Of course this is only a conjecture, but we do have some evidence.

There is a fundamental passage in *Cybernetics* in which Wiener informs us about his cooperation in the creation of the modern computer. He states:

> At this time, the construction of computing machines had proved to be more essential for the war effort than the first opinion of Dr. Bush might have indicated, and was progressing at several centers along lines not too different from those which my earlier report had

indicated. [...] There was a continual going and coming of those interested in these fields. We had an opportunity to communicate our ideas to our colleagues, in particular to Dr. Aiken of Harvard, Dr. von Neumann of the Institute for Advanced Study, and Dr. Goldstine of the Eniac and Edvac machines at the University of Pennsylvania. Everywhere we met with a sympathetic hearing, and the vocabulary of the engineers soon became contaminated with the terms of the neurophysiologist and the psychologist.

At this stage of the proceedings, Dr. von Neumann and myself felt it desirable to hold a joint meeting of all those interested in what we now call cybernetics, and this meeting took place at Princeton in the late winter of 1943–1944 (61c [48f1], 14–15).

Several epistolary sources allow us to correct the date of the "joint meeting", which took place, not in the winter of 1943–'44, but in the next one, in particular on Saturday 6 and Sunday 7 of January 1945.[25] Therefore, we have to place the "going and coming", and Wiener's conversations with Aiken, von Neumann, and Goldstine, during 1944. In the same passage, Wiener adds:

Harvard, Aberdeen Proving Ground, and the University of Pennsylvania were already constructing machines [...]. In this program there was a gradual progress from the mechanical assembly to the electrical assembly, from the scale of ten to the scale of two, from the mechanical relay to the electrical relay, from humanly directed operation to automatically directed operation; and in short, each new machine more than the last was in conformity with the memorandum I had sent Dr. Bush (61c [48f1], 14–15).

In this passage Wiener shows that he knew of the Harvard Mark I and the computing machine operating at the Ballistic Research Laboratory (BRL) at the Aberdeen Proving Ground of the US Army, where in 1944 there was a differential analyzer, an IBM multiplier, and at least one BTL relay computer [Goldstine 1973 (1980, 201). Regarding Wiener's visit to see the Harvard Mark I and his discussion with Aiken, Wiener says in his autobiography:

---

[25]Checking the days of the week with the perpetual calendar we discover that 6 and 7 January 1945 were a Saturday and Sunday, a strange fact for Anglo-Saxon habits, but perhaps consistent with the idea of an informal and largely secret meeting is that everyone was in a hurry at that time, especially von Neumann. On this date, the literature is quite unanimous [Cf. among other Aspray 1990a, notes 46, 47, 48 at p. 315; Heims 1984 [1980], 185; Hellman 1981, 229–230; Piccinini 2003, 79] and it can itself be placed on a solid documentary basis: the summary of the meeting prepared by von Neumann and sent to the participants, dated 12 January 1945 (VNLC) refers to a "meeting on 6–7 January in Princeton." von Neumann sent Wiener two copies of the summary, with a cover letter dated 12 January 1945 (VNLC). The dates of all letters concerning the conference (VNLC, WAMIT, and HGAP) relate to the turn of 1944 and 1945 and are consistent. In any case, Wiener also failed to correct the indication of winter 1943–44, even in the second edition of (61c [48f1], 15), even though many of the errors that plagued the first edition were amended. In addition, Wiener proposes the same date in the autobiography, anticipating by one year—compared to the epistolary sources—a series of events described in the letters, such as his trip to Mexico to attend the conference of the Mexican Mathematical Society in Guadalajara, documented by various letters to von Neumann sent during the course of 1945, and which took place between 28 May and 2 June 1945 (Science, vol. 102, No. 2647, 21 September, 1945, 296–297), a meeting that Wiener stubbornly places a year before (64g [56g], 276 ss.); so Wiener does also for the told of his crisis of conscience as a result of the atomic bombing of Japan, starting from the end of 1944 (Cf. 64g [56g], 293 ss).

It was at Harvard, under the supervision of Howard Aiken, that I found the first of the newer switching computers dependent on relays. Aiken was developing them under a government grant. [...] I was surprised to find that Aiken was completely committed to the relatively slow mechanical relay as the mechanical computer's first tool and that he did not put any enormous value on the speed which could be derived by the use of electronic relays (64g [56g], 266).

As we have seen, Aiken adopted relays, but not to implement the counting devices, which had remained mechanical and decimal. This made the machine slow. One must consider that all the computing machines Wiener visited were based either in military structures like the US Army Aberdeen Proving Ground, or worked for military purposes, like the Harvard Mark I, working for the US Navy,[26] and in some cases we know that they were classified like the ENIAC, which up to the spring of 1945 was "confidential", and after that "restricted" (Goldstine 1980 [1973], 215–216). Therefore Wiener would not have had the opportunity to see them without permission. Taking into account the fact that von Neumann's request to Weaver in January was made in the context of the AMP, one could suppose from Wiener's visits to the various computers operating in that period, that in the spring of 1944 he was asked by the AMP to monitor the same calculators as Weaver had mentioned to von Neumann. It is known that the AMP carried out research on "the development of techniques adapted to the solution of special problems", and on "the nature and capabilities of computing equipment", (Weaver 1946, *Foreword*) and several reports were prepared concerning a *Survey of Computing Machines*.[27] In one report by Stibitz, we find the name of Wiener among the addressees of the distribution lists.

In 1944 Wiener began to work again with Julian Bigelow who, after the end of the collaboration on prediction theory, had joined the Applied Mathematics Group at Columbia University. In April 1944 Weaver turned to Bigelow to ask him advice on how "to apply the much respected talents of Norbert Wiener as a critical check on the quality of mathematical studies—to no avail". Bigelow replied that Wiener had no interest in problem-solving as such, whether it concerned him or others, if this had happened "simply on the basis of utility, particularly if it lacks the qualities suggestive of an elegant, general, formal solution".[28] The only way to engage him

[26]The Harvard Mark I "was taken over by the US Navy Bureau of Ships" (Cohen 2003, 1079).

[27]About the "AMP Study 171" Weaver speaks of a "Survey of Computing Machines" (Weaver, 1946, *Foreword*, VIII). Owens (1996 n. 29, p. 41) says that it is in the archive (NARA), RG 227, OSRD, Records of the Applied Mathematics Panel, Box 62. We know that "AMP Study 171" collects various reports. In particular, about Eckert J.P. et al. 1945, and its presence in "AMP Study 171" was discussed in the trial Honeywell vs Sperry Rand (Cf. Verdict 1973). Stibitz 1945a and 1945b (Cf. Stibitz 1986). It is interesting to note that, among the recipients of the distribution list in Stibitz 1945, we find John von Neumann, Norbert Wiener, E[dmund] C. Berkeley, Howard Aiken, Warren Weaver, Herman Goldstine, and J.G. Brainerd. (Cf. Hook, Norman and Williams 2002, 457).

[28]Bigelow a Weaver, 22 April 1944, (NARA), AMP, cit. by Owens 1988, 295.

—Bigelow advised—was to present him with the problem as "an emergency of catastrophic consequences requiring his decision immediately".[29]

It is also clear that, in the following years, Warren Weaver formed an impression of Wiener as an "expert on computers" (Weaver 1949).

7.9 The ENIAC Before von Neumann

After a first phase in which Wiener must have given advice to von Neumann about numerical methods and also about the possibilities of existing computers, there was a second phase, begun in August 1944, just when von Neumann discovered the existence of the ENIAC (Electronic Numerical Integrator and Computer Analyser) at the Moore School, the only electronic computer then under construction in the USA.

The Moore School of Electrical Engineering at the University of Pennsylvania had a long working relationship with the BRL, on a project concerning a Bush Differential Analyzer to calculate ballistic and bombing tables. In Autumn 1942, the BRL had named Lieutenant (then Captain) Herman Goldstine, a mathematical expert on external ballistics, as a liaison officer for these computing tasks between the laboratory and the school. The latter instead entrusted the contacts with the BRL to Professor John G. Brainerd, a good organizer with some experience in analog and digital computing (Goldstine 1980 [1973], 104).

The ENIAC project was born with the main purpose of creating an electronic digital computer to do the same work, at very high speed, as the Bush's Differential Analyser already present in the school. The first idea came from John W. Mauchly, a physicist who in 1941 had discussions on the subject with John V. Atanasoff, inventor of a small digital electronic computer to calculate systems of linear equations.[30] After a training course in electronics, he had been appointed by the Moore School.

Together Mauchly and the engineer J. Presper Eckert prepared a project which merged most of the experience gathered in the field from 1939 onwards: studies by Bennet Lewis on electronic counters for nuclear detection, recounted in the book Lewis 1942, the thesis for the M.S. by Perry Crawford, (Cf. Eckert J.P. 1980, 531)

---

[29]Ibid.

[30]The name of John Vincent Atanasoff emerged during the last of the legal battles over patents for EDVAC, which began in 1945. The court recognized the priority of invention to him for creating, between 1937 and 1942, with his student Clifford E. Berry, a small electronic computer, the Atanasoff-Berry Computer (ABC), designed to solve large systems of linear equations, based on a device for addition and subtraction with 300 vacuum tubes, derived from the technology used for nuclear computations. The entire dispute over EDVAC and ABC is summarized in detail by Morelli 2001.

that summed up the four years of research, from 1938 until 1942, made at MIT on the Rapid Arithmetical Machine,[31] and some reports about the projects for digital electronic computers by the National Cash Register and Radio Corporation of America (Cf. Goldstine 1973 [1980], 148–149).

From this emerged a project for a machine equipped with a pulser operating at 100,000 pulses per second and 18,000 electronic tubes dissipating 140 kilowatts of power. The numerical data were calculated and stored in decimal format. Considering that the ENIAC had to calculate ballistic and bombing tables, the system setting up the mathematical problems was extremely rudimentary (Cf. Burks 1947).

The project was highly appreciated by the Moore School. The school saw in it a way to profit, like many other small institutions, from the river of funding which flowed from the government in times of war. Previously, they had had two contracts with the NDRC, one for the MIT Radiation Lab, in which Eckert had worked, and another for the Signal Corps, in which Mauchly and Burks had worked.

Despite some concerns about the reliability of an aggregate of 18,000 valves expressed by T.H. Johnson, BRL's electronics expert BRL (Goldstine 1980 (1973), 149), the idea won over both the latter and the Ordnance Department,[32] pressured by the need for ballistic and bombing tables coming from the front. Goldstine explains that a ballistic trajectory required about 750 multiplications. A Bush Analyser did the task in 10–20 min. Since a ballistic table contained about 3,000 trajectories, it had taken a month to complete it. In contrast, the ENIAC, which promised a computing speed 1,000 times faster, would take only a couple of hours. (Cf. Goldstine 1980 [1973], 138).

It is often thought that the main divide to be overcome to reach the modern computer concerned the analog-digital dualism. Actually, this was only partially so. The main dualism was between electromechanical digital and electronic digital.

During 1943, both the Moore school and the Ordnance Department insistently asked the NDRC Division 7 to help fund the ENIAC (Cf. Brainerd 1976; Stern 1981, 18–19). Harold Hazen, director of Division 7, after consulting Caldwell, Weaver and Stibitz, among others, refused to fund the project and expressed extremely negative opinions towards it.[33]

---

[31]Ceruzzi observes that "it appears that this thesis was partially the inspiration for work that Eckert later did at the Moore School." (Ceruzzi 1998, note 88, p. 322).

[32]J.G. Brainerd, "Report on an Electronic Diff. [sic] Analyzer," Submitted to the Ballistic Research Laboratory, Aberdeen Proving Ground, by the Moore School of Electrical Engineering, University of Pennsylvania, First Draft, April 2, 1943. Cit by Goldstine 1973 (1980, 149).

[33]Cf. Harold L. Hazen, Diary of Hazen, 14 April 1943, (NARA) OSRD GP, Project#62, Box 46 and Caldwell to Warren Weaver, 15 May 1943 (NARA) OSRD GP, Ballistics, General Correspondence Folder, Box 80. Cit by Mindell 1996, 455 and 456.

Caldwell's advice was that the Rockefeller Differential Analyzer was about to come into operation and would be able to provide the tables needed by the US Army. In addition, he reminded him that the same Center for Analysis already had another project for an electronic computer, even better than ENIAC, "packed away in camphor balls"[34] for the postwar period.

Actually, from 1940 to 1945, the NDRC's Division 7 supported digital systems, but also the electromechanical digital machines of the BTL and IBM. Funding was denied for every project involving electronic digital computing and there was unrelenting criticism against them. We may mention Wiener's machine for PDEs, the MIT Rapid Arithmetical Machine, two digital electronic designs at the RCA, one a predictor and one a computer *stricto sensu*, and finally, the ENIAC. It is incredible to note that in 1945 the section at the Rockefeller Foundation directed by Weaver refused on advice from the former staff at Division 7 to fund the project for a general purpose computer that von Neumann was planning to create at the Institute for Advanced Study.

The reasons expressed by the NDRC's Section D-2, after becoming Division 7, remained those given during the Conference "Electronic Fire Control Computers" held in New York in April 1942. There were in fact two groups of reasons. On the one hand, there were those who like Harold Hazen preferred analog systems such as "network analyzers" and "Bush's integrators". On the other hand, there were those who preferred the digital approach, but were hostile toward an electronic implementation (Cf. Mindell 1996, 451–2).

One reason shared by all was that there was no need for very fast digital computers. According to Hazen, predictors did not need the higher accuracy provided by the digital electronical approach, because the input signals were coarse and the output would be used by machines that did not require very accurate signals. Those who defended the electromechanical digital approach argued that an aggregate of many electronic tubes would be very unreliable. In contrast, a computation by means of electromechanical relay systems without failure would in the end be faster.[35]

This argument did not convinced members of the military like Colonel Gillon, who replied that an improvement of 1,000 times in the computation speed would largely offset in terms of time the defects in the less reliable electronic systems.[36]

The suspicion that the different approaches were influenced by industrial interests is legitimate. The BTL computers used the basic component of automatic telephone exchanges, namely, relays. The RCA computers used the basic component of radio and TV sets, namely, vacuum tubes. The BTL was an industrial giant, while RCA weighed much less on the stock market.

---

[34]Memorandum by S.H. Caldwell to Harold L. Hazen, 23 October 1943, NDRC. Cit by Wildes e Lindgren 1986, 231. Cf. also Goldstine 1980 [1973], 151.

[35]Cf. Memorandum from G.R. Stibitz to W. Weaver, 6 November 1943, cit. by Goldstine 1980 [1973], 151–2.

[36]Gillon to Hazen, 7 October 1943, cit. by Goldstine 1980 [1973] 152–3.

The US Army did not succeed in drilling through the granite opposition deployed by Division 7. In the end the project remained under the sole responsibility of the Ordnance Department. Work began in mid-May 1943 and the definitive contract was signed on 5 June 1943. Brainerd obtained the position of Administrative Director, while Eckert was chief engineer, and Mauchly become the principal consultant, at least until the arrival of von Neumann (Cf. Brainerd 1976 and Goldstine 1980 [1973], 150).

In Caldwell's opinion the ENIAC project was worse than those previously proposed by National Cash Register, Radio Corporation of America, and the Massachusetts Institute of Technology. This view coincided also with those of some RCA researchers. Rajchman in particular judged it "extraordinarily naïve" (Stern 1981, 44).

Actually, one of the main limitations of the ENIAC was common to all other decimal computing machines. The basic technology of seventeenth century mechanical calculators by Wilhelm Schickard, Blaise Pascal, and Gottfried Wilhelm Leibniz (Cf. Williams 1997, 118–136) consisted in a succession of wheels provided with 10 possible positions. Each of these wheels represented respectively units, tens, hundreds, and so on, and were coordinated in such a way that a complete turn of the units wheel induced one step in the tens wheel, and so on.

In the decimal electronic calculators there were devices used either as arithmetic devices or as storage devices for the numerical data, called "totalizers" in the MIT machine and "accumulators" in the ENIAC. They followed the same seventeenth century method, but with electronic technology. An accumulator was composed of a succession of rings, each comprising 10 flip-flops (coupled triodes), indicating the digits from 0 to 9, in order to represent units, tens, hundreds, and so on (cf., e.g., Bush 1940, 338). With this way of representing a 10-digit number they needed 200 triodes, plus a few more triodes to represents the sign.

Several special kinds of tubes were invented in the previous decimal electronic computers to reduce the number of triodes, e.g., the MIT created the Digitron, a tube containing ten flip-flops [Cfr. Desch 1942a, 8]. This decreased the number of tubes, but it did also increase costs.

The ENIAC designers introduced some smart tricks. First of all, instead of creating special tubes, they adopted a commercial triode, mounted in such a way that it could be easily changed when broken. To increase reliability, they avoided turning the machine on and off, to eliminate the peaks due to transients. Furthermore, they realized that a commercial triode used in a couple to form a flip-flop could work well below the voltage needed in a radioset. Consequently, the breakdown rate did not exceed 2 or 3 tubes per week, a number slightly higher than that of the relay computers, but which, considering the high speed of calculation, meant a much lower failure rate for operation (Cf. Burks 1947 and Goldstine 1980 [1973], 144–145).

At the beginning of July 1944, two accumulators began to run on the machine, which performed addition and subtraction, and managed to solve differential equations of the second order. In Goldstine's opinion, "this point, roughly one year after its inception, was the water-shed of the project. In that time the project had gone from a highly speculative idea to an engineering status that ensured its success." (Goldstine 1980 [1973], 164).

7.10   The ENIAC from von Neumann's Arrival

In January 1944 Weaver had made no mention of the existence of this controversial project, of which he had been aware at least since 15 May 1943.[37] There was no word about the ENIAC, even at the meeting at the end of July 1944, at the Ballistic Research Laboratory in Aberdeen, where Herman Goldstine was also present (Cf. von Neumann 1944).

By chance, according to Goldstine 1973 (1980, 182), von Neumann heard about the ENIAC during a casual conversation at the Aberdeen train station, perhaps just after returning from the meeting at the BRL, a conversation which von Neumann did not mention to Oppenheimer in the report of 1 August 1944. Von Neumann immediately asked for permission to see the ENIAC, and was already at the Moore School on 7 August 1944, with clear ideas on how to adapt an electronic digital computer to solve PDEs.[38] On 11 August 1944 Goldstine sent a memo to Leslie E. Simon, director of the BRL, in which he stressed the "paucity of high speed storage devices", and proposed a new contract with the BRL "with the object of building ultimately a new ENIAC of improved design".[39]

Throughout the month of August 1944, von Neumann had frequent conversations with Eckert, Mauchly, Burks, Goldstine, and his wife Adele (Cf. Goldstine 1980 [1973], 186). On 21 August 1944 Goldstine wrote to the Assistant Director of the BRL, Paul N. Gillon:

> von Neumann is displaying great interest in the ENIAC and is conferring with me weekly on the use of the machine. He is working on the aerodynamical problems of blast [...] (Goldstine 1980 [1973], 199).

---

[37]On 15 May 1943 Caldwell had written to Weaver: "There is a certain amount of agitation, coming primarily from Brainerd at the University of Pennsylvania, for the development of an electronic differential analyzer to do high-speed numerical integration. This is a huge undertaking. I doubt that it can be finished until five years after the war is over." (NARA) OSRD GP, Ballistics, General Correspondence Folder, Box 80, cit. by Mindell 1996, 456.

[38]On 5 September 1944, von Neumann asked the Moore School for official permission to see the ENIAC for two months from 7 September 1944 (HSRC). But from a letter by J.G. Brainerd to Paul N. Gillon, on 13 September 1944 (HGAP), we can argue that he had been already to the Moore School before having "extensive discussions" with the ENIAC staff about scientific problems with the computer. To do this von Neumann had the permission from the BRL at the Aberdeen base. Cf. Aspray 1990a, note 50, pp. 262–3.

[39]Goldstine to Simon, "Further Research and Development ENIAC," 11 August 1944. Cit. by Goldstine 1973 (1980, 185).

The ENIAC required two kinds of alteration. On the one hand, after an interview with S.B. Williams, i.e., the BTL engineer collaborating with Stibitz they would have to replace the device for manual reconfiguration of problems using a relay system controlled by a teletypewriter tape. On the other hand, it was planned to provide:

> [...] a more economical electronic device for storing data than the accumulator. Inasmuch as the accumulator is so powerful an instrument, it seems foolish to tie up such tools merely to hold numbers temporarily. Eckert has some ex-cellent ideas on a very cheap device for this purpose.[40]

In none of the electronic digital computer projects hitherto proposed, apart from the 1940 Wiener machine, had the idea of "a more economical electronic device for storing data than the accumulator [...] to hold numbers temporarily" been introduced. The other suggestion instead was independent of Wiener and referred to the conversations that von Neumann had had with Stibitz and S.B. Williams.

On 2 September 1944 Goldstine wrote again to Gillon about the device for storing data, explaining that "to solve a quite complex partial differential equation of von Neumann's", the ENIAC would have required 30 min, of which 28 would serve

> just in card cutting and 2 min for computing. The card cutting is needed simply because the solution of partial differential equations requires the temporary storage of large amounts of data. We hope to build a cheap high-speed device for this purpose.[41]

The point came back on 13 September 1944 in a letter from Brainerd, administrative director of the ENIAC, to Gillon. He wrote:

> The progress of work on the ENIAC has led to some rather extensive discussions concerning the solution of problems of a type for which the ENIAC was not designed. In particular, these discussions have been carried out with Dr. von Neumann, who is consultant to BRL on hydrody namical and aerodynamical problems associated with projectile motion. Dr. von Neumann is particularly interested in mathematical analyses which are the logical accompaniment of the experimental work which will be carried out in the supersonic wind tunnels. [...]

> It is not feasible to increase the storage capacity of the ENIAC [...] to the extent necessary for handling non-linear partial differential equations on a practical basis. The problem requires an entirely new approach. At the present time we know of two principles which might be used as a basis. One is the possible use of iconoscope tubes, concerning which Dr. von Neumann has talked to Dr. Zworykin of the R.C.A. Research Laboratories, and another of which is the use of storage in a delay line, with which we have some experience. Such a line could store a large number of characters in a relatively small space, and would [...] enable a machine of moderate size to be constructed for the solution of partial differential equations which now block progress in certain fields of research at the BRL.[42]

---

[40]Goldstine to Gillon, 21 August 1944, cit. by Goldstine 1980 [1973], 199.

[41]Goldstine to Gillon, 2 September 1944, cit. ibid, 198.

[42]Brainerd to Gillon, 13 September1944 (HGAP). Cit. by Goldstine 1980 [1973], 187, integrated with the quotation from Aspray 1990a, 37.

In order to meet the requirements of Wiener's *Memorandum on PDEs*, the new ENIAC would be modified for the same purposes, containing large, fast, and rewritable data storage for a huge quantity of temporary data. This device could be obtained by means of iconoscope tubes, according to another of the technical suggestions in Wiener's *Memorandum on PDEs*.

Shortly afterwards, at the Moore School, two groups were created, one that continued to work on the construction of a modified ENIAC, introducing the improvements required by von Neumann; and another that focused on a machine designed ex novo, which would soon be called EDVAC (Electronic Discrete Variable Calculator). Its construction would begin on 1 January 1945, but on a smaller scale, would already be initiated on 1 October 1944. (Cf. Goldstine 1980 [1973], 187)

Coming back to the question of data storage, on 31 August 1944 Brainerd sent Goldstine a report in which Eckert and Mauchly proposed to implement the storage device requested by von Neumann using mercury delay lines. In these delay lines, the electrical signal is cyclically transduced into acoustic signals and vice versa (Cf. Goldstine 1980 [1973], 185). Brainerd (1976, 485) made it clear that Eckert's main contribution in adopting already existing delay lines had essentially consisted in replacing water by mercury. This was an innovation that Eckert had introduced for the contract between the Moore School and the MIT Rad Lab.

Maybe for an engineer the fact of inventing a new device is enough to deserve credit. However, the real novelty with the mercury delay lines consisted in the new purpose they were used for, i.e., as data storage devices. This was a different purpose from the one they had in radar technology. In the neural networks of McCulloch and Pitts, it had been expected, at the suggestion of Pitts, that closed circuits of neurons could serve as memories. Although the issue may look rather like the question of which came first, the chicken or the egg, a passage in *Cybernetics* seems to justify our thinking that the adoption of the delay line as memory for the ENIAC was suggested by reading *A logical calculus* (McCulloch and Pitts 1943), if not by Pitts himself. Actually, Wiener writes:

> To return to the problem of memory, a very satisfactory method for constructing a short-time memory is to keep a sequence of impulses traveling around a closed circuit until this circuit is cleared by intervention from outside. There is much reason to believe that this happens in our brains during the retention of impulses, which occurs over what is known as the specious present. *This method has been imitated in several devices which have been used in computing machines, or at least suggested for such a use* (61c [48f1], 121–122. Italics added; cf. also McCulloch 1974).

The reference to the mercury delay lines adopted as memories for the ENIAC and EDVAC is evident.

7.11   Meanwhile, Wiener with Pitts, Bigelow, and von Neumann …

Wiener's stamp is evident in the proposed changes. And von Neumann cannot only have used his memorandum on PDEs. It should be noted that in *Cybernetics*, Wiener does not give a simple summary of the content of his *Memorandum on PDEs*, but a list of "suggestions", which is slightly but significantly different compared to 1940. In particular, Wiener recommends:

1. That the central adding and multiplying apparatus of the computing machine should be numerical, as in an ordinary adding machine, rather than on a basis of measurement, as in the Bush differential analyzer.
2. That these mechanisms, which are essentially switching devices, should depend on electronic tubes rather than on gears or mechanical relays, in order to secure quicker action.
3. That, in accordance with the policy adopted in some existing apparatus of the Bell Telephone Laboratories, it would probably be more economical in apparatus to adopt the scale of two for addition and multiplication, rather than the scale of ten.
4. That the entire sequence of operations be laid out on the machine itself so that there should be no human intervention from the time the data were entered until the final results should be taken off, and *that all logical decisions necessary for this should be built into the machine itself.*
5. That the machine contain an apparatus for the storage of data which should record them quickly, hold them firmly until erasure, read them quickly, erase them quickly, and then be immediately available for the storage of new material (61c [48f1], 4. Italics added).

Here we find something more than what the *Memorandum on PDEs* contains. In the latter we find the draft for a real machine design, where these rules are actually applied, but without explicitly describing the inspiring principles, which we find instead in the list quoted above. Making an accurate comparison, Wiener's machine of 1940 can be considered as an application of these principles. There is only one substantial difference: the recommendation "*that all logical decisions necessary for this should be built into the machine itself*" is only very partially implemented in the *Memorandum on PDEs*.

One might think that, if the list of suggestions was ever given to someone in such a didactic form, the recipient was not Bush in 1940, but von Neumann in 1944. However, as we have seen in the previous section, those suggestions were precisely the guidelines that were adopted practically by von Neumann from 1 August 1944 onwards, when suggesting his changes for the ENIAC and then when designing the EDVAC, although the latter, in the form it takes in the *First Draft for EDVAC*, certainly represents a more advanced step than what is proposed in Wiener's *Memorandum on PDEs*. And we also know that there were extensive conversations with its author. Wiener himself states:

At this time, the construction of computing machines had proved to be more essential for the war effort than the first opinion of Dr. Bush might have indicated, and was progressing at several centers along lines not too different from those which *my earlier report* had indicated. [...] There was a continual going and coming of those interested in these fields. We had an opportunity to communicate our ideas to our colleagues, in particular to [...] Dr. von Neumann of the Institute for Advanced Study, and Dr. Goldstine of the Eniac and Edvac machines at the University of Pennsylvania (61c [48f1], 14–15. Italics added).

We are now in 1944. As we have already seen, Wiener in the context we know concerning automatic calculation, had already resumed contact with Bigelow. We know precisely from Bigelow that Wiener took a copy of *A logical Calculus*, published in December 1943, to von Neumann (Cf. Aspray 1990b, note 13, p. 297). During the second half of 1944, Wiener reconnected his working relationship with Pitts, although the latter was committed in New York to the Manhattan Project, directly under General Leslie Groves, who was directing the entire project.

In early August of 1944 Wiener, Bigelow, and Pitts met in New York. Bigelow in this period was in charge of the AMP station c/o Columbia University, where he was working on a small computing system. Bigelow wrote to Wiener a few days later to express his desire to cooperate with the MIT Center of Analysis after the war, with a view to building a computer that could be at the height of MIT prestige. On 17 October 1944 Wiener wrote to Pitts: "I have a lot to talk over with you [...] concerning computing machine theory."[43] He also asked von Neumann if they could meet Pitts in New York [Cfr. Piccinini 2003, p. 77]. Two days later, Wiener was actually in New York.[44] Regarding the content of this collective work, we do not know many things. There is precious evidence, but it remains fragmentary.

Resuming his work in autumn 1945, Wiener wrote to Giorgio de Santillana:

In conjunction with Walter [Pitts] and von Neumann, I have very definite ideas as to the relation between computing machines and the brain, mainly that both are in essence switching devices where the conjunction of the particular complexities of open and closed channels offer new channels to be open or closed as the case may be.[45]

And suggesting Bigelow as his chief engineer for the IAS computer project in 1946, von Neumann states:

He [Bigelow] has a profound interest in automatic computing and control which is clearly a very important asset in this work. Part of his war work was done in one group with Norbert Wiener from M.I.T. whose ideas in this field are very significant.[46]

In addition there is a letter from von Neumann to Wiener, which we shall discuss in more detail later on, from which it is clear how von Neumann's conception of the general purpose digital computer had sprung from his close collaboration with Wiener and Pitts, to reach a very high level of theoretical generality. Von Neumann wrote to Wiener:

---

[43]Wiener to Pitts, 17 October 1944 (WAMIT) Box 4, ff. 66, cit. by Piccinini 2003, 77.

[44]Wiener to Rosenblueth, 19 October 1944 (WAMIT), Box 4, ff. 66. Cit. by Piccinini 2003, 77.

[45]Wiener to Giorgio de Santillana, 16 October 1945 (WAMIT), box 2, folder 69. Cit. by Hellman 1981, 242.

[46]Von Neumann, "Report of Computer Project," 16 March 1946. Cit. by Goldstine 1980 [1973], 252.

Our thoughts – I mean yours and Pitts' and mine – were so far mainly focused on the subject of neurology, and more specifically on the human nervous system, and there primarily on the central nervous system. Thus, in trying to understand the function of automata and the general principles governing them, we selected for prompt action the most complicated object under the sun – literally. In spite of its formidable complexity this subject has yielded very interesting information under the pressure of the efforts of Pitts and McCulloch, Pitts, Wiener and Rosenblueth. Our thinking—or at any rate mine—on the entire subject of automata would be much more muddled than it is, if these extremely bold efforts—with which I would like to put on one par the very un-neurological thesis of A. Turing—had not been made.[47]

However, this collaboration must have occurred, not only during the "comings and goings" of 1944 that preceded the meeting in Princeton on 6 and 7 January 1945, but also during and after it.

---

[47]Von Neumann to Wiener, 29 November 1946 (VNLC), General Correspondence, Box 7.

Chapter 8
From the Princeton Meeting to Hiroshima

8.1 The Preliminary Phase of the Princeton Meeting

Concluding the passage in *Cybernetics* which tells us about the comings and goings for consultations with Aiken, von Neumann, and Goldstine, Wiener asserts:

> At this stage of the proceedings, Dr. von Neumann and myself felt it desirable to hold a joint meeting of all those interested in what we now call cybernetics, and this meeting took place at Princeton in the late winter of 1943–1944 [sic, true date: 1944–1945] (61c [48f1], 15).

The early idea from which the Princeton meeting sprang was a conversation between Wiener and Aiken. On 17 October 1944, Wiener wrote to von Neumann:

> I saw Aiken the other day and I am much impressed by the coincidence of his point of view with mine on the future of computing machines. We are thinking of trying to get some meeting of the American Society of the Advancement of Science after the war is over for the discussion of the whole complex problems relating to computing machines, communication engineering, prediction theory, and control engineering. I would like to get your point of view on that.[1]

Two days later Wiener also acquainted Arturo Rosenblueth with his discussion with Aiken and the hypothesis of the conference at the American Society of the Advancement of Science, to deal with "the complex of subjects in which you and I are interested". He added:

> If we do that we shall try to make it a matter of invited addresses of perhaps a half an hour each by a mathematician, physiologist, electrical engineer, statistician, etc. Certainly, this wants to be done in such a way that you can participate in it.[2]

---

[1]Wiener to von Neumann, 17 October 1944 (WAMIT) Box 4, ff 66. Cit. by Piccinini 2003, 77.
[2]Wiener to Rosenblueth, 19 October 1944 (WAMIT) Box 4, ff 66. Cit. ibid., 78.

© Springer International Publishing AG 2017
L. Montagnini, *Harmonies of Disorder*, Springer Biographies,
DOI 10.1007/978-3-319-50657-9_8

If we compare the two letters, there appears a duality in the way of understanding the new field that Wiener called "cybernetics". In the first letter it is
intended as a generalized form of communication engineering, centered on information, according to the ideas of the *Yellow peril* (1 February 1942). It was a theory
that now firmly included both analog and digital computing machines. We could
perhaps speak of "cybernetics on a large scale", based on a rigorous and highly
theoretical approach. In the second letter, another kind of cybernetics appears, in
which physiology plays an important role, in line with the model represented by
Behavior, Purpose and Teleology. Here we could speak of "cybernetics on a small
scale", based on a meticulous experimental approach. Wiener probably hoped that
the two approaches could be brought together. For the moment he preferred not to
clarify this duality.

However, as we read in *Cybernetics*, both von Neumann and Wiener "felt it
desirable to hold a joint meeting of all those interested in what we now call
cybernetics, and this meeting took place at Princeton" (61c [48f1], 15). Indeed, von
Neumann asked to increase the number of guests, but required a very low profile.

On 4 December 1944 a letter of invitation was sent for a meeting to be held at
Princeton University. It was actually written by Wiener, but also in the name of von
Neumann and Aiken. Its recipients were Walter H. Pitts, Warren S. McCulloch,
Rafael Lorente de Nó, Samuel S. Wilks, Ernest H. Vestine, W. Edwards Deming,
and Leland E. Cunningham. A few days later Herman Goldstine was added.[3]
Wiener's letter explained that:

> Owing to the war, it is not yet the time to call together a completely open meeting on the
> matter, because so many researches developed in the war effort are concerned, but it seems
> highly desirable to summon together a small group of those interested to discuss questions
> of common interest and make plans for the future development of this field of effort, *which
> as yet is not even named.*[4]

The meeting concerned what I have called "cybernetics on a large scale". As
Wiener writes in the letter:

---

[3]Letter from Aiken, von Neumann and Wiener to S.S. Wilks, W.H. Pitts, E.H. Vestine, W.E.
Deming, W.S. McCulloch, R. Lorente de Nó e Leland E. Cunningham, December 4, 1944
(WAMIT). Cit. by Hellman 1981, 229–230. The official senders were Wiener, Aiken, and von
Neumann, but it was written by Wiener (Cf. Heims 1984 [1980], 185). By letter on 16 December
1944, von Neumann asked Wiener to add the name of Goldstine (WAMIT, Box 4, ff. 66, cit. by
Piccinini 2003, 79), which explains why, in the Goldstine Archive (HGAP), there is no copy of
Wiener's letter of 4 December (Cf. Aspray 1990a, note 46, p. 183). Another letter dated 22
December 1944 was sent to Goldstine (Cf. Goldstine 1980 [1973], 275). von Neumann meanwhile
had written to Goldstine on 10 December 1944 (in HSRC. Cf. Aspray 1990a, note 62, p. 263].
[4]Aiken, von Neumann and Wiener to S.S. Wilks, W.H. Pitts, E.H. Vestine, W.E. Deming, W.S.
McCulloch, R. Lorente de Nó e Leland E. Cunningham, December 4, 1944 (WAMIT). Cit. by
Hellman 1981, 229–230. Italics added.

Communication engineering, the engineering of computing machines, the engineering of control devices, the mathematics of time series in statistics, and the communication and control aspects of the nervous system, [...] have developed to a degree of intimacy that makes a get-together meeting between people interested in them highly desirable.[5]

It was stressed that this was a "field of effort, which as yet is not even named".[6]*Cybernetics* would return to this issue, speaking of an "essential unity of the set of problems centering about communication, control, and statistical mechanics, whether in the machine or in living tissue", but hampered by "the lack of unity of the literature concerning these problems, and by the absence of any common terminology, *or even of a single name for the field*" (61c [48f1], 11; italics added), a problem, he adds, that had then been resolved by entitling the book "Cybernetics" (61c [48f1], 11).

Wiener raised the issue of the name for the field again in a letter dated 28 December 1944, proposing to discuss it during the meeting. For the moment he suggested establishing a permanent society that could be called the "Teleological Society", while "Teleology" or "Teleologia", could be the title of the associated scientific journal.[7]

8.2   The Princeton Meeting

The meeting was held on 6 and 7 January 1945. Aiken was unable to participate, but does not seem to have withdrawn because he was assigned to one of the four working groups formed at the meeting. On 24 January 1945 Wiener wrote to Rosenblueth:

[It] was a great success [...]. The first day von Neumann spoke on computing machines and I spoke on communication engineering. The second day Lorente de Nó and McCulloch joined forces for a very convincing presentation of the present status of the problem of the organization of the brain. In the end we were all convinced that the subject embracing both the engineering and neurology aspects is essentially one, and we should go ahead with plans to embody these ideas in a permanent program of research [...]. While we are going to meet again in the spring, we have not organized in a formal permanent society.[8]

We can interpret that Wiener must have provided a unified basis for discussion when presenting his idea of a generalized communication theory, with a focus on

---

[5]Aiken, von Neumann and Wiener to S.S. Wilks, W.H. Pitts, E.H. Vestine, W.E. Deming, W.S. McCulloch, R. Lorente de Nó, and Leland E. Cunningham, December 4, 1944 (WAMIT). Cit. by Hellman 1981, 229–230.

[6]Ibid.

[7]Aiken, von Neumann and Wiener to Goldstein [sic], 28 December 1944 (WAMIT). Cit. by Hellman 1981, 231.

[8]Wiener to Rosenblueth, 24 January 1945 (WAMIT). Cit. by Heims 1984 [1980], 185–186; by Hellman 1981, 231; by Aspray 1990a, note 53, p. 316.

the ideas of the *Yellow peril*, and their extension to computers, in short what I have called "cybernetics on a large scale". McCulloch must have spoken about the properties of the neural net of *A logical calculus,* explaining their similarity with networks formed by relays in the digital computers. It remains to explain the presence of Lorente de Nó. An eminent neuro-anatomist and pupil of Ramon y Cajal, he had studied in vivo the reverberating circuits in the brain, demonstrating their importance in the vestibular nystagmus (involuntary rhythmic oscillation of the eyes) and in other neurological situations (Lorente de Nó 1938). We have already mentioned the relation between the supposed memory function of closed circuits of neurons, and the mercury delay line of ENIAC and EDVAC.[9] Therefore we can suppose that Lorente de Nó spoke about this theme.

In this period von Neumann had taken the entire field of *"computing machines"* firmly in his hands. We know nothing directly of his talk. We can form a draft idea of what he said, considering the content of seminars and informal conversations he gave around that time in Los Alamos, as related by Nelson and Metropolis. He regularly use to discuss the latest developments in computing, including the Harvard Mark I, BTL relay computers, and the ENIAC, and had formed a personal view of the whole subject (Metropolis and Nelson 1982, 351). In particular, he had become to see:

> [...] the technical links between the separate independent developments. He also described his ideas on the computer of the future, outlining his single-address architecture, later implemented in the IAS computer, in the IBM 701, and in other computers. [...] In addition to information about digital-computer development, von Neumann told us about conversations he had had with McCulloch and Pitts, who were investigating brain functioning using Boolean algebra. Their ideas stimulated his thinking about advanced, digital concepts (Metropolis and Nelson 1982, 352 and 354).

Wiener was very satisfied with the meeting, which he considered "a great success".[10] "This meeting", he would state triumphantly in his autobiography, "I may consider the birthplace of the new science of cybernetics, or the theory of communication and control in the machine and in the living organism" (64g [56g], 269).

Regarding the "Teleological Society", its establishment was postponed until after the war. "The reasons are", Wiener explained to Rosenblueth, "that owing to the control of different and by no means unified government departments over parts of our program, it is best not to stir up a fuss until the military situation makes matters of classification less important".[11] Furthermore, the choice of name for the new field and the public communication of research in it had also been postponed. The conference was held in a strictly confidential manner at the behest of von

---

[9]See the above section "Meanwhile, Wiener with Pitts, Bigelow, and von Neumann ...".
[10]Wiener to Rosenblueth, 24 January 1945 (WAMIT). Cit. by Heims 1984 [1980], 185–186; Hellman 1981, 231; Aspray 1990a, note 53, p. 316.
[11]Ibid.

Neumann. "However", Wiener assured Rosenblueth, "we definitely do have the intention of organizing a society and a journal after the war, and founding at Tech or elsewhere in the country a center of research in our new field".[12]

Possible funding sources for the project were also discussed: the J.S. Guggenheim Memorial Foundation, through one of its leaders, Henry A. Moe and the Rockefeller Foundation through Warren Weaver. In addition, Wiener had "heard from von Neumann mysterious words concerning some 30 megabucks which is likely to be available for scientific research. von Neumann is quite confident that he can siphon some of it off". In general, Wiener was enthusiastic and confident about the future of the project, whereas "McCulloch and von Neumann are very slick organizers".[13]

Von Neumann prepared a "Memorandum Summary of the meeting", of which a copy was sent to each participant with the date 12 January 1945.[14] He focused on the division of several tasks and the future agenda, referring to the letter of invitation for the more general view regarding the new field.[15] The working groups were the following: "1" on "Filtering and prediction problems", comprising Pitts and Wiener; "2" on "Application of fast, mechanized computing methods to statistical problems", with statisticians Deming, Vestine, Wilks, and if possible, von Neumann himself; "3" on "Application to differential equations (astronomy, ballistics, hydrodynamics, etc.)". It seems to me that here von Neumann took for implicit, without repeating it again, the specification "Application of fast, mechanized computing methods to". It is no accident that the components of this group were experts in computer design, such as Aiken of the Harvard Mark I, Leland E. Cunningham, astronomer in charge of the BRL automatic computing section (Cf. von Neumann 1944), Goldstine, and von Neumann. Finally, there was a group "4", concerning the "Connected aspects of neurology", comprising Lorente de Nó, McCulloch, and Pitts.[16]

Each group had the task of drawing up "preliminary draft memoranda", within 3 months, i.e., by the end of March or the beginning of April 1945. After 6 months, so by the end of June or the beginning of July, there would be a new general meeting,

---

[12]Ibid.

[13]Ibid.

[14]Von Neumann sent Wiener two "extra" copies of the document. The sender was only von Neumann and the recipients Howard Aiken, Leland E. Cunningham, W.E. Deming, H.H. Goldstine, W.S. McCulloch, Walter H. Pitts, E.H. Vestine, Norbert Wiener, and S.S. Wilks, 12 January 1945 (VNLC) General Correspondence, Box 7, Wiener. Hellman 1981, 231–2, citing the copy contained in WAMIT calls it "Memorandum Summary of the meeting"; in the original I received from the Library of Congress there is no reference to this title.

[15]It contains a description of the field in which we are interested, only in a rather oblique manner. von Neumann to Wiener, 12 January 1945, (VNLC), General Correspondence, Box 7, Norbert Wiener.

[16]Von Neumann to Messrs. Howard Aiken, Lelnd E. Cunningham, W.E. Deming, H.H. Goldstine, W.S. McCulloch, Walter H. Pitts, E.H. Vestine, Norbert Wiener, and S.S. Wilks, 12 January 1945, (VNLC), General Correspondence, Box 7, Wiener.

similar to the January one, where "a joint final memorandum of the research program will be undertaken and a small group put in charge of writing it".[17]

8.3   The Work of the Four Groups

What results did the four groups produce? The usual thesis among historians of cybernetics and computer science is that the project ran aground. Considering the secrecy in which the groups worked, the fact that there is a shortage of documents (not an absolute absence, as a matter of fact) does not seem to authorize such a drastic conclusion. Bigelow, asked by Aspray about the fate of the teleological society, "attributes this failure in part to a clash of personalities among the organizers".[18]

The fact is irrefutable, but the clashes—we have to use the plural because there were at least two—which actually provoked a cataclysm within the new field of cybernetics occurred after August 1945, i.e., after the nuclear bombing of Japan. As we shall see later, there was a clash, or at least a strong cooling of their relations, between Wiener and von Neumann in the Fall of 1945, after Hiroshima and Nagasaki. A second clash, with Wiener on one side and McCulloch and Pitts on the other, happened between December 1951 and January 1952 (Cf. Conway and Siegelman 2005, 213–234). However, from the amount of letters in our possession, no sign of a clash emerges at least before August 1945, the date of the nuclear bombing of Japan. Indeed, the 7 months between the conference in Princeton and the end of the war were marked by a genuine honeymoon between Wiener and von Neumann, two men forming a sort of hub around which the activities of the others were spun enthusiastically.

Nothing suggests that the program proposed by von Neumann in the *Summary* was not applied in the seven first months of 1945. After reading it, Wiener quickly replied to von Neumann, stating:

> I found one thing missing in your assignment of topics: namely, there was no single place where the problem of transition from the computing machine to the control machine was discussed. I think this is one of the most important aspects of our project and as it is closely related to the prediction and filtering problems, I have assumed that it goes to our sub-committee for a report.[19]

---

[17]Ibid.

[18]Aspray (1990b, 302) quotes a personal communication with Bigelow of 6 November 1987. In another place he states that "plans for the Teleological Society fell apart" (Aspray 1990a, 185).

[19]Wiener to von Neumann, 24 January 1945 (WAMIT). Cit. by Hellman 1981, p. 233 note 17. See also Aspray 1990a, note 52, 315–6.

On 1 February 1945 von Neumann answered Wiener, saying he was "most interested to see the memorandum by you and Pitts as soon as it is ready".[20] He also claimed to have "thought a good deal more about transitions and intermediate forms between counting and continuous processes, and I have some ideas on the subject I should like to discuss with you".[21]

Moreover, von Neumann informed him: "after some discussions with Cunningham and Goldstine I think it would be desirable to add Eckert of the University of Pennsylvania and Chandrasekhar to our group, and possibly also G. Stibitz".[22] On 23 January 1945, von Neumann wrote to Deming expressing his wish to include Harold Hotelling (1895–1973) in the group of statisticians.[23] It therefore appears clear that the groups were working.

I do not know of letters between Wiener and von Neumann from 1 February to late March. On 24 March 1945 Wiener wrote to von Neumann: "our meeting last Monday was a grand success, and hope we shall have many more of them".[24]

It seems to me that these words could not refer to a face to face meeting. It is easier to imagine a conference, in which Wiener would also have taken part in a active way. Through the perpetual calendar we discover that the previous Monday was 19 March 1945. On that date a conference was held at the IAS in Princeton on "Shock-waves and Supersonic Flow", where von Neumann had presented a paper on "Refraction, Intersection and Reflection of Shock Waves".[25] Wiener also tells of having spoken with Oswald Veblen, professor at the IAS in Princeton, and one of its main fathers, along with ballistic research in the United States. Veblen had also been a real mentor for von Neumann in the United States. On 26 March 1945 von Neumann also sent Veblen a memorandum entitled "Use of variational methods in hydrodynamics", in which he discussed nonlinear PDEs, especially of hyperbolic or mixed type, that is, those PDEs that one typically meets in the problems discussed in the paper he gave at the conference.

In the memorandum to Veblen, von Neumann concluded that he would make a systematic attack on hydrodynamics, both by introducing new analytical perspectives, and by using the new high speed computing machines. He began to question the inner nature of the numerical difficulties posed by hydrodynamic problems.[26] Metropolis and Nelson call von Neumann's new approach "experimental mathematics" (Metropolis and Nelson 1982, 348). It implied that computational work should not just be a shortcut to get solutions to mathematical problems for which

---

[20]Von Neumann to Wiener, 1 February 1945 (VNLC) General Correspondence, Box 7.

[21]Ibid.

[22]Ibid.

[23]Von Neumann to Deming, 23 January 1945 (VNLC), cit. by von Neumann 2005, 95–96.

[24]Wiener to von Neumann, 24 March 1945 (VNLC), General Correspondence, Box 7.

[25]Von Neumann gave a paper on "Refraction, intersection, and reflection of shock waves". In von Neumann 1963, 300–308.

[26]Von Neumann, *Use of Variational Methods in Hydrodynamics*, Memorandum from J. von Neumann to O. Veblen, 26 March 1945. In von Neumann 1963, 357–359.

analytical solutions were not available. On the contrary, von Neumann thought that, using results obtained numerically, one could find new analytical methods. At about the time of the memorandum to Veblen, von Neumann began to think, as he had then written to Commodore Strauss in October, that:

> An electronic machine of the most advanced conceivable type should be constructed, not for use on specific applied mathematical or physical or engineering problems, but with the purpose of experimentation with the machine itself in order to develop new approximation and computing methods, and generally to acquire the mathematical and logical forms of thinking which are necessary for the really efficient operation of such a device, with the methods it will have brought into existence.[27]

Von Neumann believed that this line of research would best be conducted within an academic environment, rather than in military or industrial laboratories "which have definite, and necessarily relatively narrowly defined, applied problems to which they must devote all or most of the time of their equipment".[28]

Wiener immediately welcomed von Neumann's ideas in an encouraging way. On 24 March 1945 he wrote to von Neumann: "Veblen told me about your post-war plans for hydrodynamics. I think your balance between pure and applied maths is the right one and our little control scheme fits perfectly into the picture".[29]

Wiener considered that their common plan for the Teleological Society would be perfect for this project. However, he thought that the IAS was not the right place to carry out von Neumann's ideas, since they would require laboratories, and noted that "labs don't grow in ivory towers".[30]

Wiener informed also von Neumann that he had spoken about him with George R. Harrison, dean of the MIT Faculty of Science, in view of an appointment of von Neumann as director of the Department of Mathematics, to replace Henry B. Phillips who was about to retire.[31] Wiener also proposed to von Neumann that MIT should be the headquarters of the Teleological Society. However, he considered it a disadvantage that, on the physiological side, they would have to support the Harvard Medical School, who in his opinion, now without either Cannon or Rosenblueth, had become a primarily clinical institution. Another option would be to convince MIT to establish a small physiological studies program.[32]

On 21 April 1945 von Neumann told Wiener he had been at MIT to discuss his employment (Cf. Heims 1984 [1980], 189). He had been fascinated most of all by the meeting with Richard Taylor. Von Neumann wrote:

---

[27]Von Neumann to Strauss, 20 October 1945, in von Neumann 2005, 234.

[28]Ibid.

[29]Wiener to von Neumann, 24 March 1945 (VNLC), General Correspondence, Box 7.

[30]Ibid.

[31]Cf. Wiener to von Neumann, 24 March 1945 (VNLC), General Correspondence, Box 7.

[32]Cf. Wiener to von Neumann, 24 January 1945 (WAMIT). Cit. by Hellman 1981, p. 233 note 17.

I was in Cambridge and saw R. Taylor and the Tech mechanico-electronic analyzer, we spent two days together. It was very interesting, particularly considering what Taylor might do in the future. We should by all means have a talk 'à trois.' I think there is more to learn from Taylor than from Aiken—and not on analyzers only or mainly.[33]

The letter confirms once again the special intimacy that built up in this period between von Neumann and Wiener, around the themes of cybernetic and the idea of a future joint collaboration at MIT.

8.4 The "Dark Side" of Cybernetics

Von Neumann was a great scientist, and like Wiener, aimed to get general results. In his hands, although with the help of others, and in particular Wiener, the design of computers became a science. But he was engaged in a military project of extreme strategic relevance. In Los Alamos throughout 1944, theoretical and computational work had focused mainly on the plutonium bomb and the implosion problem. In early 1945 Oppenheimer scheduled the various stages leading to the July deadline for the completion of the uranium and plutonium bombs. By then, it was expected that the industrial production centers would have reached a sufficient amount of uranium for Little Boy. It was also believed that a sufficient amount of plutonium would be available for Fat Man, and for a real scale test. The latter was carried out on 16 July 1945, at the Alamogordo air base, New Mexico, at the so-called Trinity site, producing as expected a destructive power of 21 kilotons (1 kiloton is the equivalent explosive power of 1000 tons of TNT). This test alone guaranteed the effectiveness of the method and allowed physicists to evaluate the consequences of the explosion (Cf. Frank, Cochran and Norris 1996).

In the aforementioned letters to Wiener of 1 February 1945, von Neumann informed Wiener: "I am leaving on February 4 for Aberdeen, and on February 6 for the *West*, and expect to be back in the first days of March".[34] He concluded: "I hope that we shall see each other very quickly thereafter. I am sure it would be very profitable for me at least if we could have another conversation".[35] The "West" was of course Los Alamos; in fact, on 12 February 1945 it was from there that von Neumann wrote to Goldstine, offering a number of suggestions about the development of the computers at the Moore School.[36]

---

[33]Von Neumann to Wiener, 21 April 1945 (WAMIT). Cit. by Aspray 1990a, note 6, p. 51.

[34]Von Neumann to Wiener, 1 February 1945 (VNLC), General Correspondence, Box 7. Italics added.

[35]Ibid.

[36]We know this for sure because Goldstine informs us that on 12 February 1945 von Neumann sent him a letter from Los Alamos. (Cf. Goldstine 1980 [1973], 195–196).

Being a great scientist, just like Wiener, von Neumann would certainly also have been motivated by the discovery of new knowledge in the emerging field. However, in the course of 1944 and 1945 von Neumann had been engaged with the activities at Los Alamos and his main interests were the theoretical and computational needs of that laboratory, in charge of the most important military project in the United States. Therefore the Princeton meeting in January 1945, like the talks before it and the activities following it, must have been strictly designed, in his mind, to cater for those more immediate interests.

It is rather clear that the changes von Neumann had asked to introduce in the ENIAC and the new EDVAC project were connected to the computing needs at Los Alamos. It is in my opinion possible to hypothesize that the contribution of the working group formed at Princeton in January, especially Group 3, was related to those projects.

On the one hand, von Neumann needed analytical and numerical tools, and on the other, he needed to know how best to build the new electronic digital general purpose computer.

In this regard it is interesting to consider the co-optation of the astrophysicist Subrahmanyan Chandrasekhar into Group 3. Von Neumann had long been in close contact with him. They were both speakers at the 4th Annual Conference on Theoretical Physics on "The Problem of Stellar Energy", held on 21–23 March 1938, focused on nuclear fusion in stars. Among the other speakers, there was also Teller, the father of the H-bomb.[37] In 1943 he had published with von Neumann's blessing an important essay on "Stochastic Problems in Physics and Astronomy".[38] In that same year, Chandrasekhar had become a consultant for the BRL, for which he dealt mainly with shock waves, his first encounter with hydrodynamics, in which he would later make his career.[39] It does not seem unreasonable to assume that von Neumann's interest in involving him concerned the mathematical demands at Los Alamos which related to the problems of hydrodynamics. In particular, he would have been able to help in connection with nuclear fusion.

On 12 February 1944, from Los Alamos, von Neumann told Goldstine: "I am also working on the problem of formulating a two-dimensional, non-stationary hydrodynamical problem for the ENIAC".[40] Metropolis and Nelson add that "in early 1945" von Neumann asked Frankel and Metropolis "to perform the very complex calculations involved in hydrogen bomb design" on the ENIAC (Metropolis and Nelson 1982, 353).

---

[37]Cf. Chandrasekhar, Gamow and Tuve, 1938; Gamow and Tuve, 1938; and also Chandrasekhar 1939.

[38]Chandrasekhar (1977) states: "my interest in the field of Brownian Motion was to use it as a basis for the theory of stellar encounters". It was with von Neumann's blessing that he published *Stochastic Problems in Physics and Astronomy* (Chandrasekhar 1943).

[39]Chandrasekhar (1977) states: "[The BRL] was my first serious introduction to hydrodynamics. I learned hydrodynamics at that time, but it did not have any immediate effect on the work I was doing in astrophysics".

[40]Von Neumann to Goldstine, 12 February 1945, cit. by Goldstine 1980 [1973], 195–196.

It seems unlikely, however, that the purpose of this computation concerned the H-bomb as such. This bomb was definitely not a priority at that stage. Regardless of the real moral reasons for a bomb thousands of times more powerful than a fission bomb, the low priority placed on it by Oppenheimer is easily explained by the fact that a fusion bomb needs to be triggered by the explosion of a fission bomb, and this fact remained uncertain until the Alamogordo test.

On the other hand, Goldstine assures us that "this problem was of great importance, since it was to test out a dramatic new idea for Los Alamos and an equally dramatic one for the Moore School, the ENIAC". (Goldstine 1980 [1973], p. 246). In an interview, Metropolis clarified: "We tried to run a set whose ensemble would enable us to make certain inferences about what the prospects were for the possibilities of *thermonuclear ignition*" (Metropolis 1987).

In my opinion it is more likely that the computation aimed to assess the consequences of the explosion of an atomic fission bomb, and in particular the verification of the hypothesis of the "ignition of the atmosphere". Actually, since 1942, physicists had been considering the risk that the very high temperatures and pressures produced by the explosion of an atomic fission bomb could trigger a chain reaction of thermonuclear fusion in the light nuclei of the atmosphere. Only very accurate computations could solve the question. Frankel and Metropolis went to the Moore School in the Spring of 1945 (Metropolis 1987 and Goldstine 1980 [1973], 214–5), and various computations were carried out under the supervision of Teller up to the first months of 1946 (Goldstine 1980 [1973], 258). In August 1946, a report by Teller and others on "Ignition of the atmosphere with the nuclear bombs"[41] was distributed at Los Alamos. No doubt the same computations could have been a prelude to the construction of the H-bomb itself (Fitzpatrick 1999, spec. 118), but it seems quite possible that, prior to the Trinity test, consisting in the explosion of a real plutonium nuclear bomb, the ENIAC and all the computational and intellectual resources available at that time, were being used to describe in detail the entire course of the nuclear explosion, from its triggering through gun or implosion methods up to its final effects on the atmosphere.

8.5 Wiener and Rosenblueth Study Cardiac Conduction

In the letter to von Neumann of 24 March 1945, Wiener also announced that on 31 March he would leave for Mexico City, going as invited speaker to the Congress of the Mexican Mathematical Society at Guadalajara, scheduled from 28 May to 2 June 1945. He would in fact stay from 9 April at Instituto Nacional de Cardiologia.[42]

---

[41]Cf. Emil Konopiski, Cloyd Heck, Marvin, Edward Teller, *Ignition of the atmosphere with the nuclear bombs*, Los Alamos National Laboratory, 14 August 1946.

[42]Wiener to von Neumann, 24 March 1945 (VNLC), General Correspondence, Box 7.

On 27 April he informed von Neumann that in Mexico City he had teamed up with Rosenblueth on the "mathematics of nerve conduction".[43] In another letter, Wiener described this as "the study of the applications of the theory of networks of sensitive tissues to the flutter and fibrillation of the heart".[44] The provisional results of the research were presented at the Guadalajara Congress and then in a definitive form in [46b].

It was in effect the first true experience of "cybernetics on a small scale". The research was discussed by Wiener and Rosenblueth together with Walter B. Cannon himself, taking advantage of his visit to Rosenblueth. Cannon would die in October 1945. On the one hand the research can be understood as the continuation of an experimental study, begun by Rosenblueth and Cannon in the year 1941 (cf. Rosenblueth and Cannon 1942), based on studies of the response to stimulation of the central nervous tissue in some animals, observed by oscilloscope. Between 1942 and 1944 Rosenblueth also continued his experimental studies of rhythmic responses in the nervous tissue and the striated muscles.

In his research with Wiener, Rosenblueth took into account two pathological forms of heart rhythm, flutter and atrial fibrillation, diseases of the heart rhythm that are both characterized by a much higher frequency than the normal rhythm (the so-called "sinus rhythm"). Fibrillation is characterized by variable and substantially random periods. The main purpose of the research was not really to study the two diseases, but rather to develop methods that could be useful for studying the neural net, taking advantage of the fact that, in the heart muscle, stimuli spread in a similar manner to the way they do in the cerebral cortex, but without the complex situation created there by synapses and the on/off behavior of the neurons.

Wiener introduced a strong mathematical component into the research. In particular, the equations to explain fibrillation were based largely on the ideas of a paper on "Discrete Chaos" (43a), written with A. Wintner. In addition, Wiener states: "The statistical technique used in the study of heart-muscle nets has been extended to the treatment of neuronal nets by Mr. Walter Pitts" (61c [48f1], 17). This would become the theme of Pitts' doctoral thesis: from the collaboration between Pitts and Wiener, the project of a brain based on stochastic neuronal nets had sprung.

8.6   At the Height of Enthusiasm

Both von Neumann's "experimental mathematics" and Wiener and Rosenblueth's work on "cardiac conduction" can be considered in the framework of post-war plans. At that point, the end of the war seemed very close. On 28 April 1945, Mussolini had been killed, while on 30 April the Soviets had taken the Reichtag and

---

[43]Wiener to von Neumann, 27 April 1945, (VNLC), General Correspondence, Box 7.

[44]Wiener to Santillana, 16 October 1945, (WAMIT) box 2, folder 69. Cit. by Hellman 1981, 242.

Hitler had committed suicide. Between 17 July and 2 August, the last inter-Allied conference was held at Potsdam, where on 26 July the new President Truman, former vice-president of Roosevelt, who had become president on the latter's death on 12 April 1945, had ordered the surrender of Japan.

At the beginning of July, Wiener went to talk to Henry Moe of Guggenheim Foundation about the plans for the new field, and also for a scholarship to allow Pitts to complete his Ph.D. at MIT, since his relationship with the Kellex Corporation was nearing its end.[45] Moe showed interest in the project and submitted the forms for Pitts' scholarship.[46] In the same month, Wiener and Pitts went to visit McCulloch at the University of Illinois. McCulloch wrote to Rosenblueth that they "were full of your experiments and Wiener's calculations on the theory of flutter and fibrillation and they showed me swell manuscripts under way".[47]

Wiener's enthusiasm arose from the prospects opened up by this "field yet unnamed", so full of promise. An important aspect for the success of the projects in this field was certainly von Neumann's appointment at MIT, which was now considered as the most probable place for the "Teleological Society" or whatever it would be called. On 1 July 1945 Wiener wrote to Rosenblueth: "I have had several consultations with von Neumann [...] and it really looks to me now as if the appointment and his acceptance were in the bag".[48] On 11 July he added: "it is quite clear that if the appointment comes through, all of our ideas concerning an organized collaboration between physiological and mathematical subjects will follow as a matter of course".[49] The same day he wrote to another person: "Johnny has been and gone and it looks as though he is in the bag. Everybody is delighted and we are going to go places".[50]

Wiener was ready, together with his collaborators, to begin to work with athletic spirit on the new field which had emerged during the war. On 22 July 1945 he wrote to ask Rosenblueth to consider coming back to Mexico. Meanwhile, their research manuscript had also been subjected to von Neumann's criticism before being sent to Rosenblueth for publication.[51]

Further evidence of Wiener's enthusiasm consists in a curious letter of 26 July, in which he replied at length to a young student who had asked for advice on what were the most promising fields of study for the future. Wiener spoke of his research with Rosenblueth, concerning "the region in which physiology and mathematics come together. In particular, both in the nervous system and in such muscular

---

[45]I assume that Pitts, who was born in 1923, had carried out his activities at the Kellex Corporation of New York as a conscript. In fact, after Pearl Harbor, compulsory military service was established for men between 18 and 45 years old, lasting 18 months.

[46]Letter from Wiener to Rosenblueth, dated July 11, 1945 (WAMIT), Box 44, ff. 68. Cit by Piccinini 2003, 84.

[47]McCulloch to Rosenblueth, 20 September 1945 (MCAPS), ff. Rosenblueth.

[48]Wiener to Rosenblueth, 1 July 1945 (WAMIT). Cit. by Heims 1984 [1980], 188.

[49]Wiener to Rosenblueth, 11 July 1945 (WAMIT). Cit. by Aspray 1990a, note 4, p. 267.

[50]Wiener to Gretel, 11 July 1945 (VNLC). Cit. by Aspray 1990a, note 5, p. 267.

[51]Rosenblueth to Wiener, 3 September 1945 (WAMIT) Box 44, ff. 68. Cit. by Piccinini 2003, 86.

systems as the heart".[52] This was clearly cybernetics on the small scale, but bound up with cybernetics on the big scale. In fact he added:

> Closely related to the problem of the analysis of organization in living tissue is a problem of the synthesis in organization in such devices as computing and control machines",[53] requiring "a revision of statistical theory in which the procession of events in time is fully considered.[54]

8.7   The Nuclear Bombing of Japan and Wiener's Crisis

Wiener could not have dreamt how the political and military leadership of the United States had chosen to put the word "end" to the war: with the nuclear bombing of Hiroshima on 6 August 1945 and Nagasaki on 9 August. This event deeply marked the fate of cybernetics, and even our way of understanding it.

At first it did not seem that Wiener suffered that much over the still fragmentary news of the nuclear bombing of Japan. Even on 11 August 1945, he was referring to Rosenblueth with the usual refrain about von Neumann: "Johnny was down here the last 2 days. He is almost hooked".[55] During the visit to MIT, von Neumann had also reassured him of the internal consistency of the manuscript on heart conduction.[56] Three days later, a letter from Harrison reached von Neumann, offering him a mathematics professorship at MIT, with a good salary of $ 15,000, and a commitment to appoint him as Director of the Department of Mathematics, after Phillips' retirement.[57]

However, as the news about the atomic bombing of Japan became more precise, the broadly cheerful atmosphere due to the end of the war dissolved in Wiener's soul. He began to build up a better understanding of the circumstances. His mood began to darken, and finally he fell into an acute crisis of conscience. In the letter of 11 August, he had confided to Rosenblueth that "in the present almost certain to come interval between wars (and I hope to goodness it will be a long one), I think we can do an enormous amount with our new schemes".[58]

He had a presentiment, common to other enlightened spirits of the time, that the end of the war would have been only the prelude to a World War III. He just hoped that it was a long interim, in order to cultivate the new projects in the best possible way. A further point followed, however, and in response, Rosenblueth tried to comfort him:

---

[52]Wiener to Lawrence Weller, 26 July 1945 (WAMIT). Cit. by Hellman 1981, 241.

[53]Ibid., 242.

[54]Ibid.

[55]Wiener to Rosenblueth, 11 August 1945 (WAMIT). Cit. by Heims 1984 [1980], 188.

[56]Cf. Ibid.

[57]Cf. Harrison to von Neumann, 14 August 1945 (VNLC). Cit. by Aspray 1990a, note 5, p. 267.

[58]Wiener to Rosenblueth, 11 August 1945 (WAMIT). Cit. by Heims 1984 [1980], 188.

You sounded rather pessimistic in your last letter. That is wrong. The war could not possibly be going better [...] Your work seems to be going along beautifully from what you tell me. Your family is doing handsomely. Our projects, although still in the realm of the 'we shall see' are alive and kicking (or maybe I should say, and wagging their caudal appendage). Your novel is still trying to crack its shell. You have friends and they don't forget you—witness thereof, the pleasant time I'm having writing to you. What the Avernus can you crab about, anyhow? It's really a great world and a great life, my dear Norbert, notwithstanding their occasional infirmities.[59]

Before Wiener's eyes, the doors of Avernus were opening wide. As in 1942, he was experiencing a rude awakening from the enthusiasm of his tireless research. But this time the vision of reality had the apocalyptic colors of a nuclear bombing that had erased the lives of about 200,000 people, mostly civilians, with only two bombs. And maybe a vague feeling of having been an unwitting accomplice. He felt in a deep crisis of conscience.

On 21 October 1945 the physicist Daniel Q. Posin of the MIT Radiation Laboratory, illustrating a widespread uneasiness with regard to nuclear weapons among scientists, wrote to Albert Einstein:

Here, at the Massachusetts Institute for Technology, [Norbert] Wiener stands aghast—as though a man in a confused dream—and wonders what we must do, and he protests at scientific meetings the "Massacre of Nagasaki" which makes it easier, for some, to contemplate other massacres.[60]

On 18 October 1945, Wiener presented a resignation letter to the MIT rector Karl T. Compton, claiming that he wished "to leave scientific work completely and finally. I shall try to find some way of living on my farm in the country. I am not too sanguine of success, but I see no other course which accords with my conscience".[61]

Some reasons for the crisis leaked out in a letter written on 16 October to his friend Santillana. This was a long letter in which Wiener made a list of the results he had been working on, which he had been so proud of until a few months before, and which had formed the basis for all his projects after the war. Wiener explained:

Ever since the atomic bomb fell I have been recovering from an acute attack of conscience as one of the scientists who has been doing war work and *who has seen his war work a[s] part of a larger body which is being used in a way of which I do not approve and over which I have absolutely no control.* I think the omens for a third world war are black and I have no intention of letting my services be used in such a conflict. I have seriously considered the possibility of giving up my scientific productive effort because I know no way to publish without letting my inventions go to the *wrong hands.*[62]

---

[59]Rosenblueth to Wiener (WAMIT) 66, cit. by Masani 1990, 198. The dating of this letter is uncertain. Masani dates it to August 1944, without specifying the day. Based on its content, I believe the date of August 1945 to be more consistent. Moreover, Masani also dates the Princeton conference to the winter of 1943–1944, in disagreement with all the other historians.

[60]Posin to Einstein, 21 October 1945, cit. by Einstein 1963, 342.

[61]Wiener to Karl Compton, 18 October 1945 (WAMIT) box 2, folder 69. Cit. by Galison 1994, 254 and Heims 1984 [1980], 188–189. Both Galison and Heims think the letter was never sent.

[62]Wiener to Santillana, 16 October 1945 (WAMIT) box 2, folder 69. Cit. by Galison 1994, 253. Italics added.

Wiener had not directly participated in the Manhattan Project, but must have been aware that "his war work a[s] part of a larger body" had been used in a way that he did not approve, that is, of course, to destroy Hiroshima and Nagasaki. But his war work coincided with the "science still without a name", namely, cybernetics. What frightened him more about this research was the question of the "wrong hands", an expression he would use several times in future writings.

Considering all these facts, it is easy to understand the gloomy language used by Wiener to announce the birth of cybernetics:

> Those of us who have contributed to the new science of cybernetics thus stand in a moral position which is, to say the least, not very comfortable. We have contributed to the initiation of a new science which, as I have said, embraces technical developments with great possibilities for good and for evil. We can only hand it over into the world that exists about us, and this is the world of Belsen and Hiroshima (61c [48f1], 28).

Part IV
(1946–1964)

For fools rush in where angels fear to tread
Alexander Pope, *An essay on criticism*
(1717 [1709], 198)

Chapter 9
"1946"

In a series of follow-ups, Wiener took more and more distance from what I called "Cybernetics on a large scale", and first of all from the grand projects that would lead to the construction of computers. We can thus explain why the "cyber" prefix is used nowadays in close connection with computer science, while, when we think of Wiener, the father of cybernetics, we do not tend to associate him with computers.

For the time being, the crisis of autumn 1945 did not cause the sharp cut which his declarations had seemed to predict; the resignation from MIT never happened, and instead he became once more fully immersed in his research on the new science which did not even have a name. His relations with von Neumann did not seem much affected by his change of mood, even though he had taken a totally different stance toward the decision to bomb Japan, to the point of taking part as a calculus consultant in the top secret commission to find out the targets on which to drop the two nuclear bombs (Cf. Groves 1962). Moreover, in the fall of 1945, von Neumann had taken part in another secret committee, run by Teller, to evaluate the feasibility of the ambitious project Teller had been involved with at Los Alamos during the war: the hydrogen bomb, a super-bomb a 1000 times more powerful than those dropped on Japan.

In the meantime, the project of hiring von Neumann at MIT had fallen through. In my view, the reasons for this need only partially be sought in the probable cooling down of his relationship with Wiener. At MIT, von Neumann must have found those misunderstandings and resistances from the ex-members of NDRC division 7 (in particular Caldwell and Hazen) who would have blocked the Rockefeller Foundation from funding his project for the IAS computer. In fact, alternatively to the plan advocated by Wiener for his engagement at MIT, von Neumann succeeded in convincing the reluctant IAS, jealous of its reputation as an institute of pure theoretical research, to become the site of a project for the construction of a "general purpose" computer for academic purposes.

On 20 November 1945 von Neumann wrote enthusiastically about his success to Wiener. The tone of his letter was warm. He was showing him all the particular organizational miracles he had managed to pull off to build a computer to be used

© Springer International Publishing AG 2017
L. Montagnini, *Harmonies of Disorder*, Springer Biographies,
DOI 10.1007/978-3-319-50657-9_9

exclusively as a "research tool", and explaining how he had succeeded in becoming the sole director of the project. von Neumann declared that the task was "of great importance in several vital respects", so he felt that it was his "prime responsibility to see it through here".[1] He added:

> I need not tell you how much I regret that this means that I cannot join you at Tech [MIT], especially since I am sure that without the decisive encouragement that I received from you and from the Tech authorities I would have hardly had the perseverance and the strength of conviction which are the minimum requirements in this project. *On the other hand, I hope that we shall work together in the field just the same. I think that we should discuss the modalities as quickly as possible.*[2]

Thus von Neumann wanted to keep up the collaboration with Wiener in the "field with no name" and invited him to go to Princeton as soon as possible as a guest of the IAS since after a few days he would have to leave for Los Alamos (he had stopped saying "for the West"). He would bid him farewell with an affable "hoping to see you soon, and very much looking forward to it, Yours as ever John von Neumann".[3]

Von Neumann had to find a chief engineer. The possibility of getting Presper Eckert for the job had fallen through because of the litigation regarding the EDVAC patents issue. Somebody suggested the name of Wilcox Overbeck, who had worked on the MIT Rapid Arithmetical Machine, but he declined the invitation. That was why at the beginning of 1946, from Princeton, Wiener and von Neumann called Julian Bigelow to invite him to Princeton to be the chief engineer of the project (Cf. 64g [56g], 243). On 7 March 1946 Bigelow accepted the proposal from the IAS, with the promise that he would work full time on the project as soon as he had freed himself from his current commitments (Goldstine 1980 [1973], 252). In any case, even the close collaboration between Wiener and Bigelow soon came to a halt. The project for a Teleological Society had not succeeded and never would.

Shortly after the passage from *Cybernetics* with its apocalyptic tone, quoted at the end of the previous chapter, Wiener adds:

> The best we can do is to see that a large public understands the trend and the bearing of the present work, and to confine our personal efforts to those fields, such as physiology and psychology, most remote from war and exploitation (61c [48f1], 28).

Wiener therefore consciously narrowed the "science with no name" to include only physiology and psychology, hence basically to the program I called "Cybernetics on a small scale". He put forward ethical motivations, even though it is very unlikely that he was under any illusions about the military use which could be made with physiology and psychology. But his choice made a lot of sense from the emotional point of view. He found a safe haven in his research and friendship

---

[1]Von Neumann to Wiener, 20 November 1945 (VNLC), General Correspondence, Box 7.
[2]ibid. Italics added.
[3]ibid.

with Rosenblueth. He felt he had found a trustworthy friend after the 1942 "awakening", and he took shelter with him there after 1945 as well.

Wiener focused on research within the small working group with Rosenblueth and Garcia Ramos, set up to investigate through cardiac conduction the real neural nets. They were joined by Walter Pitts, who, after concluding the contract with Kellex, had obtained the Guggenheim Memorial Foundation Fellowship[4] and carried on working alongside Wiener as an assistant. Pitts also kept in touch with McCulloch, who for the time being remained at the University of Illinois.

A five year program was formalized by a research contract between MIT and the Instituto National de Cardiologia (INC), with $27,500 funding from the biological science section of the Rockefeller Foundation, whose director was Robert Morison, a friend of Rosenblueth who also knew Wiener well, since he had been one of the participants at the "dinner group" (64g [56g], 286).

From the 1947 Yearbook of the Rockefeller Foundation, we learn that the contract was for "joint research in mathematical biology under Professor Norbert Wiener and Dr. Arturo Rosenblueth" [p. 111], and it was the continuation of "collaborative work begun three years before", referring to the 1945 research on cardiac conduction, and perhaps also to *Behavior, Purpose and Teleology* (43b).[5] It was set up "to continue joint research on the application of mathematical analysis to problems of the central nervous system".[6]

It was also mentioned that "they are particularly interested in mathematical analyses of activity in the nervous system".[7] These were mainly biophysics or biomathematics studies characterized by their interest in the nervous system and by the joint application of Wiener's mathematical expertise and Rosenblueth's physiological experimentation.

Before the contract began, the group met in Mexico City in the summer of 1946 for research on muscular clonus (a phenomenon in which a muscle is characterized by uncontrolled oscillation; this "clonic phase" usually follows a previous "tonic phase", in which the muscle is rigid). Actually it can be seen as the continuation of the earlier study on muscular clonus by Rosenblueth, Bond, and Cannon 1942. Thereafter, abiding by what had been formalized in the contract, they continued to meet for a semester every two years. Wiener and Pitts still went to Mexico City in 1947, 1949 and 1951. During the intermediate years, Rosenblueth worked at MIT (64g [56g], 286).

---

[4]Wiener to Moe, 6 September 1945. (WAMIT), Box 4, ff. 68. Cit by Piccinini 2003, 86.

[5]"Massachussetts Institute of Technology. Mathematical Biology" In *The Rockefeller Foundation Annual Report 1947*, New York, 1947, 111–112. Cf. also 160.

[6]"The Cross-Breeding of Biology" in *The Rockefeller Foundation Annual Report 1947*, New York, 1947, 31–34.

[7]"Massachussetts Institute of Technology. Mathematical Biology". In *The Rockefeller Foundation Annual Report 1947*, New York, 1947, 111–112.

9.1 The MIT Rockefeller Electronic Digital Computer

Wiener and Rosenblueth's research contract became entangled with a later aborted project on which very little has been written up to now, namely, the MIT Rockefeller Electronic Digital Computer. Toward the end of 1945, the idea of building an electronic digital calculator at MIT had been taking shape. It would be structured later on, in the first few months of 1946. This is the calculator Wiener talks about in *Cybernetics*, saying:

> Harvard, Aberdeen Proving Ground, and the University of Pennsylvania were already constructing machines, and the Institute for Advanced Study at Princeton and *the Massachusetts Institute of Technology were soon to enter the same field* (61c [48f1], 14. Italics added).

It would have been built at the Center of Analysis directed by Caldwell, in collaboration with the Research Laboratory of Electronics (RLE), which, under the direction of Julius A. Stratton, had inherited the structures and part of the staff of the Radiation Laboratory, and also with the Mathematics Department, in which Wiener and his team represented the prominent element.

The Rockefeller Foundation physics research section, directed by Weaver, immediately granted MIT a 100,000 dollars funding (Cf. Wildes and Lindgren 1985, 233) , so that the project could be administratively part of the one Wiener and Rosenblueth had subscribed to with the biological science section of the foundation. In the computer project, a prominent role was given to Wiener and his team. In the memorandum attached to the funding request sent by Stratton to Weaver in April 1946, it was explained:

> I believe that the strength of this proposal from MIT rests on the collaboration that may be anticipated between the Department of Mathematics, the Research Laboratory of Electronics, and the Center of Analysis. The interest in mathematics focuses, of course, on Professor Norbert Wiener and his students. Professor Wiener's current ideas about computing processes, as well as certain physiological analogues that he has studied, are enormously stimulating. The part of the Electronics Laboratories in this program is that of translating such as those of Wiener and his group into physical reality, and supplying an intermediary stage of basic physical research between the purely mathematical concepts and the ultimate specific application to a computing machine in the Center of Analysis (Wildes and Lindgren 1985, 234).

To disprove all those who deny a connection between Wiener and the digital computer, Warren Weaver gave the utmost importance to the collaboration with Wiener, whom he held in great esteem, above all for his knowledge of computers, as we see from a letter written at the beginning of 1947, in which he turned to Wiener for advice on the possibility of automatic translation, ending the letter with the question: "As a linguist and expert on computers, do you think it is worth thinking about?"[8]

---

[8]Weaver to Wiener, in "Translation", 15 July 1949. In: Locke and Booth 1955, 15–23. Wiener replied that he was highly sceptical about the possibility of automatic translation.

Giving news of the beginning of the computer project, the Rockefeller Foundation Annual Report for 1946 concluded: "Professor Norbert Wiener and others in the Mathematics Department are collaborating on the difficult mathematical aspects of the program".[9]

However, while Wiener got more and more involved in the research with Rosenblueth, the same thing did not happen with the computer project. Wiener was feeling he ought to get back to developing the "science with no name", but without being involved in projects of military use, and the electronic digital computer had been created essentially to fulfill those aims. Moreover, it would seem that Weaver's opinion on the need for Wiener to have a prominent role in the project was not fully understood by MIT engineers like Caldwell. The latter, with his idea that a project could be "packed away in camphor balls"—without understanding the complex and dynamic nature of technological innovation—seems to be the very epitome of the engineer against whom Wiener would address some of his criticism in later years.

The lack of consideration for what Wiener represented seems to be confirmed by the proposal which came up at MIT to do away with his hesitations by replacing him with a group of mathematicians. The proposal made Warren Weaver furious. He wrote in a letter:

> I think that the question, 'What does one want a computer to do?' requires great knowledge of mathematics, great imagination, great sweep and depth of mind. One could get some inspiration, doubtless, by talking to a dozen leaders in applied mathematics. But this is where I had hoped and expected we would have the genius of Norbert Wiener. (Wildes and Lindgren 1985, 234–5)

The fact that, throughout the five war years, Wiener's boss, although often infuriated by Wiener's "moonchild" manners, would write these words shows quite clearly the important role played by Wiener especially in the research on electronic digital calculators in the biennium 1944–1945. As a matter of fact for the whole year 1946, Wiener took part sporadically in meetings on the computer at MIT, a project which in any case was dragging on wearily, regardless of whether Wiener was present or not.

It must be said that Caldwell did not represent the only line of thought in this area at MIT. Richard Taylor was also in charge of the Center of Analysis. He was the engineer von Neumann met in April 1945 and who had been the interim supervisor of that Center when Caldwell had taken a steering role in Division 7 (Owens 1986, 81). Taylor got quite excited about the Rockefeller Electronic Computer project and, from some of his reports dating back to the months between 1946 and 1947, it emerges that he had quite a different way of thinking from Caldwell. In one instance, he stated his doubts about the differential analyzer, stressing its lack of precision, and using an argument that seems to have been drawn from *The Memorandum on PDEs*, he stressed how the "mechanical inertia" of its

---

[9]The Rockefeller Foundation Annual Report 1946", New York, 1946, with the title "Massachussetts Institute of Technology Electronic Computation", 168–169, 169.

elements was posing an insurmountable limit to speed. In another instance, reasoning much like von Neumann, he considered the electronic calculator, not as a machine designed to solve routine problems on a large scale, but rather as an instrument for experimentation in the hands of the applied mathematician. He also pointed out that it was necessary to study new numerical methods specifically created for electronic calculators, which were different from traditional ones suitable for manual calculus.[10]

Wiener was not the type of man who was able to organize all the technical factors and the human resources to boost a big project. And indeed, he was never asked to do that. In any case, he did not even want to tackle something he was less and less interested in fighting for anyway. At the beginning of June 1947, Caldwell noticed a drop in the morale among the men of his project and looked at the possibility of pouring all the energy of MIT into another computer project which was just taking off, namely, the Whirlwind (Cf. Wildes and Lindgren 1985, 235) . On 25 June 1947 Compton, the dean, wrote to Weaver that MIT would close the project, giving back the 100,000 dollars to the foundation and focusing its efforts on the US Navy's Whirlwind project from the Servo Lab (Cf. Wildes and Lindgren 1985, 235). The project, directed by Jay Forrester, had initially been destined to produce an analog computer for a flight simulator, financed by the US Navy. While it was underway, it changed its nature, becoming a general purpose electronic digital technology similar to EDVAC. In any case, its origins made it an ideal system to replace the analog calculators used for the predictors. In December 1950, the US Air Force gave MIT the task of setting up a research center to draw up plans for the air defense of the whole of North America based on the Whirlwind computer. In this way the SAGE (Semi-Automatic Ground Environment for US air defense) project came into being, consisting of a network of radar stations equipped with Whirlwind computers and anti-aircraft systems (cannons and surface-to-air missiles), designed and developed in the 1950s by the Lincoln Laboratory at MIT.[11] Wiener's fears over the military use of the MIT computer had not been unjustified.

9.2   The Preparatory Phase of the Macy Conferences on Cybernetics

By the beginning of 1946, especially after von Neumann had started his computer project for academic purposes with Bigelow at IAS, it had become clear that the "teleological society" project, or whatever they would have liked to call it, had gone under. But the people who were at the core of the 6–7 January 1945 Princeton meeting still wished to discuss "the new field with no name", and an opportunity to

---

[10]Cf. Wildes and Lindgren 1985, 234, who refer to a *Progress report* by Richard Taylor of 8 February 1947.

[11]Cf. "Telecommunications Systems: Network Milestones" in *EB97*.

do it came about at what would later become known as the "Macy Conferences on Cybernetics". In his autobiography Wiener tells us about cardiac conduction, referring to his Mexican stay in 1945, during which he had worked with Rosenblueth:

> When I returned to the States I found that the interest in the sort of work that Arturo and I had been doing together, namely the application of modern mathematical techniques to the study of the nervous system as a problem in communication, had excited a spirited interest. A colleague of mine had persuaded the Macy Foundation, in New York, to organize a number of meetings devoted to this subject. The series ran for several years. Here a group of psychiatrists, sociologists, anthropologists, and the like came together with neurophysiologists, mathematicians, communication experts, and the designers of computing machines, to see if they couldn't find a common basis of thought (64g [56g], 285).

To understand this "spirited interest" toward Wiener and Rosenblueth's research, we must go back to 13–15 May 1942, when the meeting on "Mental Inhibition" had taken place, organized by the Macy Foundation in New York, where Rosenblueth had offered a preview of the ideas later summarized in *Behavior, purpose and teleology* (Cf. 61c [48f1], 12; see above § 6.7). The talk had fascinated the audience which included not only physiologists, psychologists, and psychiatrists, but also social scientists, in particular the cultural anthropologists Margaret Mead and Gregory Bateson. Such had been the intensity of the interest that Mead only realized she had broken a tooth at the end of the meeting (Cf. Mead 1968, 1). After the 1942 conference, each of the participants got fully involved with their own war commitments, but promised to return to the subject once the war was over. As Bateson said:

> In 1942, at a Macy Foundation conference, I met Warren McCulloch and Julian Bigelow, who were then talking excitedly about "feedback". The writing of Naven had brought me to the very edge of what later became cybernetics, but I lacked the concept of negative feedback. When I returned from overseas after the war, I went to Frank Fremont-Smith of the Macy Foundation to ask for a conference on this then-mysterious matter. Frank said that he had just arranged such a conference with McCulloch as chairman. It thus happened that I was privileged to be a member of the famous Macy Conferences on Cybernetics (Bateson 1975 [1972], X).

That was why, on 17 and 18 March 1946, the Macy Foundation in New York organized, under the chairmanship of McCulloch, a conference to deepen the notion of feedback and its connection with intentional behaviour. In a letter to McCulloch, dated 8 February 1946, Fremont-Smith, medical director and executive secretary of the Foundation (Tudico 2012, 26), explained that the meeting aimed to present to psychologists, psychiatrists, anthropologists, sociologists, and so on the more recent developments in mathematics and engineering, concerning computing machines and automatic self-aiming devices. He explained that these new kinds of systems promised to clarify the mechanisms underlying the self-correcting and intentional behavior of both individuals and groups studied in the fields of psychology, psychiatry, anthropology, and sociology.[12]

---

[12]Cf. Fremont-Smith to McCulloch, 8 February 1946 (MCAPS). Cit. by Segal 2003, 188.

To this end, the first day was devoted to presenting the new computing machines and devices self-seeking a goal, together with their counterparts in physiology. The second day focused on psychosomatic, psychological, psychiatric, and sociological contexts in which these concepts were applicable.[13]

In the above-mentioned letter, Fremont-Smith also suggested the title of the conference: *Feedback mechanisms and circular causal systems in biology and the social sciences.*

The meeting saw the participation of some of those who had attended the 1945 Princeton conference, in particular, Wiener and von Neumann, together with Pitts, the young and promising logician, but also the neurophysiologists Lorente de Nó and McCulloch, now joined by Rosenblueth, who had come from Mexico for the occasion. These scientists could have been considered as a more or less cohesive group, because of their shared insights and because of the projects that emerged at Princeton.

Alongside them, there were scientists coming from the social and human sciences who looked to the Macy Foundation as their organizational as well as theoretical reference: in particular, Lawrence Frank, a manager at Macy, who, together with Margaret Mead and Gregory Bateson, formed a close-knit trio, who were in turn in close connection with the social psychologist Kurt Lewin, considered as the father of the new American social psychology, and the philosopher F.S.C. Northrop; there were also the neuroanatomists Ralf Gerard and von Bonin, the ex-neurophysiologist turned psychoanalyst Lawrence Kubie, and the experimental psychologists Heinrich Klüver and Molly Harrower. The neurophysiologists who had taken part in the work at Princeton were in close connection with the Macy network and virtually formed the link between the two groups: it was no coincidence that McCulloch had been appointed chairman of the conferences. Moreover, there are also likely to have been scientists who stood further outside the two groups, like the economist Oskar Morgenstern, invited by John von Neumann, the econometrician and statistician Leonard J. Savage, and the sociologist Paul Lazarsfeld.[14]

This March '46 conference became the first of ten meetings, from 1946 to 1953, lasting two days each and taking place every six months at the beginning, but later becoming annual Wiener took part in all of them, up to the eighth. About twenty people participated regularly, apart from the occasional guests. During the sixth meeting in 1949, it was decided to publish the acts of the following meetings. These would be edited by Heinz von Förster, and it was decided on von Förster's own proposal to add the title "cybernetics" (Cf. von Förster 1991, 121), as on Wiener's book, which had appeared in 1948. From then on it became the established custom

---

[13]Cf. ibid.

[14]Cf. McCulloch to Pitts, sent care of Wiener, 2 February 1946 (WAMIT). Cit. by Heims 1984 [1980], 202. Heims (1991) understood very well this difference of points of view within the *Macy Conferences on Cybernetics.*

to refer to them as to the Macy Conferences on Cybernetics (see Heims 1991, Dupuy 1999).[15]

The interest in feedback and circular causality among the scientists orbiting around the Macy Foundation can be understood if we take into consideration the organicistic, relational, holistic, and gestaltic background they all shared (Cf. Heims 1991, spec. 52–89). Many of them could have been considered heirs of the psychological and sociological school of Chicago, having as fundamental reference points John Dewey's anti-atomistic gnoseology (see the seminal paper Dewey 1896), the symbolic interactionism of the social psychologist George Herbert Mead, and Park's urban ecology (Cf. Bulmer 1984).

Behind this tradition of thought there was the psychological and sociological anti-atomism which insisted on interaction and in particular on the refusal to consider the individual as made up of social atom regardless of his or her interaction with others; on the contrary, it was thought that individuals would become such through the very act of interacting among themselves. It would follow that a modification in any such interaction would entail a modification in the very same individuals: or that a modification in an individual would produce a readjustment of the entire network of interactions among the individuals, and hence a change in all the other individuals. This way of thinking had been used by George Herbert Mead in symbolic interactionism, a psychological theory according to which the "self" of an individual understood as an image he has of himself becomes evident through his relationship with others, reflecting the idea others have of him. It had also been used by Sullivan, the psychiatrist, to develop a relational psychiatry, which considered psychic disorder as something that does not belong to an individual, but which is born out of interactions among individuals, whence he concluded that the whole context to which an individual belongs ought to be cured. Following this way of thinking, we come to the notion of family therapy.[16]

Many anthropologists, psychologists, and psychiatrists endorsed the "Personality and Culture" movement (Cf. Greenberg-Mitchell 1983; Swick Perry 1982), which sprang from Edward Sapir's social anthropology and Harry Stak Sullivan's relational psychiatry. Lawrence Frank, a manager of the Macy Foundation (Cf. Sullivan 1964) had given a decisive contribution toward its constitution at the beginning of the 1930s. In the movement "Personality and Culture" it was thought that personality, indeed the "self", was established within a specific society, and that therefore in order to study personality it was necessary to consider the particular culture in which it was formed.

---

[15]Our main source for the Macy Conference on Cybernetics is the official proceedings (Von Förster et al. eds. 1950–1955), regarding the last five meetings. For the first three 1946–1947, there is the *McCulloch Summary*, a typescript by W. McCulloch, *To the members of the conference on teleological mechanisms*, an abstract of the first three meetings, given to those present at the fourth meeting on 23 and 24 October 1947.

[16]About Sullivan, Cf. Greenberg-Mitchell 1983.

Toward the end of the 1930s there had also been a broad confluence of the themes of the Gestalt psychology from Europe, a phenomenon which had been aided by Jewish immigration into the US due to the racial persecution. The basic idea of the "gestaltists", who were mainly interested in the psychology of perception, consists in thinking that the properties of the various elements in a perceptive field can only be explained by starting from the whole they compose; and that was how a change would cause a readjustment of the whole, which is a "dynamic whole" (Lewin 1948, 17 ff.).

The need for theoretical rigor was important among these scientists, pushing them to seek the collaboration of the natural and mathematical sciences, especially on the relational-systemic side. In this respect, in the past, the encounter with Cannon's physiology and his idea of "homeostasis" had always happened through Frank's mediation (Heims 1991, 64). In this way one can also explain the fact that physiologists could be found alongside psychologists and social scientists, and alongside people like George E. Hutchinson, one of the founder fathers of modern ecology, a science which studies the biosphere as a self-adjusting system (Cf. Slack 2010 ; Grinevald 1988), at the Macy Conferences on Cybernetics. And one can thus understand the strong attraction these scientists felt for Wiener and his collaborators. As the psychiatrist David Mc K. Rioch said at the conference held at the New York Academy of Sciences in the Autumn 1946:

> Dr. Wiener et al. have offered convincing evidence that the occult force called 'purpose,' which is implicit but hidden in the concept of 'reflex reaction to a stimulus,' may be rigorously dealt with by the operationally more precise formulation of 'interaction in time' (McK.Rioch 1948, 219).

Recalling G. H. Mead's symbolic interactionism and Sullivan's relational psychiatry, Mc K. Rioch explained that they had asked to go theoretically beyond the old concept of "individual", but he added that with Wiener's research

> we have been taken a step further. It has been demonstrated how the concepts derived from one set of operations of a 'feedback' mechanism—i.e., on interacting processes—may be tested by another set of operations on the same system, thus removing the interpretation of the significance of the data from personal opinion to consistent operational demonstration (McK. Rioch 1948, 219–220).

It should be added that in this theoretical trend it was always felt, starting with Dewey himself, that social research should be accompanied by social intervention (Cf. Madge 1962), which was thought of as family psychotherapy, social politics, and eventually social engineering.

In this way one can also explained why these scientists (Harry Stak Sullivan, Margaret Mead, Lawrence Frank) belonged to the "mental health" movement as well, a movement which organized in 1948 the London International Congress on Mental Health [Flugel 1950], where it would be argued that at that point "the goal of mental health has been enlarged from the concern for the development of healthy personalities to the larger task of creating a healthy society" [Cit. by Heims 1991, 173].

As a matter of fact there was, from the beginning, a fundamental difference in the setup: the psychologists and sociologists needed somehow to draw up "principles" to export from their own disciplines, and this was the main aim, at least in their first meeting. In contrast, the participants at the Princeton conference, and Wiener more than any, felt the need to deepen the subject which had just started at Princeton. Probably, in a period in which new restrictions and secrecy requirements were rapidly being created, these conferences could have been seen as a freer discussion space. This duplicity of intents was a contradiction that was never rectified, as is made clear in Wiener's book *Cybernetics. Or control and communication in the animal and the machine* (48f1), and what would be chosen for the official acts in 1949: *Cybernetics. Circular causal, and feedback mechanisms in biological and social systems* (von Förster et al. 1950–1955), which resurrected the title suggested for the first meeting by Fremont-Smith and McCulloch. Hence, rather than helping the theoretical clarification of the nascent cybernetics, the meeting with the social and human sciences ended up introducing further problematic knots into the already tangled and tormented cybernetics panorama.

9.3   A "Duet" Between Wiener and von Neumann

The actual program of the first Macy Conference coincided with the one arranged in the above-mentioned letter from Fremont-Smith to McCulloch. On the morning of the first day there was a short talk from von Neumann about digital computers, followed by a report from Lorente de Nó highlighting new discoveries about the brain, in particular regarding Pitts and McCulloch's neural network model.

In the afternoon it was Wiener's turn to give a talk, on machines that were able to pursue an objective, after which Rosenblueth spoke about automatic regulation mechanisms in living organisms (Cf. *McCulloch Summary*, 2–3). The following day was dedicated to discussion's about psychological, psychiatric, and sociological implications, but in the evening there was already an opportunity to hear a report by Bateson about theoretical issues in the social sciences, while the philosopher Northrop spoke about applications of circular causality to ethics.

In his paper, von Neumann spoke about the characteristics of base-2 numerical calculators, with Boolean logic, emphasizing their high degree of accuracy compared to analog calculators, and the possibility of increasing that accuracy at will by increasing the basic components, i.e., the thermionic valves used as automatic on/off devices. He stressed the general idea that those devices were able to "compute any computable number or solve any logical problem presented to them in their own language provided it had a solution" (*McCulloch Summary*, 1). He was basically pointing out that these were Turing's universal machines, that is, systems in which, just by changing the program, it would be possible to solve any problem provided with a solution.

Wiener intervened several times, interrupting von Neumann and showing how both of them had perfect mastery of the subject. Their speeches appeared so harmoniously interwoven that the *McCulloch Summary* even used musical terms to describe it. In McCulloch's text, Wiener's observations appeared as "counterpoint", and the von Neumann's paper ended with a genuine "duet" (*McCulloch Summary*, 1) between the two mathematicians on the theme of computer memories. It is interesting to note where Wiener placed one of the main counterpoint notes: he observed that, if some Russellian paradoxes had been given to one of the devices described by von Neumann, "it ought to go into a series of operations instead of coming to a conclusion, so that if it first decided that something was true it would next decide that it was false and vice versa" (*McCulloch Summary*, 1).

I like to think that here Wiener still had in mind the experience of 1940, when he had tried to make Stibitz' computer divide by zero and the printer had kept on responding without stopping. In that case, the observation contains the proviso "as long as the problem has a solution", with which von Neumann had concluded his talk on the universal possibilities of the new machines. Emphasizing the fact that the machine would respond to an impossible problem without stopping, going on forever producing an infinite sequence of "true, false, true, false, ...", was a way to clarify the role of time when it is introduced into the logic of the machine. This issue was relevant to Wienerian considerations on the nature of logic, already developed in *The role of the observer*. By now the computer had become a tangible demonstration of just how deep the intuition of an operationally interpreted temporalized logic actually was. This was a subject to which Wiener would return the following year in *Cybernetics*, where he would write: "All logic is limited by the limitations of the human mind when it is engaged in that activity known as logical thinking" (61c [48f1], 125). By applying the fundamental principle of operationalism to the computer, as he had already done for the human mind, in the tradition extending from Mach to Bridgman—the principle states that it is not possible to assign any reality to a physical concept beyond what is actually being observed—Wiener argued that: "According to this, the study of logic must reduce to the study of the logical machine, whether nervous or mechanical, with all its non-removable limitations and imperfections" (61c [48f1], 125). And a few lines later we read:

> A logical machine following definite rules need never come to a conclusion. It may go on grinding through different stages without ever coming to a stop [...]. This occurs in the case of some of the paradoxes of Cantor and Russell. [...] A machine to answer this question would give the successive temporary answers: 'yes', 'no', 'yes', 'no', and so on, and would never come to equilibrium. [...] The method by which we resolve the paradoxes is also to attach a parameter to each statement, this parameter being the time at which it is asserted. [...] We thus see that the logic of the machine resembles human logic (61c [48f1], 126).

Upon this theory of the 'concrete', 'temporalized' logic of the automaton, there was a truly deep convergence with von Neumann, to whom it was clear—as Paolo Zellini writes—that:

> It was possible to think of the calculator as "materializing" what mathematical logic, at least "constructive" mathematical logic, had invented between the beginning of the century and the 1940s. [...] However, as von Neumann specified, there is an important difference between ordinary logic and the automata representing it. *Time* never intervenes in logic, while every network or nervous system admits a *definite time* elapsing between the input

and output signal. A definite temporal sequence, von Neumann explained, is always relevant to the operations of this or that system. [...] The automaton, von Neumann concluded, contains something *more* of the logic formula it symbolizes, and this something is precisely a *time* frame (Zellini 1996, 76–77. Original in Italian).[17]

There is no doubt that von Neumann and Wiener relied upon different ways of thinking. In many ways Heims was right in insisting upon the epistemological differences between the two: von Neumann had been influenced by the cultural climate of logical neo-positivism, and in particular by the axiomatic approach of Hilbert's meta-mathematical program; Wiener on the other hand had developed his ideas in the ideal climate of the first American pragmatism, had a fallibility notion of science, and saw in nature more the chaos of clouds than the order of clocks. Nevertheless, as Pesi R. Masani keenly highlighted, the similarities between the two mathematicians and the parallelism of their scientific itineraries cannot be forgotten. While von Neumann had been a student of Hilbert, Wiener had been a student of Russell, and von Neumann himself was not totally unaware of the complex dimension of the world, especially after he had understood and digested, perhaps reluctantly, but in a very convinced way, Gödel's lesson.

In the afternoon of the same day, it was time for Wiener and Rosenblueth's long-awaited reports. Wiener introduced the subject of automata capable of pursuing an objective by giving a brief historical excursus which went from the Hellenistic period up to the steam engine. In particular, he spoke about a device made by Humphrey Potter which used positive (or "regenerative") feedback and could produce a swinging movement in the steam engine; and about Watt's centrifugal regulator, the "*governor*", which used negative (also called "degenerative" or "inverse") feedback, making it possible to maintain a constant speed whatever the load. He pointed out that automata using this last type of device differed from those in the past, because they depended on knowledge of the outside world and on their own working, whence they could reduce the gap between their actual state of operation and the one desired. He extended those concepts to reflexes and in general to intentional activities, showing how all that was needed was a circuit with adequate receptors and effectors. In particular, he stressed the importance of this kind of goal-seeking system "which had built into them computing devices, some of which might so base their action on previous information as to guess the future" (*McCulloch Summary*, 2). Rosenblueth built on Wiener's talk by giving the examples of homeostasis: regulation of blood pressure, of respiratory functions, and so on.

It is important to consider the morning and afternoon reports as an integral whole. According to Wiener, insofar as a digital calculator could well form the "brain" of an automatic control device, both von Neumann's talk and his own could appear deeply interconnected, and at the same time could be complemented by Lorente de Nó's and Rosenblueth's reports.

---

[17]Zellini refers in particular to von Neumann (1956, 1966).

9.4   The Prevalence of Physiology

Notwithstanding the fact that the Macy conferences had the stated objective of an in-depth analysis of the theoretical issues concerning circular causality, the discussion focused from the beginning on neurological issues, and that was not so surprising considering that most of the participants had something to do with neurophysiology, neuroanatomy, psychiatry, or psychology. After von Neumann and Lorente de Nó's report, the neurophysiologist Frederick Bremer called into question the Pitts-McCulloch model, objecting that neurons, under specific experimental conditions, "might detract from the all-or-none properties which he had ascribed to them" (*McCulloch Summary*, 2).

Bremer was supported in this by the neuroanatomist Ralf Gerard, who, based on his laboratory experience, stressed the importance of field properties in the central nervous system. Some doubts were also expressed about the scheme of physiological explanation making use of feedback mechanisms. In fact, after Rosenblueth's report, Gerard told the audience about a result achieved by one of his own students, Roger Sperry, according to whom there might be good motor control even without information coming from the sensory organs. It is likely that the idea of research on muscular clonus arose from this same critical observation, research that would be developed by Rosenblueth and Wiener in Mexico City the following summer, for the specific purpose of clarifying the feedback role in animal movement.

Regarding the Pitts-McCulloch model, an heir in this respect of the more classical behavioral "stimulus–response" approach, the fact that it assumed that the brain had to be conceived as a device which does not work without external stimulation was also criticized. It was suggested as an alternative to consider it rather as a "wound-up automaton within". This way of looking at things was supported by a phenomenon which was receiving a great deal of interest at the time, that is, the fact that neurons in the brain formed oscillating closed "reverberating" circuits. Lawrence Kubie had been among the first to hypothesize the existence of such circuits in 1930; they had later been observed experimentally by Lorente de Nó in 1938. On the morning of the second day, the psychologist Molly Harrower spoke about the repetitive behavior of an individual suffering from brain damage, as observed by psychological tests, after which she compared those oscillations with the repetitive behavior found in neurosis. Kubie returned to the subject, suggesting that the repetitive behavior of a neurotic individual could have a neurophysiological basis in reverberant circuits. The use he made of expressions such as "energy to be released" or "psychic tension", elicited a critical reaction from Wiener, who argued that it was incorrect to use the notion of "energy" "with respect to communication systems which worked in terms of information" (*McCulloch Summary*, 4). By such an argument Wiener was for the first time making the participants aware of his innovative ideas on the specificity of information: automata, both artificial and natural, were to be conceived primarily as information processors, as "*communication machines*"; their working characteristics had to be related to some sort of information balance rather than to the energy balance introduced in the nineteenth century by engineers building heat engines. It followed that the processes that could be found in the brain ought to be conceived as circuits in which there flowed information rather than energy; and in this specific case, resonant circuits were considered as nature's equivalent of the devices used in computers as quick and erasable memory (Cf. 61c [48f1], 121–124) and De Luca-Ricciardi (1986, 397).

9.5 Social Sciences

The English anthropologist Gregory Bateson was among the scholars who most strongly felt the need to examine the issue of circular causality. He had been a student of the English social anthropology school of Bronislaw Malinowski and Alfred Radcliffe-Brown. The latter had advocated going beyond nineteenth century evolutionary anthropology and reinstating a functionalist vision which tended to describe cultures, synchronically, as integrated organic systems. The committed organicism which inspired the whole of Bateson's intellectual research can ultimately be referred to this approach, although he was deeply dissatisfied with the reforms introduced by his teachers, which he considered inadequate to provide anthropology or the social sciences in general with a solid theoretical foundation (cf. Lipset 1982; Montagnini 2007; Peterson 2010).[18] This had been the trigger which had brought him to the Macy Conferences: "Bateson you will remember is the man who insists on the importance and lack of theory in sociology".[19]

Bateson spoke at length on the evening of the first day about this very lack of theory in the social sciences, supported by his wife Margaret Mead. In the report he introduced the main concepts developed in his anthropological research (see Bateson 1958 [1938], 171–197). In particular, he distinguished two ways in which societies tend to subdivide by gradual cumulative processes which he called "schismogenesis" (from the Greek "σχίσμα" = schism, scission): on the one hand, there was "complementary schismogenesis", in which society is divided on the basis of the modality domination/submission, e.g., the man/woman or master/servant relationship in which the comparison between individuals is made through opposite modalities; on the other, there is "symmetric schismogenesis", in

---

[18]After studying the thinking of the English anthropologist Gregory Bateson, I have always been convinced that his strong holistic-relational background must have come from his relations with Malinowski and Radcliffe-Brown, and particularly from the latter. This is a connection usually neglected by scholars who have dealt with Bateson's ideas, apart from some hints by Lipset (1982). Recently, Peterson (2010) has shown how long and close the relation between Bateson and Radcliffe-Brown actually was, lasting from the 1920s until the end of the 1930s, not to mention a less intense relation with Malinowski.

[19]McCulloch to Pitts, sent care of Wiener, 2 February 1946 (WAMIT). Cit. by Heims 1984 [1980], 202.

which individuals tend to divide themselves into two groups that confront each other according to identical modalities, by boasting to each other and engaging in an arms race or armed conflict. Bateson thought that both these processes were characterized by positive or regenerative feedback, meaning that they were gradual cumulative processes. He thought that they could work together in order to form a single larger process which, on the whole, could have been considered as a degenerative circuit (with negative feedback); he noted that stabilization seemed to be successful only when the group was numerically small. Finally, he talked about the deutero-learning theory, that is "learning to learn", according to which, when we learn, not only do we acquire notions, but we modify the learning process. However, as McCulloch reports, this observation fell on deaf ears (Cf. *McCulloch Summary*, 3).

In the debate which followed Bateson and Mead's report, there were doubts regarding the possibility of exporting the feedback theory "into domains of which we were ignorant as to what the significant variables were" and especially to the social sciences. From this a challenge was born, as McCulloch explains, "which ultimately led to the special meeting of those of us more closely related to the social sciences" (Cf. *McCulloch Summary*, 3). The proposal was also supported, when the meeting came to an end, by the sociologist Paul Lazarsfeld, who asked for a specifically sociological conference in which to discuss the mathematical means needed to study causal connections in general and circular ones in particular. The meeting in question took place on 21 and 22 October 1946, under the patronage of the New York Academy of Sciences, and was set in a larger auditorium than the one usually used for the Macy Conferences. The reports to be found in the official acts were at times submitted to heavy revisions, and curiously, just those of Bateson and Lazarsfeld, the men that most desired the meeting, were totally omitted. Fortunately, we have detailed *Preprinted abstracts* in which we find drafts of the original reports.

The *Preprinted abstracts* clarify the original theme of the conference and what was meant by the expression "circular causality". In the past, the most studied causal relations in the physical sciences, or indeed in the biological or social sciences, were those involving unidirectional causality, there being two similar modalities: more precisely, the model according to which a "variable A affects B, and B affects C" or the one in which the variables "A and B together affect C". But the theme here was a third causal model "in which variable A affects B, B affects C, and C affects A", a model exemplified by such systems as "the thermostat, the homeostatic mechanisms the body, the mechanisms underlying learning, the community in which government has sources of information about public opinion, games involving two or more players, armaments races, etc." (*Introductory Statement*).[20]

---

[20]We do not know who presented this paper, a role which, as we know from the published proceedings, was played by Lawrence K. Frank, one of the main administrators of the Macy Foundation.

In the past, the text adds, such a circular causal model had been studied especially by Clerk Maxwell, in his analysis of regulators, and by Claude Bernal, in physiology. Finally, during the war, physicists, mathematicians, and engineers working on servomechanisms and computers had made headway in the understanding of such mechanisms, dealing with "many types of self-correcting and purposive mechanism", and the main objective of the conference was to make those results available to biologists and social scientists (Cf. *Introductory Statement*).

Therefore, the *Introduction* to the meeting summarized the aims of the conference which was held in March, according to the needs of the social scientists, as shared by the Macy Foundation. It would deal with instances that had been circumvented in March and for which Gregory Bateson and Paul Lazarsfeld in particular had requested this new meeting specifically dedicated to circular causality in living and social systems. Yet, paradoxically, neither Bateson's nor Lazarsfeld's paper appeared in the published proceedings of the conference.

The final draft of the Bateson paper, entitled *"Circular causal systems in society"* (Bateson 1946a),[21] reiterated the need to learn from "physicists and mathematicians" ways of dealing with phenomena in organisms and society that were subject to circular causality. Thus it could be considered that "the logical and mathematical arguments adduced by the physicists and mathematicians are, if not sound, at least the best that can be obtained today" (Bateson 1946a). He reiterated the need to have in the social sciences "not less but rather more rigorous thinking than is usual among the physicists", considering that "the entities with which we deal are much more complex than even their computing machines, and by that token we need more complex, more flexible, and more precise concepts" (Bateson 1946a).

As a first step, for an essentially exploratory research programme, Bateson suggested a classification of the various cases in which circular causality takes place, according to more or less rudimental criteria such as the size of the system, distinguishing those included in a single individual from those "more comprehensive systems whose causal arcs pass through two or more individuals, or through the organized structures of a society", often including in itself other "sub-systems composed of organised aggregates of individuals such as [...] nations, business corporations, etc."; discerning in their intrinsic causal complexity the regenerative circuits from the degenerative ones, the systems based on single signals from those depending on the statistical properties of the signals.

We do not know if the final draft of Lazarsfeld's paper also exists. I was only able to obtain the abstract of this paper entitled *Circular processes in public opinion* (Lazarsfeld 1946).

Paul Lazarsfeld was a sociologist of Viennese origin who arrived in the US in 1933. From 1937, he ran an important research project on the effects of the radio in American society, initially funded by the Rockefeller Foundation and afterwards as part of the Bureau of Applied Social Research of the Columbia University.

---

[21]There is also a final typescript of this paper, ready for publication (Bateson 1946b), which follows the lines of the above abstract.

According to a recurrent accusation, his market research ended up being used as a tool for psychological manipulation, exploited by the commissioning companies (Cf. Heims 1991, 189), even if his main aim was to create a rigorous quantitative methodology, less concerned with the final practical results of his work.

In his paper Lazarsfeld stated that market research had shown various mutual effects, such as: "A certain type of movie has a large audience; because of this, more movies of this type are produced; thereupon, people become bored and start to avoid movies of this kind; as a result the producers reduce the supply." (Lazarsfeld 1946) According to his own style of research, very attentive to the methodological issues involved in the quantitative measurement of variables, Lazarsfeld wanted to understand better from Wiener how to use the "theory of prediction" in that context.

9.6   The Discussions on Circular Causal Mechanisms

Another curious aspect arises by comparing the published proceedings of the conference on 21 and 22 October 1946, under the patronage of the New York Academy of Sciences, with the theme of Wiener's paper. It was to be called "*Self-correcting and goal-seeking devices and their breakdown*" (46d), but the published paper was instead entitled *Time, communication, and the nervous system* (48a). Wiener's paper was actually very closely linked to Rosenblueth's on "*The control of movements of animals organisms*" (Rosenblueth 1946). As we read in Wiener's abstract: "Doctors Rosenbluth [sic], García-Ramos and Wiener have made an experimental and theoretical study of clonus especially as shown in the quadriceps extensor memories of the cat". The two papers showed the results of an experiment on the muscles of a cat, carried out with the assistance of Ramos and Pitts in Mexico City in the summer of the same year.[22] In the physiological profile, as Rosenblueth would explain, clonus was a periodical contraction of some muscles resulting from the excessive reaction of reflexes and which can be obtained by a blow on a particular tendon. It had been chosen for its relative simplicity as a first step toward the development of a theory on animal motion, which was assumed to involve feedback mechanisms, and in particular postural feedback.

Wiener showed from the mathematical and engineering point of view that the frequency measurements of the oscillations obtained experimentally could have been approximated by a theoretical model based on a two-neuron loop subject to feedback (Cf. 61c [48f1], 19–21). One of the main theoretical hurdles was the search for the exact physiological equivalent of the concepts of electrical current and voltage, and since the circuit was a highly non-linear one, for the identification of suitable linearization methods (Cf. Masani 1990, 206–207).

---

[22]The early results of the research appeared in (Rosenblueth 1946), then in *Cybernetics* (61c [48f1], 19–21). Finally, *Muscular clonus* (85c) appeared, a paper that was never published. Indeed, it had undergone a number of revisions because Wiener was not fully convinced of the rigour of its mathematical conclusions.

In general, following the servomechanism manual (MacColl 1945), Wiener assumed that, in such phenomena, formed by systems with more than one degree of freedom, stabilization was reached by multiple feedbacks which actually seemed to be found in the nervous system [46d]. However, it seems that Wiener and Rosenblueth had chosen a topic involving feedback, not just because of their particular interest in feedback, but to satisfy the curiosity of their interlocutors, attracted in a spirited way by the concept.

The report of neurosurgeon William Livingston on "*The vicious circle in causalgia*" showed analogies with the research on the clonus, even though it went into less detail regarding the mathematical aspects. It presented an explicative model of causalgia, a burning pain of neurological origin which can be due to a reverberant circuit whose efferent nervous component (the outgoing nerves) excites some damaged nerves, causing pain and increasing the temperature. The most appropriate treatment, according to Livingston, consisted in a ganglionectomy, that is, basically cutting the path of the nerve impulses, or the administration of drugs which induced a block chemically. See also the paper Livingston 1948 that was actually published.

The biologist George Evelyn Hutchinson, one of the leading exponents of ecology as a science studying the environment and what it includes as an organic whole, presented various examples of circular causal circuits that can be found in the biosphere. Influenced by M. Goldschmidt and Vladimir I. Vernadsky, Hutchinson was the founder of an important school of ecology at Yale University. His paper was entitled "*Circular causal mechanisms in ecology*" (Hutchinson 1946). In the published proceedings, that became "*Circular causal systems in ecology*". He dwelled on the two prevalent ecological approaches, *biogeochemistry*, in which the magnitudes considered are continuous, and *biodemography*, where the fundamental entities, the individuals in a population of organisms, are represented by a discrete magnitude.[23] He stressed that, in the biogeochemical approach, an "energy transferal" and a "transferal of matter" took place (Hutchinson 1946, 1948). For that purpose he presented a model for the phosphorus cycle in a lake, in which he described a circular causality relation between the amount of living organisms in the lake and the thickness of the sediment deposited at the bottom, rich in phosphor, produced by excrement and rotting organisms, and as a consequence propitious for the generation of more living matter.[24] Regarding the biodemographic aspect, he commented on the mathematical models of demographic growth and Volterra's prey-predator equations, stressing the phenomenon of saturation.

The conclusive report by Warren S. McCulloch had originally been called "Summary of theory and extension to Gestalt psychology", becoming in the proceedings "A recapitulation of the theory with a forecast of several extensions". He

---

[23] Afterwards Hutchinson was the editor of the English translation of Vernadsky's *Problems of biogeochemistry* (1944 [1924]). He wrote also on population ecology (1978).

[24] Hutchinson will publish *A treatise on limnology* (1957), i.e. on the idrobiological study of lakes as ecological systems.

took the opportunity to present a synthesis of his own theories on the brain, taking into account the most recent developments, in particular the theoretical acquisition of reverberant circuits conceived as a memory form and the theory that Pitts was then developing as his doctoral thesis under Wiener, on a statistic neural network. McCulloch also summed up the main ideas on circular causation that emerged during the conference.

The overall impression one gets is that, in this conference, automatic control and servomechanism theory seemed to promise vast extensions outside of the engineering context in which it was born. McCulloch emphasized the widespread feeling of certainty during the conference that the feedback theory could be extended beyond neurophysiology to ecology, and to the study of societies as well. As can also be seen from the summary of the first three meetings written by McCulloch, it would seem that at first, the needs of the social scientists had been welcomed by everybody, including Wiener, despite a certain perplexity (*McCulloch Summary*, 6). Later on, however, Wiener's opinion had become more critical. McCulloch's final report at the conference, after recalling those of Bateson, Lazarsfeld, and Hutchinson, hints at Wiener's hesitations, and he is compared to Leonardo da Vinci, through the anecdote according which Leonardo would have "killed the mechanic who worked with him, lest knowledge of the submarine reach the ears of some would-be conqueror" (McCulloch 1948, 264). This is reminiscent of ideas we have seen emerging from Wiener's writings during the days of his crisis conscience, when he feared that the results of his research would end up in the hands of unscrupulous technicians and businessmen; it was an idea that would come back at the end of 1946 when writing *A scientist rebels*, a couple of months after this conference, and one more time in 1947, when drafting *Cybernetics*. Indeed, we shall see in the next chapter that the latter contained numerous reservations regarding the applicability of cybernetics and especially of prediction theory to the social sciences.

9.7   Von Neumann's Doubts

The first serious theoretical difficulties within the cybernetic circle started to make an appearance at the end of 1946. It looks like it was von Neumann himself who opened the discussion and who, in a long letter dated 29 November 1946, already mentioned in a previous chapter (see above 7.11), expressed his concerns to Wiener about the profitability of using the Pitts-McCulloch model to describe the brain, and suggested setting up a totally new project on molecular biology. The letter, which was polite and friendly, anticipated the theme von Neumann was to debate with Wiener in a meeting that had already been arranged at MIT on 4 December 1946.[25]

---

[25]Cf. von Neumann to Wiener, 29 November 1946 (WAMIT). Cit. by Masani 1990, 243.

A few days earlier Wiener had said he "was extremely eager to [...] talk over developments"[26] with von Neumann.

In the letter von Neumann briefly retraced the line of research followed over the last two or three years, stressing the utility of the Pitts-McCulloch model, the Turing machine, and the research by Wiener with Pitts and Rosenblueth with a view to planning a "general purpose" computer. He then observed that, at least regarding the parallel study of computer and brain, the idea of carrying on the collaboration between engineering and physiology was destined to meet major obstacles. He even went so far as to say that they may have reached a dead end. As he explained:

> Our thoughts - I mean yours and Pitts' and mine - were so mainly focused on the subject of neurology, and more specifically on the human nervous system and there primarily on the central nervous system. Thus, in trying to understand the function of automata and the general principles governing them, we selected for prompt action the most complicated object under the sun - literally. In spite of its formidable complexity this subject has yielded very interesting information under the pressure of efforts of Pitts and McCulloch, Pitts, Wiener and Rosenblueth. Our thinking - or at any rate mine - on the entire subject of automata would be much more muddled than it is, if these extremely bold efforts - with which I would like to put on one par the very un-neurological thesis of A. Turing - had not been made. Yet, I think that these successes should not blind us to the difficulties of the subject, difficulties, which, I think, stand out now just as - if not more - forbiddingly as ever: the complexity of the human nervous system, and indeed of any nervous system. What seems worth emphasizing to me is, however, that after the great positive contribution of Turing - cum - Pitts - and - Mc-Culloch is assimilated, the situation is rather worse than better than before. Indeed, these authors have demonstrated in absolute and hopeless generality, that anything and everything Brouwerian[27] can be done by an appropriate mechanism and specifically by a neural mechanism - and that even one, definite mechanism can be "universal". Inverting the argument: Nothing that we may know or learn about the functioning of the organism can give, without "microscopic", cytological work any clues regarding the further details of the neural mechanism. I know that this was well known to Pitts, that the "nothing" is not wholly fair, and that it should be taken with an appropriate dose of salt, but I think that you will feel with me the type of frustration that I am trying to express. [...] After these devastatingly general and positive results one is therefore thrown back on microwork and cytology - where one might have remained in the first place.[28]

The Pitts-McCulloch model and the Turing machine had proved to be of great utility on the road to the "general purpose" computer, able to run a universal Turing machine that could deal with any "Brouwerian" concept, that is, any logical or mathematical concept that could be represented by a finite number of instructions. In this way, Russellian paradoxes and the like were excluded. But, all in all, von Neumann reflected, by reaching this result it had also been shown that the Pitts-McCulloch model of the brain could have been put into practice by any

---

[26]Wiener to von Neumann, 25 November 1946 (WAMIT). Cit. by Heims 1984 [1980], 204.

[27]By the term "Brouwerian", von Neumann refers here to any mathematical concept to be admitted from a finitistic point of view, according to an established use in the school of Hilbert, who equated the intuitionistic restrictions on mathematics introduced by L.E.J. Brouwer with those finitistic ones, more properly Hilbertian. For this suggestion, I acknowledge Corrado Böhm.

[28]Von Neumann to Wiener, 29 November 1946 (WAMIT). Cit. by Masani 1990, 243.

network of on/off switching devices, independently of whether it was a relay, a thermionic valve, or a neuron. Therefore, ipso facto, that model did not say anything more about the human brain than it had at the beginning, and now it did not have any further heuristic use for an in-depth analysis of the phenomena of the central nervous system.

In fact Pitts and McCulloch had started by assuming that the neuron was a simple black box, able to function like an on/off device: now, in order to learn more, it would have been necessary to open the box and look inside, hence going down to the "cytological" level.

Von Neumann had serious doubts that one could deal with the complexity of the brain using the Gestalt theory, or whatever other "verbal theory". He had instead an alternative proposal and was inviting Wiener to join the venture, that is, to completely abandon the parallel study between machines and the nervous system and engage on a different terrain which seemed to him less complex: the one represented by the single cell. Considering the phage, a microorganism that was then the focus of the first research in molecular biology, he estimated that it was made of "six million atoms", but probably only a few thousand "mechanical elements", considering that each element could not have been made of fewer than ten or more atoms. On the other hand, he considered that the number of "elements" in a locomotive would be of the order of tens of thousands of pieces, so on balance, the degree of complexity of a cell did not necessarily go beyond human understanding.[29] He presented Wiener with a long term program on the molecular biology of cells, conceived as self-reproducing automata, with a detailed prospectus including the study of everything then known about viruses, phages, the gene/enzyme relation, protein-based structures, and existing or yet to be invented microscopy. He suggested preparing a bibliography and consulting experts. He felt himself capable of finding a logical model for self-reproductive mechanisms, as he explained to Wiener:

> I did think a good deal about self-reproductive mechanisms. I can formulate the problem rigourously, in about the style in which Turing did it for his mechanisms.[30]

As a matter of fact, the way von Neumann would have solved the problem would have stuck very closely indeed to Turing's theory.

In order to prevent a negative reaction from the letter's recipient, von Neumann took care to say that the proposal was not addressed against the research undertaken with Rosenblueth.[31] The proposal can be seen as being in line with von Neumann's epistemological approach, which tended to attack the phenomenologies associated with natural and artificial automata (in fact, cybernetics, although he would never use the term proposed by Wiener) by trying to develop a complete theory of the automaton, either natural or artificial, understood as a computational theory.

---

[29]Cf. ibid. 245.

[30]ibid, 246.

[31]Cf. ibid, 247.

Moreover, he had great confidence in reductionism and thus thought that, in order to build this theory, it would suffice to pinpoint the basic elemental mechanisms, the building blocks. As a consequence, he tended to move towards the study of a single cell, approaching it as a sort of microcalculator.

In this way cybernetics was already reaching out towards molecular biology. In the letter, von Neumann expressed the intention he would subsequently realize of inviting the geneticist Max Delbrück to the Macy conferences. The latter's project on the fagus would be a milestone on the road to the discovery of the double helix.

Von Neumann's proposal also highlights how easily the cybernetics of calculators could be tempted away from neurology, abandoning the problem of the functioning of the human brain and confining itself to a set of building and planning techniques for calculators. The artificial intelligence program was there from the beginning. It consisted in asking how to simulate, using calculators, the mental functions of the brain, but avoiding the question of how the brain could provide those same functions. But this was not the case with von Neumann, who went on asking himself questions on the human brain right up to the end of his life. He died in 1957.

We do not know what Wiener and von Neumann talked about on that 4 December 1946, but it is a fact that Wiener did not agree to follow him in the new venture. We should consider the possibility that Wiener could have found the proposal interesting, even if it might have meant interrupting the research with Rosenblueth and his collaborators. Wiener was attracted by the mathematical study of those phenomena subject to strong intrinsic randomness over a certain time span, and especially those involving communication. Indeed, the attempt to describe their complexity led him to produce mathematics of great depth and sophistication: it was actually what he was doing in his research with Rosenblueth. This interest of his could well have extended to molecular biology as well, with a significant contribution from his point of view. It is no coincidence that during the 1950s, he too would be attracted by the issue of the self-reproduction of automata and of the relations between genetics and cybernetics. This is demonstrated in "On learning and self-reproducing machines", the IX chapter added to the second edition of *Cybernetics* (61c, 169–180) and *God & Golem, Inc.* (64e, 39–48): using a stochastic model taken from electronics, he proposed a demonstration of an automaton able to self-replicate and treated natural selection as a self-learning process like trial and error, once again as a typical communication and stochastic phenomenon.

Therefore a common ground could have been found, but this time Wiener did not follow von Neumann. Quite correctly, Pesi Masani observes that there are no documents on the break-up between the two mathematicians. However, behind von Neumann's no doubt real theoretical concerns, he was very likely also concealing military interests, about which, in his usual way, he was careful not to let on to Wiener.

In fact von Neumann's latest ideas had developed after his participation in the Ninth Washington Conference of Theoretical Physics, 18 November 1946 (Cf. Kay 2000, 106–107). The meeting, entitled "The physics of living matter", (ibid, 106) had dealt with "problems of heredity and the mechanisms by which the almost fantastic gene is able to imprint its characteristics on the cell constituents in a hereditary fashion" (*News Release* 1946).[32]

At the meeting Max Delbrück spoke about the fagus project. From an exchange of letters between von Neumann and the mathematician and future molecular biologist Sol Spiegelman, we know that von Neumann gave a paper on the role of negative feedback on the coupling of chromosomes, and about a theory of self-duplicating machines. On 3 December, Spiegelman invited von Neumann to write a note on these themes, and one week later, von Neumann had indeed begun to do so (Cf. Abraham 2000, 73–74).

It is interesting to learn about von Neumann's interest in feedback, which seemed to be very new for him. He was introducing this concept in a context that anticipated the micro-cybernetics of Jacques Monod (1971) by about twenty years. It is also important to note that this elitist conference was the first after the war. The last two, in 1938 and 1939, had been dedicated to nuclear fusion and fission. In addition to biochemists and biologists, prestigious physicists and mathematicians were present, like Tuve, Gamow, Szilard, Teller, and Hermann Weyl, some of whom had worked at Los Alamos. Considering that, after Hiroshima and Nagasaki, one of the main problems was to investigate the effects of ionizing radiation on living matter (Cf. Yockey, Platzman, and Quastler 1958; Creager and Santesmases 2006), we can easily imagine that at least one of the motives of the conference was of a military nature.

As usual, Wiener had no indication of this from von Neumann. But the cooling in their relations in some sense preserved the former from any further sad adventures. Then arrived the publication of *A scientist rebels*, a letter dated 2 December 1946, which better clarified the different moral positions of the two mathematicians.

9.8 A Scientist Rebels

Regarding the cooling relations between Wiener and von Neumann, a further recrudescence of conscience-based issues probably contributed, of the same kind that, in December 1946, prompted Wiener to send a clamorous statement which appeared under the title *A Scientist Rebels* (47b) to the famous Boston periodical *The Atlantic Monthly*. It was an open letter—dated 2 December 1946—in which he publicly answered a Boston engineer, involved in rocketry for Boeing, who had asked him if he could borrow a copy of the book *Yellow peril*, by then sold out.

[32]Cf. also Year Book 1946 of the Carnagie, 77–78; Gamow, G. and Abelson, P.H. 1946.

Wiener suggested that he forward his request directly to the government, explaining as follows:

> The policy of the government itself during and after the war, say in the bombing of Hiroshima and Nagasaki, has made clear that to provide scientific information is not a necessarily innocent act, and may entail the gravest consequences. [...] The interchange of ideas which is one of the great traditions of science must of course receive certain limitations when the scientist becomes an arbiter of life and death (47b, 748).

Wiener reiterated all the arguments expressed during the crisis of the previous year. Moreover, he lashed out at the missile confrontation strategy, considering it as a self-destructive kamikaze-like act for the US. Finally, he committed himself formally to no longer cooperate on military projects: "I do not expect to publish any future work of mine which may do damage in the hands of irresponsible militarists" (47b, 748).

The autobiography notes: "The moral consequences of my act were soon to follow" (64g [56g], 297). The letter prompted Wiener to a coherent non-cooperation behavior and to distance himself from military projects. He decided not to attend a conference organized by Aiken at Harvard, but he attended the Princeton University bicentennial conference on problems of mathematics, on 17–19 December 1946. There, during a special session devoted to the "new scientific fields", chaired by von Neumann, someone in the audience noted that while Wiener was doing his presentation, von Neumann was reading the *New York Times* as ostentatiously and noisily as he could, almost as if he wanted to annoy him. And at subsequent Macy meetings, according to Heims, "there was a noticeable coolness and even friction between the two men" (Heims 1984 [1980], 208 and cf. note 15, 476). However, one cannot thus argue that the relation between them had permanently broken up.

Chapter 10
Cybernetics

10.1   The genesis of the book Cybernetics

When he left for Europe in the spring of 1947, to go to a congress on harmonic analysis in Nancy, the clamor raised by *A scientist rebels* (Cf. 64g [56g], 332) was still ringing in Wiener's ears. In Paris, a colleague from MIT introduced him to Freymann from the publisher Hermann (Cf. 64g [56g], 315). As Wiener tells us, he was interested in a book in which "I would present my ideas concerning communication, the automatic factory and the nervous system" (Cf. 64g [56g], 316); that's how *Cybernetics* was born. Wiener, on the other hand, had just been waiting for such an offer to express his own ideas publicly. We read from the introduction of the book:

> Those of us who have contributed to the new science of cybernetics thus stand in a moral position which is, to say the least, not very comfortable. We have contributed to the initiation of a new science which, as I have said, embraces technical developments with great possibilities for good and for evil. We can only hand it over into the world that exists about us, and this is the world of Belsen and Hiroshima. We do not even have the choice of suppressing these new technical developments. They belong to the age, and the most any of us can do by suppression is to put the development of the subject into the hands of the most irresponsible and most venal of our engineers. The best we can do is to see that a large public understands the trend and the bearing of the present work, and to confine our personal efforts to those fields, such as physiology and psychology, most remote from war and exploitation (61c [48f1], 28).

The passage testifies how the publication of Cybernetics was strictly connected to the author's choice after *A scientist rebels*: the decision to write the book marks the change from a line of maximum discretion to one of maximum publicity, aiming to provide public opinion with a sufficient understanding of the new technologies and their possible social impact, in the hope that a democratic control could then be exercised over them (Cf. 64g [56g], 308). His decision to limit his own personal commitment to physiology and psychology, the fields "most remote from war and

© Springer International Publishing AG 2017

L. Montagnini, *Harmonies of Disorder*, Springer Biographies,
DOI 10.1007/978-3-319-50657-9_10

exploitation", may seem naïve, and to some extent it is, because even in these fields one can contribute to war. However, it accords well with the following years of Wiener's life, during which he continued to work on cybernetics, first with Rosenblueth and afterwards with others, from the viewpoint that could be defined as bioengineering, with a strong insistence on humanitarian objectives: studies on the heart, leukemia, the brain, implants like artificial arms equipped with sensors, or to make up for a lost sensory organ; research which was always carried out in small groups, with limited funding (Cf. Masani 1990, 223–238).

Most likely, the decision to stay as far away as possible from "war and exploitation" conditioned the very definition of cybernetics. Wiener had by then come to see in its fully extended version the potentialities of a technology of communications, understood in a statistical and stochastic sense, including both analog and digital communications and automatic controls, in machines, in animals, and in society. This picture, which we have seen evolving from *Yellow peril*, already reached its maximum extension and maturity in *Time, communication, and nervous system,* which would appear in the acts of the conference of the New York Academy of Sciences, in the place of the report on the clonus; and it had probably assumed its quasi-final form by the beginning of 1947. In it, to indicate this science of production, elaboration, preservation, and transmission of information, the locution *"theory of communication"* is used, defined as follows:

> This theory covers what is classically known as communication engineering and a number of other fields as well. The theory of the telephone is, of course, communication engineering, but the theory of the computing machine belongs equally to that domain. Likewise, the theory of the control mechanism involves communication to an effector machine and often from it, although the machine may not be watched by any human agent. The neuromuscular mechanism of an animal or of man is certainly a communication instrument, as are the sense organs which receive external impulses. Fundamentally, the social sciences are the study of the means of communication between man and man or, more generally, in a community of any sort of being.

> The unifying idea of these divers disciplines is the message, and not any special apparatus acting on messages. In particular, communication engineering is not in any essential way a branch of electrical engineering (48a, 202).

We would expect that, by introducing the word "cybernetics", Wiener wanted to rename the *"theory of communication"*,[1] which as a matter of fact is undoubtedly the theme of the book, what I referred to above as "Cybernetics on large scale". However, both the introduction of *Cybernetics* and some articles of the same period

---

[1]This was the interpretation of cybernetics by Gino Sacerdote, who, in the preface to the first Italian edition of *Cybernetics* (50J) in 1953, wrote: "Cybernetics in its essence is the study of communications, and more specifically information. All of the electrical communication theory is examined from a new point of view that is probabilistic and statistical: we introduce new concepts such as the quantitative measure of information […]. From the narrow field of communication, the horizon widens to other fields of knowledge; […] In all these studies it is particularly important to consider the statistical structure." (Sacerdote 1953, 7–8).

show a certain hesitation to state this explicitly, leaving open the possibility of a more restrictive interpretation which would limit the use of the term to its sole application to neurophysiology, that is to that area of study which, in the two following decades, the scientific community would prefer to call "biocybernetics", i.e., what I have called a "Cybernetics on a small scale".[2]

As a matter of fact the introduction presents the book as the research program of the Wiener-Rosenblueth group, as explicitly stated in the *incipit*: "This book represents the outcome, after more than a decade, of a program of work undertaken jointly with Dr. Arturo Rosenblueth" (61c [48f1], 1). It is also true that, shortly afterwards, it hastens to retrace the history of this research program, matching it with the events we have narrated in the last two chapters, from the *Memorandum* of the 1940s up to the Macy Conferences, with a mention even of the 1911–13 seminar by Royce. From this account, on the other hand, it is very clear that the Wiener-Rosenblueth group did not become operational before 1945, with the research on the cardiac conduction. One circumstance that could have affected it was the fact that the book was written in the middle of his 1947 visit to Mexico City, where he was required to reside according to the clauses of the research contract between MIT, INC, and the Rockefeller Foundation. Much more likely is that Wiener sought some kind of shelter in those fields which, like physiology and psychology, were "most remote from war and exploitation".

The very same choice of the term "cybernetics" does not seem to be alien to this situation. In the autobiography, Wiener recollects: "I first looked for a Greek word signifying 'messenger', but the only one I knew was *angelos*". However, 'angel' in English has a religious connotation as 'the messenger of God', and the choice would have been misleading. Therefore Wiener began to look in another direction. He explains:

> Then I looked for an appropriate word from the field of control. The only word I could think of was the Greek word for steersman, *kubernetes*. I decided that, as the word I was looking for was to be used in English, I ought to take advantage of the English pronunciation of the Greek, and hit on the name *cybernetics* (64g [56g], 322).

He had been afraid that something like "angeletics" or "angelotics" would have been misleading because of its religious implications. However, shifting the emphasis from communications to automatic control entailed an undeniable reduction of perspective. After all, the word "cybernetics" would have been perfectly appropriate to name the program of "mechanization of teleology" alone, as drafted in 1943 with Rosenblueth and Bigelow, which considered the application of the theory of automatic control to neurophysiology, but which represented only one of the branches of the extensively branching tree that the "*theory of communication*" had now become.

---

[2]Cf. e.g., *A new concept of communication engineering* (49c, 76).

This explains how in the following years many commentators believed they could correctly define the "cybernetics of Wiener", by referring solely to *Behavior, purpose and teleology*, while at the same time, it was becoming common—without immediately linking it to Wiener—to use the term "cybernetics" and words derived from it to refer to "computer science", which in the early 1960s would be called *"informatique"*[3] by the French.

10.2 Clocks and Clouds

Cybernetics was Wiener's magnum opus, describing a lifetime of original thought. It was the text in which most of the threads he had been weaving together since he was a boy merge and take shape. On the other hand it was written somewhat hastily, in the space of just one summer or a little longer, and this resulted in an essentially "heuristic" exposition, according to the meaning mathematicians give to this term, i.e., a first draft requiring further editing in order to achieve mathematical rigor. "Heuristic" in the same sense is also the philosophy contained in the book, which is appealing, but essentially intuitive and likely to need a more in-depth analysis. The author did not make any secret of this situation:

> *Cybernetics* was a new exposition of matters about which I had never written authoritatively before and, at the same time, a miscellany of my ideas. It came out in a rather unsatisfactory form, as the proofreading was done at a period at which I could not use my eyes and the young assistants who were to have helped did not take their responsibility seriously (64g [56g], 332).

It is nevertheless a great book. It is the work which announces to the world for the first time the beginning of a new era, opening up with the end of the Second World War, an era of control and communication, a cybernetic era by definition. And this was achieved by a man who seemed to have been born and lived just for that. The overall picture is outlined in the first chapter on "Newtonian time and Bergsonian time". Wiener starts out from a reflection on the concept of time already presented in *Time, communication, and nervous system*, from which some sentences are quoted word-for-word, such as when he writes:

---

[3]The word "informatique" (computer science) was introduced for the first time in French in 1962 by the engineer Philippe Dreyfus, director of the Centre de calcul électronique of Bull corporation. This was a merger between the two French words "information" and "automatique", and the word was later defined by the Académie Française in 1967 as "the science of automatic information processing". More recently, the term "télématique" was coined, combining "télé" and "informatique", on the initiative of Simon Nora and Alain Minc in the report on informatization of society, submitted in January 1978 to the French president Giscard d'Estaing (Nora-Minc 1978).

The music of the spheres is a palindrome, and the book of astronomy reads the same backward as forward. There is no difference save of initial positions and directions between the motion of an orrery turned forward and one run in reverse (61c [48f1], 31–2).[4]

Historically, Wiener claimed, there were two models of science, one represented by the astronomy of the solar system and the other by meteorology, corresponding to two different types of phenomena, symbolized respectively by clocks and clouds. The first model is characterized by completely reversible and predictable phenomena, as described in nautical almanacs, which are valid for centuries; the second considers irreversible and less easily predictable phenomena, such as the weather.

All the other sciences, he noted, were usually placed between these two extremes. But during the nineteenth and twentieth centuries, it turned out that all of them were being shifted more and more toward the meteorology side. This happened in the biological sciences with the theory of evolution, which assumes an irreversible time. Irreversibility even found its way into astronomy, as the theory of the evolution of the moon–earth system would show, depending on the tidal friction. The big breakthrough for physics, however, arrived with the emergence of statistical mechanics, developed by Maxwell, Boltzmann, and Gibbs. And then came quantum mechanics, considered by Wiener as the final step in the process, which could be interpreted as the irruption of statistics right into the classical fortress of physics. After this last lesson, physic now turned out to be itself nothing more than

[…] a picture of the average results of a statistical situation, and hence an account of an evolutionary process (61c [48f1], 37).

Wiener reminds the reader how, many years earlier, Bergson had

[…] emphasized the difference between the reversible time of physics, in which nothing new happens, and the irreversible time of evolution and biology, in which there is always something new (61c [48f1], 38).

We know that, besides Bergson, other reminiscences were at work between the lines of Wiener's exposition, and in particular Peirce. As Popper writes:

[Peirce] conjectured that the world was not only ruled by the strict Newtonian laws, but that it was also at the same time ruled by laws of chance, or of randomness, or of disorder: by laws of statistical probability. So far as I know Peirce was the first post-Newtonian physicist and philosopher who thus dared to adopt the view that to some degree all clocks are clouds; or in other words, that only clouds exist, though clouds of very different degrees of cloudiness (Popper 1972, 215).

---

[4]In *Time, communication, and nervous system*, Wiener writes: "Now, except for the minute effects of tidal friction and similar phenomena, the methods of prediction which lead us to the nautical almanacs of the future are available with no modification whatever except that of the reversal of all velocities for the extension of the nautical almanac into the past. The music of the spheres is a palindrome." (48a, 197–8).

Peirce had never used the metaphor of clouds and clocks, and it is very unlikely that Popper was unaware that it had been introduced by Wiener in *Cybernetics* (48f1).[5] But Popper almost certainly did not know that Wiener had taken his ideas phylogenetically from Peirce himself, as we have seen, through the medium of Royce, and in particular, the intuition that the "statistical" description of phenomena is methodologically superior to the "mechanical" one.

During his lifetime, these intuitions had become for Wiener a sort of compass for a personal scientific program. He had embodied Peirce's cloudiness in his science, while the development of science around him had gradually been confirming its relevance, as he had had the perspicacity to note in *The role of the observer*, in the light of Gödel's theorem and Heisenberg's uncertainty principle. It could well be said now that physical science had become more similar to what Bergson thought it could never be; and what on the contrary Wiener had already thought possible in *Relativism*. In this sense, Newtonian time had thus given way to Bergsonian time.

10.3   Clouds and Communication

Cybernetics is not marked by the discovery of the intrinsically random and statistical character of phenomena, an objective toward which much of the physical and philosophical reflection of the twentieth century was aiming at. The main novelty was to have brought communication engineering back to a random and statistical dimension: "In doing this", Wiener writes in *Cybernetics*, referring to his research with Bigelow on prediction and filtering theory, "we have made of communication engineering design a statistical science, a branch of statistical mechanics" (61c [48f1], 10).

Here lies the deepest and most original part of cybernetic thought. Although there are two chapters in *Cybernetics* dealing with statistical mechanics and information theory ("2. Groups and Statistical Mechanics" and "3. Time Series, Information and Communication"), the link between "clouds" and "communication", that is, between statistical mechanics and information theory, could easily remain buried among mathematical formulas. As Segal 2002 stressed, it can be better understood in the more concise *Time, communication, and nervous system*, where the discussion on Newtonian time and Bergsonian time ends up returning to an exposition of the *Yellow peril* theories with the phrase: "We now propose to introduce ideas belonging to the Gibbsian statistical mechanics into the theory of communication" (48a, 202).

---

[5]For example, on 17 May 1951 Karl Popper gave a paper at the Nineteenth Ordinary Meeting of the Philosophy Science Group, held at University College, London. This was followed by a paper by Wiener about *Two industrial revolutions* (BBSHS).

The main idea taken from statistical mechanics is that "the message, to convey information, must represent a choice from among possible messages" (48a, 202), and that a message always therefore implies a set of possible messages with an associated probability distribution: a typical concept of Gibbs' statistical mechanics and also Wiener's mathematics. It should be noted that, in this discussion, prediction and information are strictly interrelated: it is enough to reflect that a perfectly predictable message by the recipient does not convey any information.

In this conceptual context, thermodynamic problems tend to merge with those of communication; according to Wiener, we could even think of a new way of expressing the second principle of thermodynamics:

> No communication mechanism, whether electrical or not, can call on the future to influence the past, and any contrivance which requires that, at some stage, we should controvert this rule, is simply unconstructible. Not only is this principle used in a way similar to the second law of thermodynamics, but it is really identical with it. Another completely equivalent statement says that, once a message has been formed, a subsequent operation on it may deprive it of some of its information, but can never augment it (48a, 203).

From the calculation of the quantity of information emerges a formally identical formula to the one found by Boltzmann for entropy, but of opposite sign. Thus Wiener interprets information as negative entropy, since "entropy measures disorder" while "information measures order" (48a, 203). "This point of view", Cybernetics adds, "leads us to a number of considerations concerning the second law of thermodynamics, and to a study of the possibility of the so-called Maxwell demons" (61c [48f1], 11). This is related to the appealing perspective of considering enzymes as micro devices capable of perform an anti-entropic task, which is analogous to the idea of the demon hypothesized by Maxwell.

We can probably detect here a further echo of a discussion with von Neumann, which took place in March 1947, in which Wiener had put forward the proposal of considering information as negative entropy (Cf. McCulloch Summary, 7–8).

In his review of *Cybernetics* for "Physics Today", von Neumann shows that he had long been dealing with the concept of information in thermodynamics, and adds:

> There is reason to believe that the general degeneration laws, which hold when entropy is used as a measure of the hierarchic position of energy, have valid analogs when entropy is used as a measure of information. On this basis one may suspect the existence of connections between thermodynamics and new extensions of logics (von Neumann 1949, 34).

The correlation between information and entropy would remain one of the most debated issues, even up to now (Cf. Settimo Termini 2006, 464; Aldo De Luca 2006, 245; Segal 2003; Continenza and Gagliasso 1998). Wiener certainly sidestepped reflection on the theoretical notion of information. Emblematic of this is his paper [65a] given at the Colloques Philosophiques Internationaux de Royaumont in July 1962, which was dedicated to "the concept of information in contemporary science", with most of the scientists and philosophers of the day. Wiener completely ignored the subject, focusing on research on human prostheses and reflection on the ethical problems raised by the new relationship between man and machine.

10.4   The Evolution of Natural and Artificial Automata

Wiener matches a series of technological changes to the epistemological evolution
we have just outlined, according to the following principle: "The thought of every
age is reflected in its techniques" (61c [48f1], 38). Thus we come to the most
substantial part of the first chapter of *Cybernetics*, in which the sequence of three
technological ages is outlined:

> If the Seventeenth and early Eighteenth centuries are the age of clocks, and the later
> Eighteenth and the Nineteenth centuries constitute the age of steam engines, the present
> time is the age of communication and control (61c [48f1], 39).

The three-way split results from the combination of two epistemological events:
firstly, the transition from time in reversible processes to time in irreversible processes, which allows us to grasp the difference between clocks and heat engines;
and secondly, the transition from a technology concerned with energy and matter to
one centred on the elaboration of information.

In the 1930s, Wiener had racked his brains over the differences among
Huyghens, Newton, and Leibniz; but now the clock became like a powerful
common denominator among these three thinkers. "A watch is nothing but a pocket
orrery, moving by necessity as do the celestial spheres" (61c [48f1], 38), hence
mirroring the deterministic cosmos of Newton. On the other hand, Huyghens had
played a fundamental role in the invention of the grandfather clock, and Leibniz, his
disciple, had conceived of the human body as made of many monads, without doors
or windows, synchronized together like so many music boxes:

> Thus Leibniz considers a world of automata, which, as is natural in a disciple of Huyghens,
> he constructs after the model of clockwork. [...] The monad is a Newtonian solar system
> writ small (61c [48f1], 41).

Even Wiener's former intuition, according to which the improvement of clocks
was to be linked to the needs of navigation to determine longitude, finds here a
natural place:

> The chief technical result of this engineering after the model of Huyghens and Newton was
> the age of navigation, in which for the first time it was possible to compute longitudes with
> a respectable precision, and to convert the commerce of the great oceans from a thing of
> chance and adventure to a regular understood business. It is the engineering of the mer-
> cantilists (61c [48f1], 38).

After the age of clocks, there followed the age of steam engines, which went
from the end of the eighteenth century at least right through the whole of the
nineteenth century, and which coincided with the success of thermodynamics in
physics, with technological research focusing on prime movers (Cf. 61c [48f1], 38):
"All the fundamental notions are those associated with energy" (61c [48f1], 42). In
economic terms the first industrial revolution took place with the advent of the
traditional factory.

In order to understand the meaning of the third age, we must turn to the second epistemological leap, which had already been caught at the time of *Yellow peril*, the one Wiener had grasped icastically by reflecting on the differences between power engineering and communication engineering:

> It is this split which separates the age just past from that in which we are now living. Actually, communication engineering can deal with currents of any size whatever and with the movement of engines powerful enough to swing massive gun turrets; what distinguishes it from power engineering is that its main interest is not economy of energy but the accurate reproduction of a signal (61c [48f1], 39).

The aim of communication engineering and automatic control (both analog and digital) is not the transport of matter or motive power, but the processing of a completely different quantity: information. It is about a new reality which is not linked to the electrical or mechanical or physiological nature of things, nor to the energy levels of the support which conveys it: information may even be processed economically at extremely low energies and then amplified at will. So that was the image of an age of control and communication, and hence of a cybernetic era by definition. The machines symbolizing this era were "communication machines", a term encompassing not only the telegraph and the telephone, but also radar, "self-propelled missiles—especially such as seek their target—anti-aircraft fire-control systems, [...], ultra-rapid computing machines" (61c [48f1], 43). We can therefore see cybernetics in its full? epochal curvature.

Cybernetics is a theory of automata applicable to both machines and animals; but in order to give a precise statement of the specificities of cybernetics as compared to the materialistic conceptions of the past, e.g., those of De La Mettrie, we must take into account the two epistemological leaps mentioned above.

Wiener shows how the evolution of machines described above was also matched by an evolution in the way animals were represented: the artificial automata "in the time of Newton, the automaton becomes the clockwork music box, with the little effigies pirouetting stiffly on top" (61c [48f1], 40), lacking any communication with the environment like Leibniz's monads, "without doors or windows", and acting like so many tiny synchronized clocks.

The most consistent of the mechanists, whom Wiener does not mention, was La Mettrie with *Man a Machine* (La Mettrie 1748), where all animals, men included, are considered as mechanical automatons. But the best known mechanist is Descartes, who, in the pages of *Cybernetics*, however, is overshadowed by Leibniz with his pre-established harmony. Descartes had restricted the mechanist model to animals, stopping short of human beings: "Descartes considers the lower animals as automata," we read in *Cybernetics* (61c [48f1], 40). Descartes thought that deterministic and mechanical processes in human beings concerned only the body (*res extensa*), these being influenced by the free will of the mind (*res cogitans*). This position posed the dualistic problem of how mind and body interact. It was a problem that found a first solution in the Occasionalism of Geulincx: the changes taking place in the soul are "occasions" for God to produce relative changes in the body and vice versa. Malebranche generalized Occasionalism and considered it operating in both the orders of reality, *res extensa* and *res cogitans*, claiming that each cause-effect relation was an occasion for God's intervention. Occasionalism reached a pitch with the pre-established harmony of Leibniz. For him, both body and soul were made up of monads, perfect watches that the creator had synchronized from the very beginning of creation. So causation was resolved into mere concurrence and simultaneity. For Leibniz, monads had neither doors nor windows and give rise to a unit, like a human being, because they behaved in the manner of carillons in sync. As Wiener writes: "but closed though they are, they correspond one to the other through the pre-established harmony of God. Leibniz compares them to clocks which have so been wound up as to keep time together from the creation for all eternity" (61c [48f1], 41). In general, Wiener shows in *Cybernetics* that he had a good grasp of this rationalistic seventeenth and eighteenth picture of thought, including the most "geometrically minded" Spinoza (61c [48f1], 41).

In the XIX century, the prototype automaton was the "glorified heat engine" (61c [48f1], 40), and the living organism was thus seen "above all [as] a heat engine, burning glucose or glycogen or starch, fats, and proteins into carbon dioxide, water, and urea." (61c [48f1], 40).

The automata of *Cybernetics* were in the end machines that came into being by virtue of the first and second epistemological turning points. An anti-aircraft station equipped with a computer and guided by radar was subject to statistical behavior, and it was a "communication machine". It was not deterministic like a clock, and nor did it depend on energy considerations alone, but it existed in Bergson's universe and it had to be studied from the communication point of view. We read in *Cybernetics*:

> In short, the newer study of automata, whether in the metal or in the flesh, is a branch of communication engineering, and its cardinal notions are those of message, amount of disturbance or 'noise'—a term taken over from the telephone engineer—quantity of information, coding technique, and so on (61c [48f1], 42).

This new way of looking at the phenomena of living matter had a much broader influence on the nascent molecular biology than had previously been hypothesized. We have already quoted von Neumann's 1946 letter to Wiener, in which he envisaged a theory of the cell as a self-reproducing automaton. Another interesting record of the influence of cybernetics on molecular biology comes from a letter which Haldane (who was one of the world's leading experts on genetics) wrote to Wiener after reading *Cybernetics*:

> I suspect that a large amount of an animal or plant is redundant because it has to take some trouble to get accurately reproduced, and there is a lot of noise around. A mutation seems to be a bit of noise which gets incorporated into the message. If I could see heredity in terms of message and noise I could get somewhere.[6]

---

[6]Haldane to Wiener, 12 November 1948 (WAMIT). Cit. by Segal (2003), 464.

Is it perhaps a coincidence that, a few years later, DNA would be conceived as a long message in which the words, the genes, are formed by a combination of four letters, and where research would rapidly become oriented towards the problem of "deciphering" the "genetic code"? The climate both created and perceived by *Cybernetics* had obviously encouraged scientists to regard the cell as a "communication machine".[7]

The classic example of the cybernetics approach applied to the study of living organisms remains the consideration of the brain as a computer and vice versa. The earliest cyberneticians, with Wiener and von Neumann leading the field, tended rather quickly to reach this conclusion, although they were prepared to admit broad architectural differences between the two. One issue is undeniable for both, namely that both the brain and the computer are primarily systems which elaborate information and whose operational problems are of a completely new class compared to those implied in heat engines. Von Neumann writes:

> An active logical organ [i.e. a vacuum tube, a transistor, or a neuron] does not by its nature do any work [...]. Consequently, the energy involved is almost entirely dissipated, i.e., converted into heat without doing relevant mechanical work. Thus the energy consumed is actually energy dissipated, and one might as well talk about the energy dissipation of such organs.

> The energy dissipation in the human central nervous system (in the brain) is of the order of 10 W. Since, as pointed out above, the order of 10^{10} neurons are involved here, this means a dissipation of 10^{-9} W per neuron. The typical dissipation of a vacuum tube is of the order of 5–10 W (von Neumann 1986 [1958], 49).

In *Cybernetics*, we find a similar reasoning. Wiener states:

> A large computing machine, whether in the form of mechanical or electrical apparatus or in the form of the brain itself, uses up a considerable amount of power, all of which is wasted and dissipated in heat. The blood leaving the brain is a fraction of a degree warmer than that entering it. No other computing machine approaches the economy of energy of the brain. [...] Nevertheless, the energy spent per individual operation is almost vanishingly small, and does not even begin to form an adequate measure of the performance of the apparatus (61c [48f1], 132).

Wiener and von Neumann understood that, unlike what happens in a heat engine or an electrical engine, the energy used by a system which elaborates information cannot in any way be related to its performance, and it is not even vaguely a measure of it. This is obvious today if we compare the relationship between the computational power and the electrical energy used by a personal computer

---

[7]The thesis of a close connection between the first cybernetics and research in molecular biology has been supported in recent years by Lily E. Kay, and I agree with her. Cf. Kay (1997) and (2000).

containing the equivalent of many millions of electronic tubes, and which can be powered by a battery, with the computational power and energy used by the ENIAC. Indeed, the latter had a computational power of 18,000 electronic tubes plus a few thousands relays, and absorbed a power of 140 kilowatt (Cf. Williams 1997, 272), that is, the equivalent of the power required by a hundred domestic boilers. Such a situation was clear to Wiener and von Neumann from the beginning, partly because they were convinced to have in their hands a computer with very low electrical consumption and very high performance, i.e., the human brain.

10.5   The Philosophy of *Cybernetics*

Wiener's aforementioned passage on the brain ends with the following famous sentence:

> The mechanical brain does not secrete thought 'as the liver does bile,' as the earlier materialists claimed, nor does it put it out in the form of energy, as the muscle puts out its activity. *Information is information, not matter or energy. No materialism which does not admit this can survive at the present day* (61c [48f1], 132. Italics added).

There are several strong and catchy statements in *Cybernetics* which could possibly have been given a more in-depth justification by the author. In this case Wiener was coming to terms with the materialist philosophers studied in his youth, and in particular the Genevan zoologist Karl Vogt (1817–1895). In order to silence spiritualists, who regarded the mind as a reality which was irreducible to the physiology of the body, the latter had provocatively claimed that thought was nothing else but a brain secretion. Wiener rejected this materialism, which was similar to the monism of Ernst Haeckel's matter/energy, in the name of the new reality consisting of information.

The mental process suggested is similar to the one put forward in 1943: in order to physicalize teleology, Wiener resorted to feedback and automatic controls, and in order to physicalize the mind, he resorted to the notion of information and the electronic computer.

It was this ambition to physicalize mental phenomena which seems to have induced Wiener to invoke Leibniz as the "patron saint" of cybernetics, since he had been among the first to attempt to mechanize reasoning using a logical symbolism: "so the calculus ratiocinator of Leibniz contains the germs of the machina ratiocinatrix, the reasoning machine" (61c [48f1], 12). Wiener insisted on the fact that the computer must be understood as resulting from the marriage between technology and logic. He pointed out that this union could be found in the biographies of many of the protagonists of cybernetics: he himself, a pupil of Russell and professor at a technological institution like MIT; Turing, "who is perhaps first among those who have studied the logical possibilities of the machine as an intellectual experiment, served the British government during the war as a worker in electronics" (61c [48f1], 13); Shannon, who had studied ways to represent Boolean logic by switches.

Wiener's autobiography would add with admiration that the marriage between technology and logic had come about precisely through Shannon's research on the implementation of Boole's algebra in an electronic circuit:

> It is through his [Shannon's] work that a training in symbolic logic, that most formal of all disciplines, has come to be one of the recognized modes of introduction into the great complex of scientific work of the Bell Telephone Laboratories (64g [56g], 179).

The fact that formal logic had become an indispensable part of the theoretical arsenal of the engineer reveals one of the less banal aspects of cybernetics, a discipline which puts together the queen of theoretical sciences, namely logic, and electrical technology, but which, even more, breaks the separation between mind and body. The computer is a "machina rationatrix", a thinking machine, an "electronic brain, a term which is not journalistic as one might think, but which can already be traced in *Cybernetics*. From this marriage, logic itself comes out modified to something that the operational considerations mentioned in the previous chapter can apply to.

At the same time, cybernetics does away with the spiritualistic barriers which, in a Cartesian way, separate the body as *res extensa* from the mind as *res cogitans*.

In the same way as he refuses the materialism of matter/energy, Wiener does not accept the vitalistic theses either (the affirmation of the irreducibility of biological phenomena to those of inanimate nature), nor spiritualistic ones (the affirmation of the irreducibility of mental phenomena to those of the body). He notes that the epistemological changes which took place in twentieth century science agreed with Bergson insofar as modern science has upheld a Bergsonian vision of the universe, that is, an evolving world not subject to determinism.

Even the machines of the cybernetic era appeared now to exist in Bergsonian time, so much so that the very term "mechanical" no longer seemed suitable to describe them. In *Cybernetics* we read:

> What is perhaps not so clear is that the theory of the sensitive automata is a statistical one [...]. Thus its theory belongs to the Gibbsian statistical mechanics rather than to the classical Newtonian mechanics [...]. Thus the modern automaton exists in the same sort of Bergsonian time as the living organism; and hence there is no reason in Bergson's considerations why the essential mode of functioning of the living organism should not be the same as that of the automaton of this type. Vitalism has won to the extent that even mechanisms correspond to the time-structure of vitalism (61c [48f1], 43–44).

To fully understand this issue, we must take into account the idea that it applies to systems which constantly deal with noise, whose functioning is of a statistical kind, which retroact to signals coming from the outside expressing a complex behavior, as might happen, for example, with an automatic radar-guided anti-aircraft station. It is very difficult to think that such systems could be clocks. However, Bergson's victory seems to him Pyrrhic, since it does not achieve the result that Bergson had very much at heart, and that was a concern above all to Royce who is not quoted, namely, as we read in *Cybernetics*, "the desire to conserve in some form or other at least the shadows of the soul and of God against the inroads of materialism" (61c [48f1], 38). Hence, we read:

This victory is a complete defeat, for from every point of view which has the slightest relation to morality or religion, the new mechanics is fully as mechanistic as the old. Whether we should call the new point of view materialistic is largely a question of words: the ascendancy of matter characterizes a phase of Nineteenth-century physics far more than the present age, and 'materialism' has come to be but little more than a loose synonym for 'mechanism'. In fact, the whole mechanist-vitalist controversy has been relegated to the limbo of badly posed questions (61c [48f1], 44).

In conclusion Wiener stays true to what he had written in the article *Vitalism* for the *Encyclopedia Americana*: the needs of vitalists and spiritualists can be accepted, provided that we stay on a physicalist plane. This was the solution suggested in 1943 for teleology, and which after all followed the line he had already identified while writing the article on *Infinity*, where the needs of absoluteness and infinity were considered at most as Kantian regulative ideas, in a world like the one we live in, which must continually come to terms with finitude.

Wiener's universe includes many vitalistic and spiritualistic instances, but what is left is a "disenchanted" universe. A world in which the absolute finality, the soul and the divine, are precluded. In a rather difficult-to-decipher passage of *Cybernetics*, he observes that vitalism had at one point claimed to incorporate the sphere of the living and also that of the non-living, within the same wall which had once divided them, that is, if we try and explain it, giving rise to a sort of "panpsychism" or "panvitalism".

The addressee of this note would seem to be Haldane, whose ideas on the quantum brain, we recall, had been compared by Wiener to Leibnizian panpsychism. But Wiener replies that:

It is true that the matter of the newer physics is not the matter of Newton, but it is something quite as remote from the anthropomorphizing desires of the vitalists. The chance of the quantum theoretician is not the ethical freedom of the Augustinian, and Tyche [the goddess of fortune] is as relentless a mistress as Ananke [goddess of destiny] (61c [48f1], 38).

Thus Wiener avoids a monistic theoretical stance, whether we consider that we are dealing with the old materialism of the single substance matter/energy or with the quantum based panpsychism of Haldane. On the other hand he also rejects the spiritualistic and vitalistic dualisms. But he holds on really hard to immanentism, reintroducing a dualism which opposes matter/energy with information. This conception is surprisingly like that of one of the philosophers the young Wiener admired most, that is, Aristoteles, the philosopher who spoke of immanence and a dualism between matter and form.

At least as regards religion, the final conclusion of *Cybernetics* is therefore a mechanism that is even more inexorably immanentist than the previous one, just because it has incorporated the *res cogitans*, the mind.

In some respects this conclusion seemed alarming to Wiener himself, since he was aware that cybernetics was not a mere philosophy, a doctrine among doctrines, but a technology. This is the point on which he would never cease to insist right up to the end: cybernetics has embodied itself into machines, and into the scientific theories required by technology. Who would prevent us from invoking the sanctity of man, the irreducibility of the *res cogitans* to the *res extensa*, as Descartes had done? But this invocation cannot protect the intangibility of the human from technological manipulation, once the dividing wall is shattered, not by a heretical philosophy, but by science itself, and precisely by applied science.

10.6   The New Industrial Revolution

Wiener devotes a large part of the introduction and the whole of the eighth chapter of *Cybernetics* (61c), the last of the first edition (48f1), to the discussion about the possibility of expanding the ideas of cybernetics to the social sciences, considering the social scientists attending the *Macy Conferences* as partners in dialogue. We ought to bear in mind that the book came out between the fifth and sixth meetings, and that it refers to the discussions that had taken place up to the fourth, on the pretext, it would seem, of suggesting that the participants should undertake their own rereading of the cybernetics program, establishing its objectives, and delimiting its field of action. In this respect it could be considered—in the terminology of Imre Lakatos—as a real "research program" for the nascent discipline. Regarding the social sciences, Wiener takes a double stance: on the one hand, he recognizes the importance of evaluating the social impact of cybernetics technologies—an aspect which left the social scientists present at the *Macy Conferences* somewhat indifferent—and suggests a sort of "informationistic" sociology of his own; on the other, he leaves *sub judice* the possibility of a full inclusion of social sciences into cybernetics, and their mathematization using the theoretical instruments of cybernetics. Here we consider first the *pars construens* of this talk and leave the criticism of the social sciences until later, along with his qualms about what seemed to him could become like a form of social engineering.

Wiener was convinced that the epistemological and technological changes presented in *Cybernetics* would inevitably be matched by a new industrial revolution, based on the fully automated factory (which would have implied replacing workers on the assembly line with more docile robots). The subject was in fact one of the major triggers which prompted him to write the book. It was something which for him had long been a cause for concern, which he had informed numerous trade unionists about, and which was finally heard by some of the senior managers in the Congress of Industrial Organizations (CIO), one of the top American trade union organizations (61c [48f1], 28). He had spoken about it with his left-wing friends— Haldane, H. Levy, and Bernal—on the occasion of his 1947 trip to Europe, and they had encouraged him to write *Cybernetics*, judging the subject "as one of the most urgent problems on the agenda of science and scientific philosophy" (61c [48f1], 23).

The crucial point of Wiener's intuition was not merely the idea of automatic control through retroaction, as generally thought, but stems from the way the computer could be used as a process control system in industrial plants. As we read in *Cybernetics*: "It has long been clear to me that the modern ultra-rapid computing machine was in principle an ideal central nervous system to an apparatus for automatic control" (61c [48f1], 26). In fact already during his crisis of conscience, on 16 october 1945, he wrote to Santilliana: "In conjuction with feedback and prediction apparatus they [the computing machines] constitute an adeguate central part for automatic control

devices such as automatic assembly lines, automatic control of chemical plants, etc."
(WAMIT, cit. by Hellman 1981, 242).

Further details about the way in which Wiener had come to this intuition are
provided in an autobiographical text written in 1958, which states:

> The thesis which I made in this book [*Cybernetics*] had implications for the sociology of the
> age of automatization. It had become clear to me that the human brain gave some sort of an
> index of what automatic machinery could do and was subjected to the same principles.
> I saw that the digital computing machine was primarily a logical rather than a numerical
> machine, and could be adapted to the control of factory processes. It was necessary for me
> to take a definite point of view with regard to the moral problems posed by this new
> industrial revolution which was clearly under way. It was in this connection that I wrote my
> book on 'The Human Use of Human Beings' (58f, 10).

Cybernetics had actually only touched on the subject in the introduction, making
several deep but rather short observations. He later followed up on this in *The human
use of human beings. Cybernetics and society*, a book written in 1949 with the express
wish of looking further into matters regarding the impact of cybernetics on society.
Here Wiener devoted a whole chapter to describing in full detail the way in which work
in a factory could be controlled by a computer, reaffirming: "I have often said that the
high-speed computing machine is primarily a logical machine, which confronts different propositions with one another and draws some of their consequences" (50j, 181).

This was a surprisingly precocious perception if we think that, in the same
period, a government commission expressed the conviction that "8 or 10 electronics
experts would have satisfied the requirements of the entire scientific community and
the few companies capable of using their unusual talents".[8] In those years, even
among the manufacturers of calculating machines for commercial use it was hard to
grasp the prospects for development inherent in the computer, firstly because it was
seen merely as another calculator, the only distinguishing feature being that it was
bigger and faster than the calculators around before the war.

In contrast, at least since the beginning of 1945, Wiener had understood the link
between electronic calculators and automatic control; just like Stibitz, he knew that an
electronic digital calculator would have been an excellent system to control an automatic shooting station, and he figured that it would have been equally suitable to
control an industrial device.

The second difficulty for the typical industrial manager in imagining the future of
computers had to do with their size and cost. The internal combustion engine, for
instance, had been born from the beginning as a light engine, and it had remained more
or less the same for at least a century. How could a mastodon like the ENIAC have been
used in a massive way in factories, with all the costs it involved? According to Wiener
this reasoning made no sense, because the human brain was for him a model of what
one could hope the "electronic brains" would become in a more or less remote future. It
was only a matter of time (Cf. 50j, 187–8). And time was what worried him the most.

---

[8]Cf. Can IBM keep up the pace?, in *Business Week*, 2 February 1963, 95 cit. by Rossi (1971), 16.

We may recall that, prior to the war, Wiener was worried about the possibility of the advent of nuclear energy as a source of cheap energy. Being a man of the Great Depression, he was worried by the phenomenon of overproduction, not so much of the economic consequences in themselves, just the excessive speed with which they would come about. Now he was worried that an accelerated and out-of-control evolution of the industrial automation process might not allow society to find a new equilibrium without even bigger upheavals than those that took place in 1929. He estimated that it would take 20 or more years to see the first fully automated factory; but he feared that a new war could bring about an unpredictable acceleration (for instance, to replace in the factory the human beings sent to the front), just as had happened for the technologies connected with radar (Cf. 61c [48f1], 26–27). In this respect he quoted a series of articles appearing in "Fortune" in 1945, one of which was titled *Radar. The industry. A clandestine business in the billions was built on the work of the physicists* [Cf. Anon 1945a and 1945b][9] in which an anonymous feature writer described the kind of multi-million dollar business which had come out of MIT's research on radar.

In any case, in his view, the final outcome was sure: just as the first industrial revolution, the one which had taken place in England from the second half of the eighteenth century, had led to the replacement of human physical energy by the energy of machines, this other industrial revolution would be characterized by the replacement of functions in which human intelligence was required. Wiener argues:

> The first industrial revolution, the revolution of the 'dark satanic mills,' was the devaluation of the human arm by the competition of machinery; [...] the modern industrial revolution is similarly bound to devalue the human brain, at least in its simpler and more routine decisions (61c [48f1], 27).

This "Second Industrial Revolution"[10] would tend to get rid of low level intellectual professions. Of course, just as the other revolution had spared skilled craftsmen, so specialized technicians would be spared, but sooner or later, in his view, the exclusion of "the average human being of mediocre attainments or less" (61c [48f1], 28) from the production process would be inevitable.

The solution Wiener suggested fell within a more or less Keynesian frame of reference. It consisted in creating "a society based on human values other than buying or selling. To arrive at this society, we need a good deal of planning and a good deal of struggle, which, if the best comes to the best, may be on the plane of ideas, and otherwise—who knows?" (61c [48f1], 28).

---

[9]Quoted in (61c [48f1], note 13 p. 27).

[10]Economic history often uses the ordinal "second" to describe the economic changes that took place between 1870 and the late nineteenth century, marking the industrial development of countries such as Germany, the United States, and Japan (cf. e.g., Landes 1969; Baracca, Ruffo and Russo 1979). The industrial revolution glimpsed from Wiener has often been recognized as a third industrial revolution. In the 1960s, the term "cyber-revolution" was in use. Cf. Rapoport (1965).

One argument deserves to be underlined, although as usual it was thrown in as though it were just a detail: Wiener observes that these developments "give the human race a new and most effective collection of mechanical slaves to perform its labor", but that "any labor that accepts the conditions of competition with slave labor accepts the conditions of slave labor, and is essentially slave labor. The key word of this statement is *competition*." (61c [48f1], 27; cf. also (50j, 189).

When we consider on the one hand the masses of dispossessed workers in poor countries, poor children weaving carpets for a mouthful of bread, masses making up cheap labor for the major globalized industries, and at the same time the half-deserted warehouses of the Ford Motor Company or FIAT, almost exclusively populated by strange robots which move up and down repetitively, we might think that they were two complementary scenes, the same as the ones glimpsed by Wiener in the summer of 1947: "human slaves" are employed until their use is no longer more advantageous than the use of "mechanical slaves".

10.7 The Information Society

Wiener arrived at a sort of "informationist" sociology whose first draft can already be found in *Time, communication, and the nervous system*, in which he wrote:

> Up to the present, we have associated a theory of communication with psychology and neurology of organisms singly. In connection with this, we have already seen that two or more communication systems, such as computing machines, may be coupled into a single, larger system. This is true whether the constituent machines are artificial or natural machines. The coupling of human beings into a larger communication system is the basis of social phenomena (48a, 217).

Perhaps he was prompted to include the theme of society in cybernetics in this way by the study of ant behavior presented by Schneirla during the second Macy conference, something he refers to explicitly in *Cybernetics*. The main theme of Schneirla's report had been about the cyclic behavior of ants, determined by the reproductive cycle of the queen. What had struck Wiener the most was that the overall organization of the anthill could be explained in terms of the more or less rich communication models existing among individuals. The resulting sociological model was rather an atomistic one, in the sense that the individuals were conceived as pre-established, regardless of society and the relationships among them. Hans Jonas was scandalized, and at a conference in 1953 would state:

> According to cybernetics, society is a communication network for the transmitting, exchanging, and pooling of information, and it is this that holds it together. No emptier notion of society has never been propounded (Jonas 1953, 191).

However, this model, undoubtedly hyper-simplified, does not lack in appeal, since it offers a simple but powerful method for sociological investigation, as Wiener himself shows in the chapter of *Cybernetics* entitled "Information, language, and society". Assuming that "*intercommunication*" is the essential factor of integration in a society (Cf. 61c [48f1], 156), he concludes that, in order to study different societies, it is enough to study their communicative structures: the more individuals have the possibility to communicate, the more evidently the possibility of integration increases and the community extends as far as the transmission of information can actually reach. It was a methodological approach which would be followed up by the Harvard psycholinguists like George A. Miller, the author of *Language and Communication* (Miller 1972), and by social psychologists like Bavelas of MIT, who was very close to Wiener. And Gregory Bateson would also accept to contemplate the phenomenon of communication, integrating it with his own holistic-relational convictions.

On the basis of his own cybernetic vision, Wiener glimpsed in its full revolutionary extent what would have been an era based on communication and information, and it is genuinely surprising how, in a clear and precocious way, he came to foresee those characteristics which present-day sociology usually considers to be the essential features of our society. Not only did he predict the rise of that process which many years later sociologists would call "post-industrialization", and which for both Bell and Touraine, the first to use this term, did not mean the advent of a society without factories, but a society without workers; but he also understood that the society of the future would be an information society.[11] As we can read in *The human use of human beings*:

> It is the thesis of this book that society can only be understood through a study of the messages and the communication facilities which belong to it; and that in the future development of these messages and communication facilities, messages between man and machines, between machine and man, and between machine and machine, are destined to play an ever-increasing part (50j, 9).

Wiener also glimpsed far in advance the new and complex issues that the information age was to entail. He even talked about the possibility of teleworking. In mid-1949 he thought it already feasible that an architect from Europe could manage the construction of a building in the United States by phone, teletype, and "ultrafax", the new device to transmit duplicates of documents which is known today as "fax", only recently developed at that time by US laboratories (Cf. 50j, 104). Wiener writes:

> In short, the bodily transmission of the architect and his documents may be replaced very effectively by the message-transmission of communications which do not entail the moving of a particle of matter from one end of the line to the other. We thus have two types of communication: namely, a material transport, and a transport of information alone (50j, 105).

In the last chapter of *The human use of human beings*, there is a sentence which could be used today as a catchphrase for an Internet service provider: "for man to be

---

[11]Kline (2015) has recently presented the figure of Wiener and the history of cybernetics, just focusing on the notion of information age. See in particular Chap. 8 "Inventing an Information Age" (pp. 202–228) for an interesting survey of the metamorphoses that the expression "Information Age" and similar phrases assumed after the Second World War.

alive is for him to participate in a world-wide scheme of communication" (50j, 217). But Wiener avoided easy optimism. He considered—and let us bear in mind that this was 1949—that the age of communication had at its core "two opposite and contradictory tendencies": on the one hand, there was "a network of communication, intranational and international, of a degree of completeness never before found in history", and on the other, a witch-hunting climate was gaining ground, one that would soon lead to McCarthyism, while the practices of secrecy introduced during the war, both inside and outside science, remained unchanged, so that "we are approaching a frame of mind paralleled in history only in the Venice of the Renaissance" (50j, 123).

He saw hints of a world that would have to deal with more and more information, in which we would have to get to know its peculiar characteristics, understanding to begin with that information is a reality to which the ways of thinking developed to deal with matter and energy cannot be adapted. He took it for granted that "the fate of information in the typically American world is to become something with a price which can be bought or sold" (50j, 125). He did not comment on whether this was right or wrong, but he felt it necessary to warn the world that information is not a commodity like any other: to be like any other means "essentially, that it can pass from hand to hand with the substantial retention of its value, and that the pieces of this commodity should combine additively in the same way as the money paid for them" (50j, 128). There must be a conservation law for a commodity, but unfortunately, unlike what happens for energy and matter, "information and entropy are not conserved, and are equally unsuited to being commodities" (50j, 129).

"This matter of the consideration of time is essential in all estimates of information" (50j, 136). Information tends to lose its value with time. Hence, the function of the patron is not just to acquire and save works of art, but rather to encourage the creation of new ones. "The idea that information can be stored in a changing world without an overwhelming depreciation in its value is false" (50j, 133). Since "information is even more a matter of process than a matter of storage", it must be understood "as a stage in the continuous process by which we observe the outer world, and make our acts effective upon it" (50j, 134). In the same way, to guarantee US military and scientific supremacy, the so-called scientific and military know-how ought not to be jealously kept "in static libraries and laboratories" (Cf. 50j, 133); this know-how would end up like the weapons stored in arsenals, which in time lose their value, becoming leftovers from the war (Cf. 50j, 132–133).

10.8 Criticism of Social Engineering

In *Cybernetics*, Wiener is highly critical about one of the things that most interested the social scientists attending the Macy conferences, that is, the possibility of transferring to the social sciences the same mathematical tools he was using in his experimental research with Rosenblueth on the human body. This is a question of interpretation about which there has been much discussion and which is still open. I have dealt with it at length elsewhere (Cf. Montagnini 1996 and 2000–2001); here I would like to show mainly how the objections expressed by Wiener in *Cybernetics* were not made casually, but were closely connected to his reflections in the 1930s, and indeed to his whole epistemology.

In *Cybernetics*, Wiener denied that society was able to self-regulate, and he did that by criticizing the prototype of all the deus ex machina supposed by social scientists to be factors in the social homeostasis, that is, the "invisible hand" of Adam Smith, the market understood in the free-trade sense. Here we may imagine an echo of the discussions with Haldane on the shortcomings of capitalism. On the other hand, from Wiener's informationist standpoint, this "social homeostasis" would have to come to terms with communication in the first place. In this respect, he thought that information on the state of the system, which would allow it to self-regulate, is constantly manipulated and distorted in the large mass societies. In fact, he pointed the finger at those statisticians, sociologists, and psychologist serving the military and industry, thinking only of their opinion polls and approval ratings, built ad hoc precisely to manipulate public opinion. For him, these were typical, intentionally misleading activities producing distortions in the communication channels and therefore preventing a healthy social homeostasis (Cf. 61c [48f1], 159–160). To remedy these dysfunctions, was it necessary then to activate some "social therapies" or even some real form of "social engineering", as Bateson and Mead seemed to wish? Wiener showed himself to be sympathetic to their humanitarian motivations, but he thought that the project was not feasible (Cf. 61c [48f1], 24). The possibility of a boomerang effect had to be taken into account:

> It is the mode of thought of the mice when faced with the problem of belling the cat. Undoubtedly it would be very pleasant for us mice if the predatory cats of this world were to be belled, but—who is going to do it? Who is to assure us that ruthless power will not find its way back into the hands of those most avid for it? (61c [48f1], 162).

That is to say, suppose we build a good social theory able to explain and predict collective phenomena: could it not then be used to control individuals to an even greater extent by a power which, in a sick society, is not in the hands of the best people? It is the subject of the "wrong hands" which comes back. The only serious choice in a dangerous and untrustworthy world is caution; and the cautious must in the first instance be the scientist, including the social scientist. In this invitation to prudence, we notice a change in Wiener's stance toward the request for unlimited freedom in research, as expressed in *Limitations of science* in 1935 (35b; see above § 4.1): the experience of the war and the atomic bomb had made him change his opinion. Therefore it is safe to assume that Wiener's fear of a bad use of social engineering had an influence on the way he assessed the application of mathematics to the social sciences. In practice, however, this "methodological qualitativism", to use a sociologists' term (Cf. Ferrarotti 1989), was rooted in Wiener's overall epistemology, as Pesi Masani has put it: "Wiener's thoughts on sociology [...] bore the impress of ideas he had acquired in the course of his deepest and hardest mathematical research" (Masani 1990, 291).

Actually, Wiener had already shown scepticism about the possibility of mathematizing the social sciences, in is discussion in *Limitations of science* (35b), as we have seen in Chap. 4. And most difficulties we found in *Cybernetics* are the same as those he expressed in the 1930s.

In *Cybernetics*, the question becomes more specific, because it deals with his prediction theory, which was the mathematical "core" of cybernetics: so could it be applied to society? The answer is, however, substantially the same: he points out that such a theory requires long historical series, but "under essentially constant conditions" (61c [48f1], 25), and this is not the case in the social sciences, where the longer the series, the more difficult it is to uphold this stationarity condition, due to the great variability of social phenomena. Furthermore, in *Cybernetics*, he concentrates on the effects of the observer, an argument which refers to another paper from the 1930s: *The role of the observer*. In sociology, the situation is not like the one in astronomy, where the observer is infinitely smaller than the observed object and has no significant influence on it; it is rather analogous to the quantum mechanics, where the effect of the photons on the particles that are being observed cannot be neglected, because they are of the same order of magnitude. He refers to the case of ethnographers who modify the lives of human groups they study, or the stock market analyses, which influence its outcome.

In 1964, returning to this issue in *God & Golem, Inc.*, he claimed to have "deliberately refrained" the social scientists in order to avoid "a flood of superficial and ill-considered work", and insisted on the particular characteristics of social phenomenology, which meant that the social sciences should not have excessive expectations about mathematics.

It seems useful to suggest a comparison with an equally great mathematician of the previous generation, namely, Vito Volterra. In a talk in 1901 "Sui tentativi di applicazione delle matematiche alle scienze biologiche e sociali" [*On attempts to apply mathematics to the biological and social sciences*], he had highlighted the importance of mathematics in the social sciences, firstly in the economics of Jevons, Walras, and Pantaleoni, stating:

> The concept of *homo economicus* [...] appears so natural to our scholar in mechanics that he is genuinely surprised when faced with the diffident wonder awakened in others by this idealized and schematic being. For he is accustomed to idealizing surfaces, considering them frictionless, assuming wires to be inextensible [...], both in his own science and in economics, it all comes down to a game of trends and constraints, the latter limiting the action of the former, which by reaction generate tensions. From this system arises sometimes balance, sometimes motion, hence also static and dynamic situations, and in both, science [Volterra 1990, 15–6, original in Italian].

Volterra believed that the method of infinitesimal calculus could be extended to every situation. It was needed to build pure conceptual forms, regardless of aspects such as friction and other effective irregularities, thought of as inessential and to be removed for an adequate scientific discussion. On the other hand, all of Wiener's research can be summed up by the discovery that irregularities in the universe are not inessential. In such a perspective, the adoption of ideal concepts like those used by a mathematized science cannot be justified, except with great prudence and extreme attention with regard to the adopted procedures. Therefore, Wiener's views on social science, and above all on economics, could no longer be identified with those of Volterra. Wiener wrote, in 1964:

> The mathematics that the social scientists employ and the mathematical physics that they use as their model are the mathematics and the mathematical physics of 1850. An econometrician will develop an elaborate and ingenious theory of demand and supply, inventories and unemployment, and the like, with a relative or total indifference to the methods by which these elusive quantities are observed or measured. Their quantitative theories are treated with the unquestioning respect with which the physicists of a less sophisticated age treated the concepts of the Newtonian physics (64e, 90).

Later on in the same text, Wiener drew attention to some studies Mandelbrot was doing:

> He [Mandelbrot] has shown that the intimate way in which the commodity market is both theoretically and practically subject to random fluctuations *arriving from the very contemplation of its own irregularities* is something much wilder and much deeper than has been supposed, and that the usual continuous approximations to the dynamics of the market must be applied with much more caution than has usually been the case, or not at all (64e, 92. Italics added).

Due to this intrinsic irregularity and chaos, the social scientist wishing to adopt mathematical models "must begin with a critical account of these quantitative notions and the means adopted for collecting and measuring them". (64e, 90).

10.9 Wiener and the USSR

The fact that Wiener had chosen to publish *Cybernetics* outside the United States, in France, could raise a suspicion that he wanted the results of his research to be conveyed beyond the Iron Curtain as soon as possible. Be that as it may, the reception given to his ideas by the Soviets was appalling. Stalin's philosophers accused *Cybernetics* and the subsequent works of "pure mysticism", "idealistic pseudo-science", and "idealistic epistemology" (Mikulak 1966, 92), labels which amounted to an excommunication. Moreover, they feared that *Cybernetics* was a Trojan horse that capitalism had come up with to replace "the struggling proletariat by docile robots" (Losano 1978, XI). Actually, they may have had a point in the light of how things have turned out, since, in the words of sociologist Anthony Giddens:

> Information and knowledge have now become media of production, displacing many kinds of manual work. Marx thought that the working class would bury capitalism, but as it has turned out, capitalism has buried the working class. (Hutton and Giddens 2000, 22).

But interestingly enough, rather than a critical confrontation with Wiener as the first to have grasped these social consequences with due concern, and the first to make them known, Stalin's ideologues denigrated him.

When Stalin died in 1953, it was recognized that the ostracism given to cybernetics had been detrimental, not only because the circulation of Wiener's work had been banned, but because such disrepute had affected research on mathematical logic and information theory, branded as "obscurantist sciences" (Elias, P. 1994. Cf. Mikulak 1966 and Ford 1966). It is true that—as Russian cybernetics historian Slava Gerovitch has recently observed—with a healthy realism and while criticizing cybernetics ideologically, the Soviets also attempted to nurture a "de-ideologized" version of it, if only for practical calculational purposes (Gerovitch 2002). On the other hand, their delay was conspicuous from the start since, as Gerovitch adds, in 1946 the main Russian mathematics journal "Advances in Mathematical Sciences" published a double issue devoted to "mathematical machines" which dealt primarily with Bush's analog analyzers, and it was not until 1949 (maybe it is no coincidence that this was after Wiener's *Cybernetics* had come out) that it published an overview of Western research on digital calculators equipped with a program (Cf. Gerovitch 2001, 262). In fact it looks as though they never caught up with it, if it is true that the transition from electromechanical relay methodologies to electronic ones dates to 1950 (Cf. Losano 1978, XVIII), that is, ten years after Wiener's *Memorandum*, and if it is true that, at the beginning of the 1960s, the USSR had to import computers from the United States and the United Kingdom. It is possible that the anti-cybernetics boycott was not even the cause for this delay, which in my view was rooted in the Soviet developmental model itself, but it at least illustrates why Khrushchev himself acted to rehabilitate cybernetics. Eventually a formal apology was offered to Wiener, and in 1960 he was officially invited to give a lecture in Moscow. It is interesting to note that, on that occasion, Wiener gave a paper in which he insisted on the importance of the free circulation of ideas. In addition, he remarked that both the liberalism of Adam Smith and the historical materialism of Marx and Engels were ideological consequences of the first industrial revolution, now unsuitable to manage the changes produced by the new industrial revolution relating to cybernetics (Cf. 61b).

After its rehabilitation, cybernetics was taken over by the soviet leadership, so much so that a scholar like Howard Gardner, a historian of cognitive sciences, has been led to see in Soviet cybernetics the most authentic continuation of the earliest cybernetics, which in the West had been evolving within the more unrestricted and liberal cognitivist approach (Gardner 1985, 21).[12] Jean-Pierre Dupuy has rightly reassessed this interpretation, pointing out the absence of any break in continuity in the subject matters of cognitive science and cybernetics, between the Hixon Symposium and the Macy Conferences, between the "second" and the "first" cybernetics (Cf. Dupuy 1999). But it must be added that Soviet scholars did with cybernetics something that Wiener would never have wanted to do: they made of it a closed theoretical system. Moreover, the cybernetics that was included in the PCUS program in 1961 as "the fundamental tool for the construction of

---

[12]Note that the expression "cognitive science" was coined only in the 1980s, precisely by the research carried out by Gardner for the Alfred P. Sloan Foundation (Cf. Elias, 1994, 24).

communism", was a theory which tended to apply the methods of automatic control to economic and social planning, one of the very aspects about which Wiener had expressed most of his reservations. Cyberneticists beyond the Iron Curtain were well aware of his stance. Still in 1963, the Czechoslovakian social scientist Viktor Knapp wrote: "Cybernetics is therefore, without doubt, the science of control and communication in society (though Wiener, for some of his idealistic conceptions, seems to put society out of the reach of cybernetics)" (Knapp 1978 [1963], 18. Trans. from Italian). On the other hand, the same Soviet scholars, attempting to build the social engineering Wiener had left *sub judice*, fearing an illiberal use, tended to focus more on automatic control than on the communicationist aspects whose importance Wiener had underlined for the social sciences. Even after the rehabilitation of cybernetics, Soviet intellectuals were unable to fully accept the informationist and stochastic vision that was the most genuine "core" of the Wienerian cybernetics (Cf. in particular Ford 1966). An ideology which had crystallized Engels dialectic materialism, a theory in turn sculpted in the granite of German materialism in the second half of the nineteenth century, could not stomach the sentence "Information is information, not matter or energy. *No materialism which does not admit this can survive at the present day*" (61c [48f1], 132. Italics added. Cf. Mikulak 1966, 100 on this point).

Chapter 11
After *Cybernetics*

> *I do not blame the American intellectual for a hostile attitude to science and the machine age. [...] I do blame him for a lack of interest in the machine age. [...] He shows a willingness to accept the trends of the day as disagreeable but inevitable. In fact, he reminds one of the refined creatures in a fable of Lord Dunsany. These delicate and refined beings have become so used to being consumed by a grosser and more brutal race that they accept their fate as natural and proper, and welcome the axe which takes their heads off.*
> Norbert Wiener, *The human use of human beings* [50j, 163].

11.1 Wiener and von Neumann in the Cold War

In order to understand Wiener's stance in the years spanning the end of the war up to the date of his death (1964), it is particularly useful to compare with the diametrically opposite stance adopted by von Neumann, a comparison whose importance was first glimpsed by Heims 1980. As we already noted, there was never a real disagreement of a scientific nature between the two, as is evident from a letter of 4 September 1949, in which von Neumann wrote to Wiener:

> I hope I need not tell you what I think of 'Cybernetics', and, more specifically, of your work on the theory of communications: We have discussed this many times, I hope we shall discuss it many more times, and I have even published my 'appraisal' (as a book review in 'Physics Today'). Hardly any two people ever agree 100% on hardly anything, but I think that we agree more-than-average on this subject.[1]

A few days earlier Wiener had written to him: "I am very much interested in what you have to say about the reproductive potentialities of the machines of the future".[2] It almost seemed like a way to take up their discussions from where they

---

[1]Von Neumann to Wiener, 4 September 1949, cit. by Aspray (1990a, 327), note 155.
[2]Wiener to von Neumann, 10 August 1949 (WAMIT). Cit. by Heims (1984 [1980], 212). Wiener added "As Grey Walter in England has just made a machine with ethics of its own, it may be an opportunity for a new Kinsey report". The "Kinsey report" was a book on the sexual behavior of males by Kinsey, Pomeroy, and Martin 1948, which had just become a bestseller.

© Springer International Publishing AG 2017
L. Montagnini, *Harmonies of Disorder*, Springer Biographies,
DOI 10.1007/978-3-319-50657-9_11

left off in 1946. On the other hand, von Neumann replied sharply that he did not think the issue was something one could joke about, and admonished him for making no further mention of it in the media.[3] Their personal contacts then seem to have come to a stop. It was an interruption that did not stem from theoretical differences, but from life choices. Still in the same period, when writing *The human use of human beings*, Wiener noted:

> This brings up a very interesting remark which Professor John von Neumann has made to me. He has said that in modern science the era of the primitive church is passing, and that the era of the Bishop is upon us. Indeed, the heads of great laboratories are very much like Bishops, with their association with the powerful in all walks of life, and the dangers they incur of the carnal sins of pride and of lust for power. On the other hand, the independent scientist who is worth the slightest consideration as a scientist, has a consecration which comes entirely from within himself: a vocation which demands the possibility of supreme self-sacrifice (50j, 226).

In the first centuries of Christianity we witness a transformation from the poor and evangelical church to the institutionalized Constantinian church of bishops. This transition is taken as a model by Wiener and von Neumann to outline the evolution of science as a transition from its poor origins, driven by creative independent scientists, to the "*big science*" of the major laboratories chaired by coordinating scientists. It is remarkable to note how in those years Wiener and von Neumann embodied these two antithetic models of the scientist in a paradigmatic fashion.

In 1948, in *A rebellious scientist after two years* (48d), which appeared in the "Bulletin of Atomic Scientists", the periodical of critical atomic scientists, Wiener restated that "in every case in which my policy can be of any effect, I intend to act on what seems to me to be the most responsible basis on which I can" [48d, 750]. This aspect of "responsibility" was the guiding thread that ran right through his commitment to science. In 1960, when speaking to Soviet academics, he would state:

> The scholar durst not achieve personal and unlimited freedom of thought at the cost of losing his moral responsibility, which is all that makes this freedom significant. To this needful combination of freedom and responsibility, there is no safe and riskless external guide. (61b, 51)

Von Neumann takes a basically opposite path. The physicist Richard Feynman, his young collaborator at Los Alamos would say:

> And von Neumann gave me an interesting idea: that you don't have to be responsible for the world that you're in. So I have developed a very powerful sense of social irresponsibility as a result of von Neumann's advice. It's made me a very happy man ever since. But it was von Neumann who put the seed in that grew into my active irresponsibility (Feynman 1992 [1985], 132).

---

[3]Cf. von Neumann to Wiener, 4 September 1949 (WAMIT). Cit. by Heims (1984 [1980], 212–213).

While Wiener became a scientist-writer, von Neumann chose a strictly reserved line of communication, and in 1948 he refused to participate in the scientific committee of the "Bulletin of atomic scientists" on the grounds that as a matter of principle

> [...] I would prefer not to join the Board, since I have throughout the last years avoided all participation in public activities, which are not of a purely technical nature, and it seems to me that the objectives of the Bulletin are defined somewhat more broadly, than would be the case for a technical publication.[4]

Wearing this uniform of a neutral technician was actually one of the most esteemed consultants of industrial and military groups, a man with a meteoric career, crowned by his appointment as a member of the Atomic Energy Commission, the supreme institution for civil and military nuclear policy. Gordon Dean, chairman of the Atomic Energy Commission, acknowledged him as "one of the best weapons men in the world" (Heims 1984 [1980], 251), where "weapons" should be understood as "nuclear weapons". A career interrupted by his death from cancer in 1957.

Von Neumann was one of the patrons of the hydrogen bomb project and the intercontinental ballistic missile equipped with the corresponding nuclear warhead. Even the aims of the "IAS computer" (its "*general purpose*" computer would be known simply as "IAS"), were only partly represented by the weather forecast (and control) calculations for which it was officially known: at least the calculation requirements for the hydrogen bomb must be added too.

While the IAS was under construction, he was supervising the realization of many other clones of the IAS scattered around the country. One of these was installed at the Rand Corporation, the think tank for strategic studies, named JOHNNIAC in his honor (Cf. Poundstone 1993).

It should be recalled that, at the same time, Wiener's technological ideas were fully achieved in the biggest project of the Cold War: the SAGE network (Semi-Automatic Ground Environment for US air defense), a computer network connected to anti-aircraft stations equipped with radar, scattered right across North America to ensure anti-aircraft protection. In this network, which would become operational during the 1950s, we can see the achievement of Wiener's intuition in the most plastic way, for he glimpsed in the computer, not only a "*computing machine*", but also a "*control machine*" and a "*communication machine*".

The SAGE project was based on the Whirlwind computer, the MIT digital computer, which Wiener shows himself to be at least informed about in *Cybernetics*. The Whirlwind was born at the beginning of 1944 as a project of MIT's Servomechanisms Laboratory, financed by the Navy, for an analog computer to be used in a flight simulator. Since December 1944, it had been put under the

---

[4]Von Neumann to Norman Cousins, 22 May 1946 (VNLC). Cit. by Heims (1984 [1980], 235). Replying to a second invitation, von Neumann stated: "I do not want to appear in public in a not primarily technical context", von Neumann to Hyman Goldsmith, 5 October 1948 (VNLC). Cit. by Heims (1984 [1980], 235).

direction of the engineer Jay Forrester. As he pointed out around mid-1946, the initial analog approach was abandoned and it was decided to turn it into a "general purpose" computer that was both digital and electronic. The motivation for taking that step probably came from reading *First draft of a Report on the EDVAC* by von Neumann and Goldstine, as well as the "Conference on Advanced Computation Techniques", held at MIT between 29 and 31 October 1945 (Cf. Archibald 1946), where the developments stemming from the research on ENIAC were presented in detail.[5]

In many ways similar to other "general purpose" computers based on von Neumann's architecture, the specificity of the Whirlwind consisted in being suitable for real-time automatic control, and because of this, costs were about ten times higher compared to other computers: in the end it would cost 5 million dollars, while the cost of the other computers was roughly around 300–600 thousand dollars (Cf. Edwards 1997, 90–1). These figures give us some idea of how much money Wiener could have relied upon, had he adopted a more collaborative stance.

In the same year, precisely because it was suitable for automatic control, the Whirlwind was chosen for the SAGE network construction project.[6] It became the first practical arena where the first programmers were trained, along with the second generation of cyberneticists. And from this arose the Artificial Intelligence paradigm, as a demonstration of how close the connections were between the two generations of scholars.

11.2 Cybernetics as a "Grammar of the Cold War"

While cybernetics had been the daughter of World War II, it was all the more the undisputed protagonist of the Cold War, since "cybernetics" became part of both the defensive and offensive systems. Not only that, but cybernetics became a more or less coherent sort of "grammar of the Cold War".[7] *Cybernetics* was a very successful book, and turned the spotlight of the mass media on the Macy Conferences, where many illustrious scholars would come and go, when they were still at their fifth meeting. In fact, the word "cybernetics", though not without contradictions, was a conglomeration of various ideas and theories which very soon became a shared issue, a real "grammar", through which the protagonists of the

---

[5]Wiener appears on the list of those present at the conference, but he does not seem to have presented any paper (Cf. Archibald 1946).

[6]Cf. Telecommunications systems: network milestones, in [EB 1997].

[7]This fact, although with some differences, has also been understood in *The closed world* (Edwards 1997). I had come to similar conclusions independently, in my article *Cibernetica e guerra fredda* (Cybernetics and the Cold War) [Montagnini 2000]. However, the main actor in the "discourse" glimpsed by Edwards in Cold War policy, was it seems the United States, and its aim was merely "control" in a negative sense. On the contrary, in my opinion, the USA and the USSR played a sort of long table tennis match, creating a typical loop.

Cold War, not only in the United States, tended to self-represent the ongoing conflict, devising strategies not only for war, but also for peace.

In fact, in no time, the world had just got itself into the biggest "schismogenetic" whirlpool, to use a concept introduced by Bateson, since the beginning of human history, where the members of the first cybernetic circle, just like the cyberneticists of the second generation found themselves playing different roles, although never marginal.

The typical circular models of automatic control were extremely well suited to represent the ongoing military "escalation" process. In a letter dated November 1951, addressed to Rear-admiral Strauss, his political patron, von Neumann declared:

> The preliminaries of war are to some extent a mutually self-excitatory process, where the actions of either side stimulate the actions of the other side. These then react back on the first side and cause him to go further than he did 'one round earlier', etc. [...] each one must systematically interpret the other's reactions to his aggression, and this, after several rounds of amplification, finally leads to 'total' conflict [...] I think, in particular, that the USA–USSR conflict will probably lead to an armed "total" collision, and that a maximum rate of armament is therefore imperative.[8]

This clear-headed awareness of the ongoing "schismogenetic" process left von Neumann with no opportunity to transcend it. On the contrary, it suggested to him the sort of behavior which added more fuel to the fire: "I think, in particular, that the USA–USSR conflict will probably lead to an armed 'total' collision, and that a maximum rate of armament is therefore imperative".[9]

Strategically speaking, at the beginning, von Neumann shared the thesis of the "first strike" party, that is, the conviction that, in order to win the Cold War, a preemptive strike was necessary (Cf. Heims 1984 [1980], 246). In his view, the earlier was the USSR bombed the better. Apparently, he declared: "If you say why not bomb them tomorrow, I say why not today? If you say today at 5 o'clock, I say why not 1 o'clock?" (Heims 1984 [1980], 247). But neither Truman nor Eisenhower ever accepted this theory, and the proof is that we are still alive.

As early as 1945, von Neumann took part in Teller's meetings to discuss the hydrogen bomb. The project was discontinued because of the unfavorable opinion of the General Advisory Committee chaired by Robert Oppenheimer. As we know, the latter had been the director of the Manhattan project. He believed that the development of the H bomb would exacerbate the arms race and he considered it more appropriate to build up a deterrence strategy by increasing the "tactical" arsenal, that is, atomic bombs of the type used in Japan.

As we can see, even Oppenheimer's idea belonged to cybernetics and unfortunately the feared military escalation took place: on the 29 August 1949, the Soviets obtained the first nuclear explosion with a 10–20 kiloton plutonium bomb. The party which was against it was weakened and was wiped out afterwards when the Korean war broke out. Von Neumann did not believe in the disarmament

---

[8]Von Neumann to Strauss, 21 November 1951, cit. by Heims (1984 [1980], 287).
[9]Ibid.

agreements. He argued that they would shift the area of conflict to a less favorable terrain for the United States: being a democracy, it would have had more difficulty than the USSR to run black bag operations. In his view, America needed a rapid and massive ability to retaliate as a valid deterrent (Cf. Heims 1984 [1980], 266). This is how, as an alternative to the *First strike*, the *Biggest Bang* strategy was born: the idea was to provide America with the maximum response capacity, in terms of power and rapidity. This explains the H bomb and long-range missile option. The United States rushed to build the new bomb and it was successfully experimented in November 1952. The following year the Russians did the same.

After the election of Eisenhower, in 1952, in full McCarthyism, as Oppenheimer was being tried for having obstructed the H bomb (on that occasion, von Neumann defended Oppenheimer), the option of a thermonuclear retaliation attack began to gain a foothold in response to violent attempts to alter the territorial status quo, even without resorting to atomic bombs from the Soviets. So, while nuclear experiments were repeatedly carried out as a mutual display of power, but also to verify the effective tolerance of populations and armies to the effects of the bombs, the so called von Neumann Committee was set up to investigate the feasibility of an intercontinental ballistic missile (ICBM) carrying an H bomb. The final report of the committee at the beginning of 1954 informed the President of the United States that the project was feasible and militarily effective. In fact, it urged the president to go ahead with it as "the highest national priority", suggesting that he should cut down on the ordinary procedures and waiting times and simultaneously carry out development, production, and deployment. In the summer of 1955, the ATLAS ICBM project was launched (Cf. Heims 1984 [1980], 253).

The Rand Corporation used von Neumann's game theory (see von Neumann and Morgenstern 1944) to elaborate strategic models. It is not known in detail how von Neumann himself may have used such a theory in his role as a government contractor. There are no doubts, however, that the best way to play Cold War according to von Neumann was to stay one step ahead and raise the stakes, maximizing secrecy and minimizing trust in one's opponent.

Wiener and Bateson were not to be outdone by von Neumann in their cybernetic reasoning skills, and they reached a different conclusion from the nuclear holocaust that von Neumann thought was inevitable. Both Wiener and Bateson were very worried by the use of game theory in the strategic field. In the meantime, Gregory Bateson had been carrying out a deeper psychiatric investigation of the dimension of circularity in conflicts. He wrote this to Wiener in 1952:

> What applications of the theory of games do, is to reinforce the players' acceptance of the rules and competitive premises, and therefore make it more and more difficult for the players to conceive that there might be other ways of meeting and dealing with each other [...]. The theory may be "static" within itself, but its use propagates changes, and I suspect that the long term changes so propagated are in a paranoidal direction and odious. I am thinking not only of the propagation of the premises of distrust which are built into the von Neumann model ex hypothesi, but also of the reflexion or corollary of the fact that the original theory was set up only to describe the games in which the rules are unchanging and the psychological characters of the players are fixed *ex hypothesi*. I know as an

anthropologist that the "rules" of the cultural game are not constant: that the psychology of the players is not fixed; and even that the psychology can be at times out of step with the rules.[10]

It is no coincidence that Bateson had been the creator of the deutero-learning theory: as we learn something we also learn to learn. In the same way, when war strategies are played by game theory, we tend to impose on the opponent the inflexibility of those rules, so that its use remodels the relations with the opponent tightening them.

Wiener agreed with Bateson's preoccupations and used his status to "play" Cold War with an alternative rationality to that of von Neumann. In the letter published in December 1946 by *The Atlantic Monthly*, he had already lashed out at the strategy of missile confrontation, considering it a self-destructive and kamikaze-like act for America (Cf. 47b). In 1950, still in *The Atlantic Monthly*, he strongly defended the thesis of negotiation, referring ironically to the blind rationality driving the opinions of the hawks:

This military version of 'Another little drink won't hurt us' shares much of the unconvincingness of the alcoholic assertion on which it is patterned. Another little drink is going to hurt us, and this is the reason why. [...] Fundamentally, they can have no more desire than we to celebrate a nominal victory by a universal funeral pyre of both sides. They share with us a healthy hope for a longer span of life on this earth [50i, 50 and 52].

"Voices of rigidity", the last chaper of *The human use of human beings* (50j), is a singular and very important text. It was deliberatedly omitted (Cf. Calimani, Riccardo and Lepschy 1990, 155) in translating the Italian editions (see edns by the publishers Einaudi, in 1953, and Boringhieri, in 1966). However, it would also disappear from the second English edition (54d), under the advice of Jason Epstein of Doubleday. Wiener argued that, after the revolution, the USSR had progressively become more and more like a strict, dogmatic church, not very different from the medieval Christian church. In addition, the competition with this "rigidity" was even making the United States lose that democratic flexibility which was inherent in its constitutional roots: the inquisitorial practices on one side of the Iron Curtain induced a witch hunt climate on the other.

The US, in his view, would have come out victorious from the Cold War just by rediscovering the importance of broad communications and a free exchange of opinions, values inscribed in their most fundamental democratic traditions, and also guiding principles of what constituted cybernetics according to Wiener (50j, 228–9 in particular).

The vicious circle would not be broken by continuing with the process of accumulating "know how" and defending it with secrecy, but on the contrary by suspending the secrecy practices inherited from the WWII and exacerbated by competition with the Soviets, thus eliminating the climate of suspicion and intimidation into which America was falling.

---

[10]Bateson to Wiener, 22 September 1952, cit. by Heims (1977, 148).

Wiener was as aware as von Neumann of the "stickiness", so to speak, of the grammar of the Cold War. On the one hand, for example, it seemed necessary to him to ride on innovation, including military innovation, while on the other, he realized that any innovation would do nothing but heighten the drama:

> We are in the position of the man who has only two ambitions in life. One is to invent the universal solvent which will dissolve any solid substance, and the second is to invent the universal container which will hold any liquid. Whatever this inventor does, he will be frustrated (50j, 141).

He realized that it was not easy to escape the maelstrom in which the world was trapped, and observed disconsolately:

> We have taken unto us the devils of the age like so many Gadarene swine (Cf. Matthew 8: 28–33) and the compulsion neurosis of scientific warfare is driving us pell-mell, head over heels into the ocean of our own destruction. Or perhaps we may say that the gentlemen who have made it their business to be our mentors, and to administer the new program of science, are so many apprentice sorcerers, fascinated with the incantation which starts a devilment that they are totally unable to stop. *Even the new psychology of advertising and salesmanship* becomes in their hands a way for obliterating the conscientious scruples of the working scientists, and for destroying such inhibitions as they may have against rowing into this maelstrom of destruction (50j, 142. Italics added).[11]

Inspired by a sea current in the Norwegian sea, Edgar Allan Poe and Jules Verne had imagined that the maelstrom was a huge whirlpool that could swallow up men and ships. The word later got into the English language as a metaphor for a wildly turbulent situation, and what better word to describe the whirlpool of the Cold War!

Notwithstanding, Wiener did not believe that the ultimate holocaust was inevitable. The complexity of the situation required one to learn to think in dynamic and relational terms. It was not easy to follow such logic: to develop new weapons one had to consider that ipso facto, a similar response would always be induced from the enemy, and sooner or later the latter would come to the same findings. It was not enough to consider how much damage these new weapons could cause to the enemy; they could turn out to be even more harmful to those who first intro-duced them, if the enemy had any opportunity to turn those same weapons against them.[12] In general, a morality of prudence sprang out of this:

> If a man with this tragic consciousness of fate approaches, not fire, but another manifes-tation of original power, like the splitting of the atom, he will do so with fear and trembling. He will not leap in where angels fear to tread, unless he is prepared to accept the pun-ishment of the fallen angels (50j, 211).

The mention of angels, which would also be the inspiration for the title of a book by Bateson, is the echo of a quote from Alexander Pope: "for fools rush in where angels fear to tread", which has become proverbial in Anglo-Saxon culture.

---

[11]The notion of 'maelstrom' in human relationships has been discussed with reference to Bateson by Elias (1987 [1983]). For an attempt to mathematize this human social dynamics, see, e.g., the classic work on the arms race by Richardson (1960).

[12]Remark to Wiener by J.B. Wiesner (Cf. 50j, 141).

In this request to use the delicate movement of an angel, there is something more than what would later become known as the "precautionary principle". It is not enough just to be cautious; one ought to think carefully about the real effects one's actions.

11.3   An Anthropology for the Information Society

Cybernetics has left open many theoretical problems, and first of all the question of a consistent definition of information outside the strictly engineering context. Despite the fact that Wiener had never strictly speaking abandoned scientific research, what most interested him after *Cybernetics* was his reflection on the ethics of science and technology in the new cybernetic age.

In *Cybernetics*, the theme of usurpation by machines already arises:

> It will be seen that for the second time I had become engaged in the study of a mechanic-electrical system which was designed to usurp a specifically human function—in the first case, the execution of a complicated pattern of computation, and in the second, the forecasting of the future (61c [48f1], 6).

This is a leitmotif that accompanies all his reflection over the last two decades, up to *God & Golem, inc.*, where it is the central theme. Once the barrier between man, machines, and animals has been broken—Wiener seems to be reasoning—the actual differences between them must be re-established. He is once again questioning the value of human beings, and trying to found again its sacredness and inviolability on bases other than those proposed by vitalism and spiritualism, but within an immanent cosmos dominated by finiteness. If science and technology, with cybernetics, had managed to physicalize teleology and even the mind, there seemed to be only one way out: to physicalize ethics itself, and perhaps even theology, reconsidering them from an operational point of view, as he had done with teleology and the mind. This would appear to be the ultimate message of Wiener's discussion in all his writings from *The human use of human beings* (1950) up to *God & Golem, Inc.* (1964).

The human use of human beings has been considered as a sociology book, so much so that some sociologists have considered it to constitute undue interference.[13] And in part it is, especially when it so early on and so effectively captures the advent of the new industrial revolution based on automated factories and the communication society. But it is above all a philosophy book dealing with philosophical anthropology and ethics.

---

[13]See the survey on the reception of Wiener's ideas by sociologists, reported in (Geyer and Van Der Zowen 1994). However, Geyer and Van Der Zowen neglect e.g. the strong influence on Talcott Parsons, one of the most eminent sociologists of the day; cf. in particular (Parsons 1966, 28 ff.).

The cultural anthropologist describes humans phenomenologically, paying attention to social behavior, with the emphasis later shifted to culture, while in the past, physical anthropology had insisted on the soma and hereditary characteristics. Philosophical anthropology, on the other hand, is an ancient discipline that raises questions about the essence of man. The first, whether cultural or physical, has the stated aim of describing humans and human behavior, while the second wants to grasp what humans "are", in their essence. There is no doubt that, from the first chapter, the book comes up against the main problem of philosophical anthropology, which is to give a definition of man:

> The problem of the definition of man is an odd one. [...] What does differentiate man from other animals in a way which leaves us not the slightest degree of doubt, is that he is a talking animal (50j, 2).

To say that man is an animal that speaks, however, is not enough for him, insofar as communication is a skill that he does not deny to either animals or machines. So it becomes necessary to specify what this "to talk human" is in comparison to other communication models. The specific difference from other beings that communicate must be sought essentially in the Santayana approach (see above § 1.6). In fact Wiener does this on the basis of human phylogeny, compared with that of insects, all reinterpreted in the light of cybernetics. The chapter on "Rigidity and Learning: two patterns of communicative behavior" (50j, 59 ff.) states:

> [There are] two contrasting forms of animal mental organization: that of the ants, in which the degree of dependence on the learning process is minimal, and yet the structure of society seems to be reasonably complicated; and that of man, in which the whole individual and social organization centers around the process of learning (50j, 84).

The human behavioral model, Wiener argues, based on experience and learning, is closely linked to the use of the word, intelligence, and memory. It is the climax of an evolutionary tendency inherent in mammals, although the distance between humans and other primates is great in this respect. In contrast, in his view, the evolution of the behavior of insects has followed a somewhat different phylogenetic path: the ant as an individual has a poor ability to learn because it has too small a body, undergoing very profound metamorphosis in the course of his life, facts that radically limit its individual memory. The communication between the ants is entrusted to a few chemical signals. It can therefore be assumed, he concluded, that the ant possesses a largely genetically programmed behavior, unlike humans, where evolution has favored the ability of each individual to communicate, learn, and remember (Cf. 50j, 68).

The essence of man, derived from what we might call his "phylogenetic vocation", is that of a free and creative being, given the opportunity to gain experience and communicate with others without impediment and censure, left free to decide for himself. It is a flexible communication model. We read in the chapter on "Voices of Rigidity":

For us, to be less than a man is to be less than alive. Those who are not fully alive do not live long even in their world of shadows. I have said, moreover, that for man to live is for him to participate in a world-wide scheme of communication. It is to have the liberty to test new opinions and to find which of them point somewhere, and which of them simply confuse us. It is to have the variability to fit into the world in more places than one, the variability which may lead us to have soldiers when we need soldiers, but which also leads us to have saints when we need saints. It is precisely this variability and this communicative integrity of man which I find to be violated and crippled by the present tendency to huddle together according to a comprehensive prearranged plan, which is handed to us from above. We must cease to kiss the whip that lashes us (50j, 217).

By reflecting upon the parallel between men and ants, Wiener reaches a criticism avant la lettre of sociobiology, a discipline that has made quite a stir since the 1970s, propagated by Edward O. Wilson. The guiding hypothesis of sociobiology is that man is a biological species, and therefore human behavior is subject to the same mechanisms that regulate the behavior of other animals. It follows that ethological knowledge, that is, knowledge related to animal behavior, can also be applied to humans to explain certain social behaviors. So far the conclusions are similar to Wiener's. But Wilson concludes: since man is an animal, he is genetically programmed.

For Wiener, this view applies to a large extent to insects, but not to humans, the development of which would lead in an entirely different direction.

It should be pointed out that Wilson is an entomologist by profession and that, shortly before writing his manual on *Sociobiology* (1975), he had written another on *The insect societies* (1971), before subsequently giving birth to his monumental treatise entitled *The ants* (Hölldobler and Wilson 1990). It seems to me that he tended to conceive other animals, including humans, in the same way as insects.

According to Wiener, the organizational model of a society of men based on the ant society is what "fascists like", along with many "strong men" in the worlds of business, politics, and even science: the ideal of a society in which "such people prefer an organization in which all orders come from above, and none return" (50j, 15–16), where "the human beings under them have been reduced to the level of effectors for a supposedly higher nervous organism" (50j, 16). Against this reduction, he rises up:

I wish to devote this book to a protest against this inhuman use of human beings; for in my mind, any use of a human being in which less is demanded of him and less is attributed to him than his full status is a degradation and a waste (50j, 16).

The purpose of philosophical anthropology is openly prescriptive, that is, it serves as a basis for ethics: if the man is such and such, you need to treat him according to his rank. Wiener concludes that "it is as completely natural for human society to be based on learning as for an ant society to be based on an inherited pattern" (50j, 69).

11.4 The Golem and Its Creator

In many pages of Wiener's writings from these years, we find a religious inspiration. The discovery of the complicated relations prevailing in the Cold War, as discussed above, converged in a tragic consciousness which, suggests Wiener, is in itself salvific, since it implies a healthy note of caution:

> The sense of tragedy is the sense that the world is not a pleasant little nest made for our protection, but a vast and largely hostile environment, in which we can achieve great things only by defying the gods; and in which this defiance inevitably brings its own punishment. It is a dangerous world, in which there is no security, save the somewhat negative one of humility and restrained ambitions. It is a world in which there is a condign punishment, not only for him who sins in conscious arrogance, but for him whose sole crime is ignorance of the gods and the world around him (50j, 211).

Faced with words such as these, readers and commentators have often questioned the origin of the religious inspiration they imply (Cf. Jdanko 1994). Attempts have been made to come back to his paternal education and Jewish background, but in the first chapter we have shown that Wiener had had little religious training. An answer to these questions is given by Wiener himself in the first volume of his autobiography:

> Although there was always a strong moral implication in my father's personality and in the course of life toward which he directed me, my interest in science started with a devotion rather to the service of truth than to the service of humanity. Such interests in the humanitarian duties of the scientist as I now have are due more to the direct impact of the moral problems besetting the research man of the present day than to any original conviction that the scientist is primarily a philanthropist (64h [53h], 73).

Wiener was not even ten years old when he began to declare himself an atheist. Even in Cybernetics, as we have observed, his conception of the universe is marked by a radical immanence. The religious inspiration is the reflection of his battle for accountability: "The hour is very late, and the choice of good and evil knocks at our door" (50j, 213).

In fact, it is his conception of an immanent and intrinsically complex cosmos that prompts him to reflect even more on ethics and the "sacred". If in *The human use of human beings* (50j) he was looking for a physicalization of the idea of man and ethics, in *God & Golem, Inc.* (64e), it seems that he wants to physicalize theology itself, or at least certain demands of theologians and religious, who were perhaps the main targets of the book. To them, as regards the need to preserve the dignity, the inviolability, and the sacredness of man, Wiener seems to say: "I'm on your side, but you must realize that it is not enough to deny a soul to machinery, as did Descartes with animals, to guarantee that human dignity is not violated".

In this respect, one should think of the self-reproducing machines, or machines playing checkers or chess, which are capable of self-learning, as considered in the book. In reply, he commonly met with remarks about "pulling the plug": after all those who build the machines are men, and machines cannot exceed their constructor; and in any case, we can always pull the plug. Wiener had already expressed his opinion on this in *The human use of human beings*:

> In constructing machines, it is often very important for us to extend to them certain human attributes which are not found among the lower members of the animal community. If the reader wishes to conceive this as a mere extension of our personality as human beings, he is welcome to do so; *but he should be cautioned that the new machines will not stop working merely because we have discontinued to give them human support* (50j, 86–87. Italics added).

In *God & Golem, inc.* he gives the example of the program that plays checkers, devised at about that time by Samuel:

> As a matter of fact, for a considerable period Samuel's machine was able to defeat him rather consistently, after a day or so of working in [...]. It did win, and it did learn to win; and the method of its learning was no different in principle from that of the human being who learns to play checkers (64e, 22).

The relationship between the designer and the machine finds a parallel in the relationship between God the Creator and man created in his image: since God creates man free, man can escape his control. This is also the story of the Golem, the legendary automaton created by the rabbi of Prague—the Hebrew version of the Renaissance magician—to be his servant, but who rebelled, causing destruction. Besides the Golem, Wiener recalls numerous legends and similar stories. *The Monkey's Paw*, by the British writer W.W. Jacobs, is about a talisman, given to two elderly parents, which can make one of their wishes come true. The parents express their wish immediately, asking for 200 lb, and shortly afterwards someone knocks on the door with the money: it is the compensation for the death of their son in an accident which had just happened at work (64e, 58–9). This chilling story, like that of the Golem, seeks to warn that, when you set up an action, there's no guarantee that you will be able to foresee its consequences with certainty: this is the Hegelian heterogenesis of ends. And it applies in particular to technology:

> A goal-seeking mechanism will not necessarily seek *our* goals unless we design it for that purpose, and in that designing we must foresee all steps of the process for which it is designed [...]. The penalties for errors of foresight, great as they are now, will be enormously increased as automatization comes into its full use (64e, 63).

This art of prediction, which goes beyond the ordinary practice of the scientist and engineer, will in his opinion become a necessity in the future:

> As engineering technique becomes more and more able to achieve human purposes, it must become more and more accustomed to formulate human purposes. In the past, a partial and inadequate view of human purpose has been relatively innocuous only because it has been accompanied by technical limitations that made it difficult for us to perform operations involving a careful evaluation of human purpose. This is only one of the many places where human impotence has hitherto shielded us from the full destructive impact of human folly (64e, 64).

According to Wiener, some people may believe that the very statement that an automaton is able to learn and think is a sort of sacrilegious act, or that man himself is an automaton. But in his view, such a statement is just a statement of a reality—man's usurpation by cybernetic technology—and it cannot be ignored if we want to have an adequate representation of the level reached by science and technology and the dangers that this entails. The real problem is something quite different: it is the worship of the gadget, the device, a dangerous idolatry, which is in fact a kind of black magic. The "worshippers of devices" blindly trust automatic devices, without questioning the limits of the machines and the need for an ethics regarding their use. The machine can also be a great excuse to avoid taking responsibility for one's own choices: "Neither will he calmly transfer to the machine made in his own image the responsibility for his choice of good and evil, without continuing to accept a full responsibility for that choice" (50j, 211–2).

11.5   The Butterfly Effect and Limits to Growth

Wiener is, of course, aware of the difficulties that lie behind the problem of prediction. In fact, he introduces the reader of *The human use of human beings* to considerations of complexity and nonlinear processes. In this context, he discusses a phenomenon that would some years later become known as the "butterfly effect". A non-linear process is a phenomenon in which, for example, doubling the cause does not double the effect. What the scientist often attempts to do is to find some technical device to treat any non-linear phenomenon at least approximately as a linear one. However, there are phenomena which are difficult to "linearize". Wiener gives the following example: if we throw a small ball on a glass plate, it will bounce with an approximately proportional speed every time we increase the strength of the launch. But at a certain point, the glass will break in a complex way, subtracting from the ball an amount of kinetic energy that will always depend on the type of fracture occurring in the glass, while this in turn will depend on the actual structure of the glass, which will generally be full of defects and impurities. The problem at this point of establishing how quickly the ball will bounce, and indeed whether it will bounce, is difficult to reduce to a linear problem (50j, 33–34). Now, according to Wiener, for both human and natural facts, we are faced with similar conditions most of the time:

> For this purpose a thousand years may be a short time or a thousandth of a second a long time, depending on the particular system considered. It only takes a small modification of the impetus of the firing pin of a gun to change a misfire into the effective projection of a bullet; and in case this gun is in the hands of an assassin, this vanishingly small difference may produce the difference between a revolution and a peaceful political development (50j, 34).

Perhaps the "trigger" of the gun is less poetic than Lorenz' "butterfly", which produces a hurricane with a beat of its wings, but the idea is the same. It is worth mentioning that Wiener wrote these things in 1949 and published them in 1950, while the meteorologist Edward Lorenz, who was also based at MIT, would talk about the "butterfly effect" in the 1960s.

In *The human use of human beings*, he also criticizes a naïve idea of progress by introducing arguments on the "limits to growth" which would later be taken up by the Club of Rome, a community of scholars that, in the 60s and 70s, would have one of its main reference poles at MIT. The interdependence and fragility of large networks had to be taken into consideration. The world had become increasingly interdependent, the supplies for cities now depended on entire continents: "The very increase of commerce and the unification of humanity render the risks of fluctuation ever more deadly" (50j, 40). In general, the risk of famine and epidemics was much less remote than anyone could imagine (Cf. 50j, 41–43). There were the problems of water shortages, energy supply, and aging of populations: "We are on the verge of closing our schools and of opening homes for the aged" (50j, 48).

On the whole, according to Wiener, the modern age was the "the age of a consistent and unrestrained exploitation", as regards both natural resources and "an exploitation of conquered so-called primitive peoples; and finally, of a systematic exploitation of the average man" (50j, 35), an age characterized by the Mad Hatter and the March Hare's development policy, to draw a parallel with *Alice in Wonderland* (Carroll 1901 [1865], Chap. VII: "A mad tea-party"). Wiener argues: "When the tea and cakes were exhausted at one seat, the most natural thing for the Mad Hatter and the March Hare was to move on and occupy the next seat [...]. As time passed, the tea table of the Americas has not proved to be infinite" (50j, 35).

11.6   Fertility of Science

Wiener's criticism of the idea of progress does not lead him to yearn for a return to nature:

> What then shall we do? Our nostalgia for the 'simple life' antedates the success of the industrial revolution to which we have been subject, and must not blind us to the fact that we are not free to return to that pristine state. Our industrial progress has mortgaged our futures (50j, 56).

Technology and science are unavoidable: "We must continue to invent and to earn bread, not merely by the sweat of our brows, but by the metabolism of our brains" (50j, 57). Indeed we must take care to keep fertile the soil that allows scientific creation and invention.

The chapter on "progress and entropy" in *The human use of human beings* concludes with the following statement:

> Thus in depending on the future of invention to extricate us from the situations into which the squandering of our natural resources has brought us we are manifesting our national love for gambling and our national worship of the gambler, but in circumstances under which no intelligent gambler would care to make a bet. Whatever skills your successful poker player must have, he must at the very least know the values of his hands. In this gamble on the future of inventions, nobody knows the value of a hand (50j, 40).

It is as if to say, the survival of humanity relies on a process, that of discovery and invention, that remains substantially unpredictable. *Invention* picks up the argument where *The human use of human beings* leaves off (Cf. 93, 3). The central thesis of the book is precisely this: "The gamble of the really great discoveries is not only a gamble against tremendous odds, but a gamble against unascertainable odds" (93, 14–5). Therefore there is a totally unpredictable element behind invention and discovery, in which the individual's action is central. All one can do is to take care of the factors that influence them, which can increase the chances that such changes occur. And it is no coincidence that the subtitle means literally "the care and nurturing of ideas".

In *The human use of human beings*, Wiener devoted an entire chapter to art and culture as a whole, treating them as interrelated with the fertility of science: "nothing less than the whole man is enough to constitute the scholar, the artist, and the man of action" (50j, 162). He also maintained that:

> To replace it [the stock of discoveries exploited by the war], we need a range of thought that will really unite the different sciences, shared among a group of men who are thoroughly trained, each in his own field, but who also possess a competent knowledge of adjoining fields.
>
> No, size is not enough. We need to cultivate fertility of thought as we have cultivated efficiency in administration. We need to find some mechanism by which an invention of interest to the public may effectively be dedicated to the public. We cannot afford to erode the brains of the country as we have eroded its soil (50j, 57–8).

Invention insists that what really matters are the ideas, and that they need an adequate "breeding ground". This ground, in his view, was formed at the beginning of modern science by Humanism and the Renaissance. The highly philosophical speculations of the humanists, but also of the Talmudists, laid the foundations for modern science. The same happened in the twentieth century when a Confucian scholar, or an Indian *pandit* became scientists (Cf. 93, 32)., However, Wiener feared a drainage of the sources of discovery in the West, mainly because the pre-eminence of Europe and in particular Germany appeared to be coming to an end: "Two wars have all but removed from the lists Germany which was the chosen home of nineteenth century learning" (93, 32). The two most important countries in scientific terms had become the US and the USSR: "In each of these two countries the European tradition of scholarship has had to survive in a climate in which scholarship, for all the favor it receives, has been very greatly subordinated to other interests" (93, 32). Moreover, Wiener added, while the scholar in the USSR had to bow down in front of a political doctrine, the scholar in the United States was expected to give his "assent to the deification of the businessman and of the order for which he stands" (93, 33).

Invention reports another element of great importance in the process of invention and discovery, namely, the combination of pure speculation and practical activities. This was the merit of Alexandrian science, and this was the factor that led to the emergence of modern science: the intertwining of the speculative activity of the philosopher of nature with the practical activity of the skilled artisan (Cf. 93, 8).[14] These were reflections with a definite autobiographical background. And Wiener was thinking about his own life when he spoke of "mega dollars science", which is *big science*, "the science of the bishops", as von Neumann had called it. Its peculiar feature was excessive specialization, discouraging bold research that would cross the boundaries between different specializations:

> The great laboratory of the present day [...] is generally devoted to the accomplishment of a specific task. This task is mapped out by some sort of planning board, and is subdivided into subordinate tasks, each within the realm of some specialist. These specialists are primarily hired for their competence in certain limited fields, outside of which they are not encouraged to go, or even to satisfy their curiosity (93, 89).

Research was taking place in an atmosphere of secrecy imposed for military reasons or due to industrial competition (Cf. 93, 90). Wiener had suffered greatly during the war from the limitations imposed on inter-scientific communication, which had always been for him like a lifeblood (Cf. 50j, 123–143). After the war, instead of being abandoned, this secrecy had been consolidated, and in *The human use of human beings*, he protested against this practice which, in his opinion, by trying to stick to old know-how, would have made it difficult to create anything genuinely new.

The big laboratories, under the rule of profit, secrecy, and specialization, were the very negation a creative science; it was fine in time of war, when it was essential to use a wealth of already acquired ideas and to find all the possible implications for applying them, as had been done for radar and also for the atomic bomb; but this system, used as a standard practice of science, would soon run out of resources for creative thought. On the other hand, the independent scientist, a Heaviside or a Wiener, was poorly tolerated and severely criticized as an unruly scientist by those who directed the mega laboratories:

> To the scientist outside the great organization, the glorification of the great institution by those within it sounds like the attempt by the fox who has lost its tail in a trap to establish a new fashion of bobbed tails. One of the reproaches most commonly made against the independent scientist, both by those foxes who have already sacrificed their tails and by the would-be employers of science, is that the independent and individual scientist is undisciplined. Undisciplined he may be if undisciplined is taken merely as freedom from the shadow of external restriction and of possible external punishment. On the other hand, he must be a man of profound discipline if he is able to seize the vague and formless hints of ideas which are all he has to work with, and to reduce them to cogent and to manageable form (93, 106–7).

Wiener doubted that there could be any strategy to remove from the scene the independent scientist, creative but eccentric, someone that is not docile and conformist, but ready to criticize the structure, and to reflect on its aims; there could be

---

[14]Wiener's idea that science arises from the encounter between skilled craftsmen and scholars is supported by extensive evidence in the history and sociology of science (see, e.g., Zilsel 1942; Hall 1959; Rossi 1962).

no strategy to replace such a person by an organization composed of many more docile hyper specialized technicians, who would be able to replace the creative abilities of the brilliant scientist with the organization and use of automated systems, like the Memex described by Vannevar Bush in *As we may think* (Bush 1945), a kind of forerunner of the encyclopedic hypertext system.

Wiener notes that the ideas that actually advance science are not obvious, sensible, or consistent with the knowledge acquired. As he wrote in *The human use of human beings*:

> I should like to comment on the difference between a love for truth and a love for consistency and logic. In the long run, either the truth must be consistent, or it must be impenetrable to us. However, a consistent view of things does not necessarily mark the first stages of our intellectual penetration into new fields, but rather their final conquest and integration in a larger world, containing in an organized form all that is valid of the different and previously unresolved alternatives [...].
>
> This consistency of final result, which emerges as a consequence of the process of investigation and study, is not characteristic of all the steps by which we learn to know the truth [...].
>
> Whenever an intellectual movement becomes a militant religious propaganda, there is also a drive to relinquish those imperfectly defensible outposts of knowledge which have not yet been integrated into a tight and consistent line of defense, and to retreat into a ritual behind the security of an apparently impregnable logic. This should be recognized as the step backward which it actually is (50j, 218–9).

Note the emphasis on logic. An impregnable logic is typical of intellectual movements that withdraw and "clam up", leaving aside the less easily defensible outposts. In the specific case of the above quote, Wiener was targeting the dogmatism of Soviet science. It reminded him of a similar dogmatism in the Catholic Church, which, in the transition from "saints" to "bishops", had adopted a closed logic, exemplified by the Aristotelian logic of Thomism (50j, 219). The USSR was behaving similarly now with dialectical materialism. In these social systems, science could indeed continue to exist, "but it is science in a strait-jacket, fearing for conditions of its own continuation" (50j, 223).

A not dissimilar criticism could have been directed at logical positivism which, in the same period, was becoming the philosophy of official science, of *big science*. It is interesting to reread what *Invention* has to say about the strengths of formalized languages and mathematics: "One of the most potent tools in reanimating a science is mathematics." (93, 25); but this was mainly because the translation of a phenomenon into mathematical terms allows one to generalize it and find analogies in other fields. On the other hand, mathematization helps with conceptual clarification, and in any case, in Wiener's opinion, once we are able to give a phenomenon a mathematical treatment, the dialogue among the sciences is made easier. But beyond that, what matters are the ideas, and at this level, demanding logical rigor can simply hinder discovery. In fact, new ideas often appear paradoxical. *Invention* gives the example of the Jesuit Father Jerome Saccheri, who had questioned the postulate of parallel lines, hoping to be able to prove it by a reductio ad absurdum procedure, but who had instead—without realizing it—demonstrated many theorems of non-Euclidean geometry, rediscovered later in the nineteenth century. Something similar had happened to the young Wiener, who had used those continuous non-differentiable curves in physics, although they had previously been considered as purely "pathological" cases, to be exhibited in a museum of mathematics (Cf. 93, 134–5).

Truly original and creative ideas go against the logic of immediate profit:

> Many inventions and discoveries have it in them to be as imposing as an elephant and as dangerous as a lion, but neither a lion nor an elephant is a particularly impressive lump of flesh in utero. Yet it is precisely this uterine period during which an invention, like a fetus, might be subject to some ideal control and is, in fact, a rare and highly improbable accident. For it is one of the paradoxes of probability theory that what happens is, in fact, always enormously improbable in its full concreteness. [...] Our employers do not invest in embryos, and our capitalists are no more able to invest in embryonic ideas. [...] The point is that by the time an idea is ripe for any predictable economic or military use, a lot of people will have seen it and it cannot likely be eliminated [...]. By this time, the phenomenon of multiple invention is not merely an outside possibility, but a very definite probability (93, 104–5).

According to Wiener, insofar as scientific discovery depends on extremely improbable ideas, on which no one would bet any money, the scientist must be animated by a great faith and dedication, because it is a search for something where immediate feedback is unlikely to be seen. Wiener noted a vicious circle that the sociologist of science Robert Merton would have called the "Matthew effect": "Whoever has will be given more, and they will have an abundance. Whoever does not have, even what they have will be taken from them" (*Matthew* 13:12). The more success you have, the more will come of it, and vice versa (Cf. Merton 1968). This vicious circle, which leads to conformism and leveling, can be reversed, according to Wiener, by the activities of academies and universities that operate outside the market and can afford to bet on the improbable, encouraging solitary scientists and improbable ideas, rather than rewarding those who have already achieved success and ideas that are already acquired (Cf. 93, 107).

11.7 The Tempter

Speaking about scientists who bet on the improbable, *Invention* also cites the communication engineer Oliver Heaviside. (Cf. 64g [56g], 94–101) Wiener always had a veneration for Heaviside. In 1941 he had already sent a script outline to Orson Welles for an unmade movie (Cf. Heims 1993, 12). The story is told in detail in *Invention*. Finally, in 1959, Wiener managed to publish the novel *The Tempter*, which he wanted to come out in place of *Invention*, as he explained to his editorial consultant Epstein:

The story is really a treatment in fictional form of my ideas on invention in the modern world. I am not too enthusiastic about carrying through my plans for a purely expository book on invention. I therefore suggest that you take this manuscript in lieu of such a book.[15]

The Tempter is a fictionalized transposition of the historical account of Heaviside's life which can be found in the sixth chapter of *Invention* (93, 63 ff.). In the first years of its existence, telephony encountered important technical difficulties for long distance communications. The signal would come in garbled and this drawback was attributed to the fact that the cables had insufficient electrical capacity. Oliver Heaviside, a poor but brilliant English electrical engineer published numerous articles in the journal *The electrician*, and even a book, which remained unsold, suggesting a nonconformist thesis: the distortion over long distances was not caused by poor electrical capacity, but rather by excessive electrical capacity, whence the distortion could be alleviated by increasing the inductance of the line, giving the so-called "loaded-line".

A decade later, the correctness of Heaviside's ideas was recognized, but they had by then entered the public domain and it was no longer possible to patent them. In that period the Bell Telephone was launching long-distance telephony on the market with the company AT&T (American Telephone and Telegraph Company). In order to secure exclusivity on Heaviside discoveries, AT&T activated its scientists, in particular Campbell, and an external associate Idvorsky Michael Pupin, to flush out those aspects neglected by Heaviside, which could subsequently be patented. Finally, in 1901, a genuine legal scam was orchestrated against Heaviside: they bought all rights to exploit discoveries related to long-distance communication from Pupin for half a million dollars. A figure that high should have taken away any shadow of a doubt as to who was the true custodian of those rights. AT&T also tried to get any other rights from Heaviside, to avoid any future claim. Heaviside, however, refused the offer, reiterating the claim that he was to be recognized as the unique and authentic inventor of the "loaded line". In addition, B.A. Behrend, an executive of AT&T, visited him to offer him an honor of the American Institute of Electrical Engineers, which had only a symbolic value.

Meanwhile, the usurper Pupin wrote books—including his autobiography *From Immigrant to Inventor*, for which he won the Pulitzer Prize and in which he discredited the figure of Heaviside. *The Tempter* is presented as a memorial, a kind of spiritual testament, that Gregory James (Behrend), former chief engineer of Williams Controls Company (AT&T), leaves at the time of his death to the heir of the company. The fictional story deals, not with telephony but with automatic controls, Williams Controls corresponds to AT&T, Heaviside is called Cedric Woodbury, Pupin is Diego Dominguez, and Campbell is Watman. Both in *The Tempter* (59th, 239–240) and in *Invention* (93), Heaviside-Woodbury is Prometheus, while Pupin-Dominguez is a Doctor Faustus. We read in *Invention*:

---

[15]Wiener to Epstein, 2 August 1957, cit. by Heims 1993, XIII.

> The fact is that circumstances had joined the story of Prometheus with the story of Marlowe's *Dr. Faustus*. Heaviside may have been a very snuffy lower-middle-class Prometheus, but at least he had snatched a piece of fire for mankind. If the vultures of poverty and the sense of persecution were gnawing at his liver, he shared with Prometheus the sense of having performed a godlike feat [...]. Pupin, on the other hand, had wrapped his soul as part and parcel of a commercial bargain. When a soul is bought by anyone, the devil is the ultimate consumer (93, 76).

We can read much the same words in *The Tempter* (59e, 239–240). But who was the devil? The novel attributes this role to the representative of Williams Controls—AT&T who had hatched the immoral plot.

So why did Wiener care so much about this novel to consider exchanging it with an almost completed essay like *Invention*? The answer is probably to be found in the dedication: "To those inventors who have preferred the claims of truth to the gifts of fortune" (59e). In this period Wiener tends to identify himself with Heaviside, especially in the late 1950s, when people no longer flocked in droves to listen to his lectures, and the "father of cybernetics" must have felt very isolated. As shown by Conway and Siegelman (2005, in particular 255–271), his "rebellion" had become reason for investigations by the US secret services. Although Wiener had many friends who were definitely communists, such as J.B.S. Haldane or Dirk Struik, every attempt to show that he was himself a communist failed. And in the same way they failed to demonstrate his lack of loyalty for his homeland. However, from various documents reported by Conway and Siegelman (2005, 259), judgements emerge that would claim to deeply discredit Wiener's person. He was considered by certain informants as "extremely erratic", "politically naïve", "a complete egotist", "a screwball", and he was even "known to be nuts". These judgements, in fact more painful than those of the Soviet ideologues, helped willy-nilly to diffuse those strange anecdotes that spread among young scholars and students at MIT and beyond, according to which he was perhaps a great scientist, but also an eccentric and unreliable individual in the practical aspects of life.

In a sudden and unexpected manner, as his last assistant Giacomo Della Riccia told me, a new atmosphere arose around Wiener. He was invited to India, then repeatedly came to lecture in Italy, especially in the Laboratory of Cybernetics founded by Eduardo R. Caianiello at the Institute of Theoretical Physics of the University of Naples (Cf. Montagnini 2016). Note, however, that Wiener had always refused to go to Italy during the fascist regime.[16]

This resurgence of interest in Wiener was marked by two awards. On the one hand, an official invitation to Moscow in 1960 and, a few weeks before his death, in

---

[16]In 1932, "a couple of the Italian mathematicians broached to me the subject of an invitation to lecture in Italy. I had no sympathy with Fascism and resented the completely Fascist and official auspices of this invitation. I talked the matter over with Leon Lichtenstein, who was also a participant in the meeting, and he told me to forget the politics and accept the offer. However, I heard nothing more of the offer, and they must have come to the conclusion that my opinions would not go well in Fascist Italy" (64g [56g], 163).

Stockholm on 18 March 1964, the recognition granted to him by President John F. Kennedy, the prestigious National Medal of Science, which since Kennedy had been assassinated, was presented to him by President Lyndon B. Johnson "for marvellously versatile contributions, profoundly original, ranging within pure and applied mathematics, and penetrating boldly into engineering and biological sciences" (Rosenblith and Wiesner 1965, 3).

Bibliography

Norbert Wiener's Complete Chronological Bibliography

Reproduced below, with very slight changes, is the "Bibliography of Norbert Wiener" published for the first time in *Bulletin of the American Mathematical Society 72* (1966), 135–145, special issue dedicated to Wiener after his death. It was published again in the four volumes of CW, in Masani (1990, 377–390), and in the Proceedings of the symposium (Mandrekar and Masani 1997, 543–556).

The indefatigable editor of the CW, Pesi R. Masani, has indexed using a code comprising the last two digits of the year of publication and a letter of the alphabet for titles that refer to the same year. This coding can now be considered as the official standard to correctly and unambiguously identify Wiener's many publications, and it is in some cases imperative because it has been used in the comments in CW to accompany Wiener's writings (e.g., E. Nagel, *Comments on [14c], [14d]*. In [CW4, 67], or McMillan B. and Deem, G.S. *The Wiener program in statistical physics. Commentary on* [38a], [39b and h], [40d], [43a], in (CW1, 654–671).

It would have been possible to include only those of Wiener's works explicitly mentioned in the text of this book, but it seemed more appropriate to offer his entire bibliography, because I believe that isolating his philosophical, sociological, or cybernetic works from the others, as well as being a difficult task, would make it harder to follow his intellectual itinerary. When the document was published in CW. I even added the pagination of CW to that of the original edition. To keep account of the actual year of publication I had to change the labels in several cases, noting this each time.

[CW] Wiener, N. 1976–1985. *Collected works. With commentaries*, 4 vv, ed. Pesi Rustom Masani. Cambridge [MA]: The MIT Press.
[CW1] v. 1. Mathematical philosophy and foundations; potential theory; Brownian movement, Wiener integrals, ergodic and chaos theories, turbulence and statistical mechanics, 1976.

© Springer International Publishing AG 2017
L. Montagnini, *Harmonies of Disorder*, Springer Biographies,
DOI 10.1007/978-3-319-50657-9

[CW2]   v. 2. Generalized harmonic analysis and Tauberian theory, classical harmonic and complex analysis, 1979.

[CW3]   v. 3. The Hopf-Wiener integral equation; prediction and filtering; quantum mechanics and relativity; miscellaneous mathematical papers, 1981.

[CW4]   v. 4. Cybernetics, science, and society; ethics, aesthetics, and literary criticism; book reviews and obituaries, 1985.

Note that [CW] brings together almost all Wiener's publications, apart from works published in volume form.

[13a]   On a method of rearranging the positive integers in a series of ordinal numbers greater than that of any given fundamental sequence of omegas. *Messenger of Mathematics* 43: 97–105. [CW1], 240–248.

[13b]   *A comparison between the treatment of the algebra of relatives by Schröder and that by Whitehead and Russell.* PhD thesis, Philosophy Department, Harvard University, Cambridge [MA].

[14a]   A simplification of the logic of relations. *Proceedings of the Cambridge Philosophical Society* 17: 387–390. [CW1], 29–32.

[14b]   A contribution to the theory of relative position. *Proceedings of the Cambridge Philosophical Society* 17: 441–449. [CW1], 34–42.

[14c]   The highest good. *Journal of Philosophy, Psychology and Scientific Method* 11: 512–520. [CW4], 41–49.

[14d]   Relativism. *Journal of Philosophy, Psychology and Scientific Method* 11: 561–577. [CW4], 50–66.

[15a]   Studies in synthetic logic. *Proceedings of the Cambridge Philosophical Society* 18: 24–28. [CW1], 43–57.

[15b]   Is mathematical certainty absolute? *Journal of Philosophy, Psychology and Scientific Method* 12: 568–574. [CW1], 218–224.

[16a]   Mr. Lewis and implication. *Journal of Philosophy, Psychology and Scientific Method* 13: 656–662. [CW1], 226–232.

[16b]   The shortest line dividing an area in a given ratio. *Journal of Philosophy, Psychology and Scientific Method* 9: 56–58. [CW3], 633–635.

[16c]   Review of: Keyser, C.J., Science and religion. The rational and the superrational. *Journal of Philosophy, Psychology and Scientific Method* 13: 273–277. [CW4], 990–994.

[16d]   Review of: Robb, A.A., A theory of time and space. *Journal of Philosophy, Psychology and Scientific Method* 13: 611–613. [CW4], 973–975.

[17a]   Certain formal invariances in Boolean algebras. *Transactions of the American Mathematical Society* 18: 65–72. [CW1], 321–328.

[17b]   Review of: Keyser, C.J., The human worth of rigorous thinking. *Journal of Philosophy, Psychology and Scientific Method* 14: 356–361. [CW4], 984–989.

[18a]   Review of: Huntington, E.V., The continuum and other types of serial order. *Journal of Philosophy, Psychology and Scientific Method* 15: 78–80. [CW4], 976–978.

[18b] Æsthetics. *Encyclopedia Americana*, ed 1918–1920. 1: 198–203. [CW4], 845–850.

[18c] Algebra. Definitions and fundamental concepts. *Encyclopedia Americana*, ed 1918–1920. 1: 381–385. [CW3], 636–640.

[18d] Alphabet. *Encyclopedia Americana*, ed 1918–1920. 1: 435–438. [CW4], 933–936.

[18e] Animals, chemical sense in. *Encyclopedia Americana*, ed 1918–1920. 1: 704. [CW4], 970.

[18f] Apperception. *Encyclopedia Americana*, ed 1918–1920. 2: 82–83. [CW4], 951–952.

[18g] Category. *Encyclopedia Americana*, ed 1918–1920. 6: 49. [CW4], 944.

[18h] Dualism. *Encyclopedia Americana*, ed 1918–1920. 9: 367. [CW4], 943.

[18i] Duty. *Encyclopedia Americana*, ed 1918–1920. 9: 440–441. [CW4], 955–956.

[18j] Ecstasy. *Encyclopedia Americana*, ed 1918–1920. 9: 570. [CW4], 953.

[19a] Geometry, non-Euclidean. *Encyclopedia Americana*, ed 1918–1920. 12: 463–467. [CW3], 641–645.

[19b] Induction. logic. *Encyclopedia Americana*, ed 1918–1920. 15: 70–73. [CW4], 964–967.

[19c] Infinity. *Encyclopedia Americana*, ed 1918–1920. 15: 120–122. [CW4], 961–963.

[19d] Meaning. *Encyclopedia Americana*, ed 1918–1920. 18: 478–479. [CW4], 959–960.

[19e] Mechanism and vitalism. *Encyclopedia Americana*, ed 1918–1920. 18: 527–528. [CW4], 968–969.

[19f] Metaphysics. *Encyclopedia Americana*, ed 1918–1920. 18: 707–710. [CW4], 937–940.

[19g] Pessimism. *Encyclopedia Americana*, ed 1918–1920. 21: 654. [CW4], 954.

[19h] Postulates. *Encyclopedia Americana*, ed 1918–1920. 22: 437–438. [CW4], 957–958.

[20a] Bilinear operations generating all operations rational in a domain Ω. *Annals of Mathematics* 21 (1920): 157–165. [CW1], 250–258.

[20b] A set of postulates for fields. *Transactions of the American Mathematical Society* 21: 237–246. [CW1], 259–267.

[20c] Certain iterative characteristics of bilinear operations. *Bulletin of the American Mathematical Society* 27: 6–10. [CW1], 269–273.

[20d] see [21f].

[20e] see [21g].

[20f] The mean of a functional of arbitrary elements. *Annals of Mathematics* (2) 22: 66–72. [CW1], 435–441.

[20g] Review of: Lewis, C.I., A survey of symbolic logic. *Journal of Philosophy, Psychology and Scientific Method* 17: 78–79. [CW4], 1001.

[20h] Soul. *Encyclopedia Americana*, ed 1918–1920. 25: 268–271. [CW4], 947–950.

[20i] Substance. *Encyclopedia Americana*, ed 1918–1920. 25: 775–776. [CW4], 941–942.

[20j] Universals. *Encyclopedia Americana*, ed 1918–1920. 27: 572–573. [CW4], 945–946.

[21a] A new theory of measurement: A study in the logic of mathematics. *Proceedings of the London Mathematical Society* 19: 181–205. [CW1], 58–86.

[21b] The isomorphisms of complex algebra. *Bulletin of the American Mathematical Society* 27: 443–445. [CW1], 277–279.

[21c] The average of an analytical functional. *Proceedings of the National Academy of Sciences* 7: 253–260. [CW1], 442–449.

[21d] The average of an analytical functional and the Brownian movement. *Proceedings of National Academy of Sciences* 7: 294–298. [CW1], 450–454.

[21e] (with Hitchcock, F.L.) A new vector in integral equations. *Journal of Mathematics and Physics* 1: 20. [CW3], 646–665.

[21f] Certain iterative characteristics of bilinear operations. In *Comptes rendus du Congrès international des mathématiciens. Strasbourg, 22–30 septembre 1920* [Proceedings of the International Congress of Mathematicians (ICM 1920)], 176–178, ed. H. Villat. Toulouse: Imprimerie É. Privat. [CW1], 274–276, as [20d].

[21g] On the theory of sets of points in terms of continuous transformations. In *Comptes rendus du Congrès international des mathématiciens. Strasbourg, 22–30 septembre 1920* [Proceedings of the International Congress of Mathematicians (ICM 1920)], 312–315, ed. H. Villat. Toulouse: Imprimerie É. Privat. [CW1], 281–284, as [20e].

[22a] The relation of space and geometry to experience. *Monist* 32. 8 lectures: I. 12–30; II: 31–45; III: 46–60; IV: 200–215; V: 216–230; VI: 231–247; VII: 364–380; VIII: 381–394. [CW1], 87–214.

[22b] The group of linear continuum. *Proceedings of the London Mathematical Society* 20: 329–346. [CW1], 285–302.

[22c] Limit in terms of continuous transformation. *Bulletin de la Société Mathématique de France* 50: 119–134. [CW1], 303–318.

[22d] (with Walsh, J.L.) The equivalence of expansions in terms of orthogonal functions. *Journal of Mathematics and Physics* 1: 103–122. [CW2], 813–832.

[22e] A new type of integral expansion. *Journal of Mathematics and Physics* 1: 167–176. [CW3], 666–675.

[23a] On the nature of mathematical thinking. *Australasian Journal of Psychology and Philosophy* 1: 268–272. [CW1], 234–238.

[23b] (with Phillips, H.B.) Nets and the Dirichlet problem. *Journal of Mathematics and Physics* 2: 105–124. [CW1], 333–352.

[23c] Discontinuous boundary conditions and the Dirichlet problem. *Transactions of the American Mathematical Society* 25: 307–314. [CW1], 355–362.

[23d] Differential space. *Journal of Mathematics and Physics* 2: 131–174. [CW1], 455–498.

[23e] Note on the series Σ (+1/n). *Bulletin de l'Académie polonaise des sciences. Série des Sciences Mathématiques, Astronomiques et Physiques* A: 87–90. [CW1], 520–523.

[23f] Note on a new type of summability. *American Journal of Mathematics* 45: 83–86. [CW2], 31–34.

[23g] Note on a paper of M. Banach. *Fundamenta Mathematicae* 4: 136–143. [CW3], 676–683.

[24a] Certain notions in potential theory. *Journal of Mathematics and Physics* 3: 24–51. [CW1], 364–391.

[24b] The Dirichlet problem. *Journal of Mathematics and Physics* 3: 127–147. [CW1], 394–413.

[24c] Une condition nécessaire et suffisante de possibilité pour le problème de Dirichlet. *Comptes Rendus de l'Académie des Sciences*. I. 178: 1050–1053. [CW1], 414–418.

[24d] The average value of a functional. *Proceedings of the London Mathematical Society* 22: 454–467. [CW1], 499–512.

[24e] Un problème de probabilités dénombrables. *Bulletin de la Société Mathématique de France* 11: 3–4. [CW1], 525–534.

[24f] The quadratic variation of a function and its Fourier coefficients. *Journal of Mathematics and Physics* 3 (1924): 72–94. [CW2], 36–58.

[24g] Review of: Four books on space: Rudolf Carnap's *Der Raum [...]*, E. Study's *Mathematik and Physik and Die realistische Weltansich* [...]; Erster Teil and Hermann Weyl's *Mathematische Analyse des Raum-problems* [...]. *Bulletin of the American Mathematical Society* 30: 258–262. [CW4], 979–983.

[24h] Review of: Study, E., Denken und Darstellung: Logik und Werte. *Bulletin of the American Mathematical Society* 30: 277. [CW4], 1003.

[24i] In memory of Joseph Lipka. *Journal of Mathematics and Physics* 3: 63–65. [CW4], 1029.

[25a] Note on a paper of O. Perron. *Journal of Mathematics and Physics* 4: 21–32. [CW1], 420–431.

[25b] The solution of a difference equation by trigonometrical integrals. *Journal of Mathematics and Physics* 4: 153–163. [CW2], 443–453.

[25c] On the representation of functions by trigonometrical integrals. *Mathematiche Zeitschrift* 24: 576–616. [CW2], 60–101.

[25d] Verallgemeinerte trigonometrische Entwicklungen. *Nachrichten von der Gesellschaft der Wissenschaften zu Göttingen*. Mathematisch-physikalische Klasse, 151–158. [CW2], 103–110.

[25e] Note on quasi-analytic functions by trigonometrical integrals. *Journal of Mathematics and Physics* 4 (1925), 193–199. [CW2], 834–840.

[25f] A contribution to the theory of interpolation. *Annals of Mathematics* (2) 26: 212–216. [CW3], 686–690.

[26a] The harmonic analysis of irregular motion. *Journal of Mathematics and Physics* 5: 99–121. [CW2], 112–134.

[26b] The harmonic analysis of irregular motion (Second paper). *Journal of Mathematics and Physics* 5: 158–189. [CW2], 135–166.

[26c] The operational calculus. *Mathematische Annalen* 95: 557–584. [CW2], 397–424.

[26d] (with Born, Max) A new formulation of the laws of quantification for periodic and aperiodic phenomena. *Journal of Mathematics and Physics* 5: 84–98. [CW3], 427–441.

[26e] (with Born, Max) Eine neue Formulierung der Quantengesetze fur periodische und nicht Periodische Borganze. *Zeitschrift für Physik* 36: 174–187. [CW3], 442–455.

[26f] (with Franklin, Philip) Analytical approximations to topological transformations. *Transactions of the American Mathematical Society* 28: 762–785. [CW3], 692–715.

[27a] The spectrum of an array and its application to the study of the translation properties of a simple class of arithmetical functions. *Journal of Mathematics and Physics* 6 (1927): 145–157. [CW2], 380–392.

[27b] A new definition of almost periodic functions. *Annals of Mathematics.* Second Series, 28. 1/4: 365–367. [CW2], 455–457.

[27c] On a theorem of Bochner and Hardy. *Journal of the London Mathematical Society* 2: 118–123. [CW2], 474–479.

[27d] Une méthode nouvelle pour la démonstration des théorèmes de Tauber. *Comptes Rendus de l'Académie des Sciences.* I. 184: 793–795. [CW2], 481–483.

[27e] On the closure of certain assemblages of trigonometrical functions. *Proceedings of the National Academy of Sciences* 13: 27–29. [CW2], 842–844.

[27f] (with Struik, Dirk J.) Quantum theory and gravitation relativity. *Nature* 119: 853–854. [CW3], 580–581.

[27g] (with Struik, Dirk J.) A relativistic theory of quanta. *Journal of Mathematics and Physics* 6: 1–23. [CW3], 582–604.

[27h] (with Struik, Dirk J.) Sur la théorie relativiste des quanta. *Comptes Rendus de l'Académie des Sciences.* I. 185: 42–44. [CW3], 605–607.

[27i] Sur la théorie relativiste des quanta (Note). *Comptes Rendus de l'Académie des Sciences.* I. 185: 184–185. [CW3], 608–609.

[27j] Laplacians and continuous linear functionals. *Acta Scientiarum Mathematicarum.* Szeged (HU). 3/1: 7–16. [CW3], 718–727.

[27k] Une généralisation des fonctions à variation bornée. *Comptes Rendus de l'Académie des Sciences.* I. 185: 65–67. [CW3], 730–732.

[28a] The spectrum of an arbitrary function. *Proceedings of the London Mathematical Society* (2) 27: 487–496. [CW2], 167–180.

[28b] A new method of Tauberian theorems. *Journal of Mathematics and Physics* 7: 161–184. [CW2], 485–508.

[28c] (with Struik, Dirk J.) The fifth dimension in relativistic quantum theory. *Proceedings of National Academy of Sciences* 14 (1928): 262–268. [CW3], 614–620.

[28d] Coherency matrices and quantum theory. *Journal of Mathematics and Physics* 1: 109–125. [CW3], 456–472.

[29a] Harmonic analysis and group theory. *Journal of Mathematics and Physics* 8: 148–154. [CW2], 461–467.

[29b] A type of Tauberian theorem applying to Fourier series. *Proceedings of the London Mathematical Society* (2), 30: 1–8. [CW2], 510–517.

[29c] *Fourier analysis and asymptotic series.* In *Operational circuit analysis*, 366-379, ed. Vannevar Bush. New York: Wiley. Appendix. [CW2], 428–441.

[29d] Hermitian polynomials and Fourier analysis. *Journal of Mathematics and Physics* 8: 70–73. [CW2], 914–917.

[29e] Harmonic analysis and the quantum theory. *Journal of The Franklin Institute* 207: 525–534. [CW3], 473–482.

[29f] (with Vallarta, Manuel S.) On the spherically symmetrical statistical field in Einstein's unified theory of electricity and gravitation. *Proceedings of the National Academy of Sciences* 15: 353–356. [CW3], 622–628.

[29g] (with Vallarta, Manuel S.) On the spherically symmetrical statistical field in Einstein's unified theory: a correction. *Proceedings of the National Academy of Sciences* 15: 802–804. [CW3], 626–628.

[29h] Mathematics and art. Fundamental identities in the emotional aspects of each. *The Technology Review* 32: 129–132, 160, 162. [CW4], 851–856.

[29i] Einsteiniana. Facts and fancies about Dr. Einstein's famous theory. *The Technology Review* 32: 403–404. [CW4], 913–914.

[29j] Murder in mathematics. A brilliantly written detective story keeps buoyantly and entertainingly afloat over scientific deeps. *The Technology Review* 32: 271–272. [CW4], 1017–1018.

[30a] Generalized harmonic analysis. *Acta Mathematica* 55: 117–258. [CW2], 183–324.

[30b] The religion of the scientist, review of: Eddington, A. *Science and the unseen world.* In *The Technology Review* 33: 150. [CW4], 995.

[31a] (with Eberhard Hopf) Über eine Klasse singulärer Integralgleichungen. *Sitzungsberichte der Berliner Akademie der Wissenchaften* 696–706. [CW3], 33–43.

[31b] A new deduction of the Gaussian distribution. *Journal of Mathematics and Physics* 10: 284–288. [CW3], 734–738.

[31c1] Education and science in England (Reports from Cambridge, 1931). *The Technology Review* 34: 82–83. [CW4], 915–916.

[31c2] Cutter and tailor review (Reports from Cambridge, 1931). *The Technology Review* 34: 131. [CW4], 917.

[31c3] The shattered nerves of Europe (Reports from Cambridge, 1931). *The Technology Review* 34: 218, 220. [CW4], 918–919.

[32a] Tauberian theorems. *Annals of Mathematics* (2) 33: 1–100. [CW2], 519–618.

[32b] A note on Tauberian theorems. *Annals of Mathematics* (2) 33: 787. [CW2], 619.

[32c] Back to Leibniz! Physics reoccupies an abandoned position. *The Technology Review* 34: 201–203, 222, 224. [CW4], 76–79.

[32d] A mathematician in Europe (Reports from Cambridge, 1932). *The Technology Review* 34: 62, 74. [CW4], 920–921.

[32e] Review of: Besicovitch, A.S., Almost periodic functions. *The Mathematical Gazette* 16 (220): 275–277. [CW4], 1011–1013.

[32f] (with Paley, R.E.A.C.) Analytic properties of the characters of infinite Abelian groups, In *Verhandlungen des Internationalen Mathematiker-Kongresses Zürich 1932* [Proceedings of the International Congress of Mathematicians, Zürich 1932 (ICM 1932)], 95. [CW2], 636.

[33a] (with Paley, R.E.A.C., and Zygmund, A.Z.) Notes on random functions. *Mathematische Zeitschrift* 37: 647–668. [CW1], 536–557.

[33b] A one-sided Tauberian theorem. *The Mathematical Gazette* 33/476: 787–789. [CW2], 624–626.

[33c] (with Paley, R.E.A.C.) Characters of Abelian groups. *Proceedings of the National Academy of Sciences* 19: 253–257. [CW2], 637–641.

[33d] (with Young, R.C.) The total variation of g $(x + h) - $ f (x). *Transactions of the American Mathematical Society* 35: 327–340. [CW2], 643–656.

[33e] (with Paley, R.E.A.C.) Notes on the theory and application of Fourier transforms. *Transactions of the American Mathematical Society* I, II, 35: 348–355; III, IV, V, VI, VII, 35: 761–791. [CW2], 919–949.

[33f] Putting matter to work. The search for cheaper power. *The Technology Review* 35 (1933): 47–49, 70, 72. [CW4], 692–696.

[33g] Review of: Bohr, H. Fastperiodische Funcktionen. *The Mathematical Gazette* 17: 54. [CW4], 1014.

[33h] Paley, R.E.A.C., In memoriam. *Bulletin of the American Mathematical Society* 39: 476. [CW4], 1032.

[33i] *The Fourier integral and certain of its applications* [Reworking of 15 lectures given at the University of Cambridge during the second term of the academic year 1931–'32]. Cambridge [UK]: Cambridge University Press.

[33j] Review of: Titchmarsh, Edward Charles, The Fourier integral and certain of its applications. *The Mathematical Gazette* 17: 129; *Science* 132: 731.

[34a] see [35g].

[34b] A class of gap theorems. *Annali della Scuola Normale Superiore di Pisa* (1934–1936) 1–6. [CW2], 845–850.

[34c] Quantum mechanics, Haldane and Leibniz. *Philosophy of Science* 1: 479–482. [CW4], 80–83.

[34d] (with Paley, R.E.A.C.) Fourier transforms in the complex domain. In *American Mathematical Society Colloquium Publications*. New York: American Mathematical Society.

[35a] Fabry's gap theorem. *The Science Reports of National Tsing Hua University. Series A: Mathematical, Physical and Engineering Sciences* 3: 239–245. [CW2], 851–857.

[35b] Limitations of science. The holiday fallacy and a response to the suggestion that scientists become sociologists. *The Technology Review* 37 (1935): 255–256, 268, 270, 272. [CW4], 697–701.

[35c] (with Bridenbaugh, C.) The student agitator. Is he accepting radicalism as an opiate? *The Technology Review* 37 (1935): 310–312, 344, 346. [CW4], 767–772.

[35d] Mathematics in American secondary schools. *Journal of the Mathematical Association of Japan for Secondary Education Japan Society of Mathematical Education*. 17: 1–5. [CW4], 893–895.

[35e] The closure of Bessel functions, Abstract 66. *Bulletin of the American Mathematical Society* 41: 35: [CW4], 1026.

[35f] Once more ... the refugee problem abroad. *Jewish Advocate* 5 February: [CW4], 927.

[35g] Random functions. *Journal of Mathematics and Physics* 14, 1–4 (1935): 17–23. [CW1], 564–570 as [34a].

[36a] A theorem of Carleman. *The Science Reports of National Tsing Hua University. Series A: Mathematical, Physical and Engineering Sciences* 3: 291–298. [CW2], 884–891.

[36b] (with Mandelbrojt, Szolem) Sur les séries de Fourier lacunaires. Théorème direct. *Comptes Rendus de l'Académie des Sciences*. I. 203: 34–36. [CW2], 858–860.

[36c] (with Mandelbrojt, Szolem) Sur les séries de Fourier lacunaires. Théorème inverse. *Comptes Rendus de l'Académie des Sciences*. I. 203: 233–234. [CW2], 861–862.

[36d] Gap theorems. In *Comptes rendus du Congrès international des mathématiciens,* Oslo, 1936 [Proceedings of the International Congress of Mathematicians (ICM 1936)], 284–296. [CW2], 863–875.

[36e] A Tauberian gap theorem of Hardy and Littlewood. *Journal of Chinese Mathematical Society* 1: 15. [CW2], 876–883.

[36f] Notes on the Kron theory of tensors in electrical machinery. Abstract. *Journal of Electrical Engineering China*. 7: 3–4. [CW3], 740–750.

[36g] The role of the observer. *Philosophy of Science* 3: 307–319. [CW4], 84–96.

[37a] (with Martin, William T.) Taylor's series of entire functions of smooth growth. *Duke Mathematical Journal* 3: 213–233. [CW2], 721–731.

[37b] (with Levinson, Norman) Random Waring's theorems. Abstract. *Science* 85: 439.

[38a] The homogeneous chaos. *American Journal of Mathematics* 60: 897–936. [CW1], 572–611.

[38b] (with Pitt, H.R.) On absolutely convergent Fourier-Stieltjes transform. *Duke Mathematical Journal* 4-2: 420–436. [CW2], 657–673.

[38c] (with Wintner, Aurel F.) Fourier-Stieltjes transforms and singular infinite convolutions. *American Journal of Mathematics* 60: 513–522. [CW2], 677–686.

[38d] (with Martin, William T.) Taylor's series of entire functions of smooth growth in the unit circle. *Duke Mathematical Journal* 4: 384–392. [CW2], 732–739.

[38e] The historical background of harmonic analysis. In *Semicentennial address of the American Mathematical Society*, 56–68. New York: American Mathematical Society. [CW2], 794–806.

[38f] (with Widder, David V.) Remarks on the classical inversion formula for the Laplace integral. *Bulletin of the American Mathematical Society* 44-8: 573–575. [CW2], 952–954.

[38g] The decline of cookbook engineering. *The Technology Review* 41: 23. [CW4], 896.

[38h] Review of: Hogben, L., Science for the citizen. *The Technology Review* 41: 66–67. [CW4], 1007–1009.

[39a] The ergodic theorem. *Duke Mathematical Journal* 5: 1–18. [CW1], 672–689.

[39b] The use of statistical theory in the study of turbulence. *Nature* 144: 728. [CW1], 641–650.

[39b1] The use of statistical theory in the study of turbulence. In *Proceedings of the Fifth International Congress for Applied Mechanics. Held at at Harvard University and the Massachusetts Institute of Technology, Cambridge, Ma, September 12–16, 1938*, 356–358. New York: Wiley.

[39c] (with Wintner, Aurel) On singular distribution. *Journal of Mathematics and Physics* 17-4: 233–246. [CW2], 680–700.

[39d] (with Cameron, Robert H.) Convergence properties of analytic functions of Fourier-Stieltjes transforms. *Transactions of the American Mathematical Society* 46: 97–109. [CW2], 705–717.

[39e] (with Pitt, J.R.) Generalization of Ikehara's theorem. *Journal of Mathematics and Physics* 17-4: 247–258. [CW2], 742–753.

[39f] Review of: Burlingame, R., March of the iron men. In *The Technology Review* 41: 115. [CW4], 1005–1006.

[39g] Review of: George, W.H., The scientist in action. In *The Technology Review* 41: 202. [CW4], 1004.

[39h] (with McMillan, B.) A new method in statistical mechanics. Abstract 133. *Bulletin of the American Mathematical Society* 45: 234 [CW1], 652; *Science* 90. November: 3.

[40a] Review of: Fukamiya, M. On dominated ergodic theorems in L_p (p 1). *Mathematical Reviews* 1: 148. [CW4], 1019.

[40b] Review of: Fukamiya, M. The Lipschitz condition of random functions. *Mathematical Reviews* 1: 149. [CW4], 1020.

[40c] Review of: De Donder, T., L'énergetique déduite de la mécanique statistique générale. *Mathematical Reviews* 1: 192. [CW4], 1021.

[40d] A canonical series for symmetric functions in statistical mechanics. Abstract 133. *Bulletin of the American Mathematical Society* 46: 57. [CW1], 653.

[41a] (with Wintner, Aurel) Harmonic analysis and ergodic theory. *American Journal of Mathematics* 63: 415–426. [CW1], 693–704.

[41b] (with Wintner, Aurel) On the ergodic dynamics of almost periodic systems. *American Journal of Mathematics* 63: 794–824. [CW1], 705–735.

[42a] (with Polya, George) On the oscillation of the derivatives of a periodic function. *Transactions of the American Mathematical Society* 52 (1942): 249–256. [CW2], 956–963.

[42b] *Wiener's Final Report.* Final report on Section D2, Project # 6. Typescript 8 pages, Confidential, signed by Wiener and sent to Warren Weaver. 1 December 1942. Enclosed to "Report to the Services no. 59, division 7 (Fire Control), Statistical method of prediction in Fire Control" [hereinafter referred to as the Report to the Services no. 59], followed by an "Initial distribution" to 64 members. Now unclassified with number AD800206 in DTIC. Accessed in 2010.

[42c] *Response of a non-linear device to noise* (8 pages). Radiation Laboratory, MIT, # 129 Report V-165. April 6, 1942. "Restricted", now unclassified number ADA800212. DTIC. Accessed in 2010.

[43c] Wiener to Weaver, 15 January 1943. 5 pages letter, enclosed in Report to the Services no. 59, AD800206.

[43a] (with Wintner, Aurel) The discrete chaos. *American Journal of Mathematics* 4 (1943): 279–298; *Mathematical Reviews* 4 (1943): 220. [CW1], 614–633.

[43b] (with Rosenblueth, Arturo, and Bigelow, Julian) Behavior, purpose and teleology. *Philosophy of Science* 10: 18–24. [CW4], 180–186.

[43c] Elements of calculus. In *Practical mathematics*, vol. 2, n. 8, 459–499, ed. R.S. Kimball. June 1943.

[43d] (with Wintner, Aurel) Ergodic dynamics of almost periodic systems. *Mathematical Reviews* 4 (1943): 15. Wiener, Norbert, Ergodic dynamics of almost periodic systems (with Wintner, Aurel). *Mathematical Reviews* 4 (1943): 15 (same as reference [41b] above).

[44a] Mits's education. Review of: Gray Liber, H., and Lieber, L.R., The education of T.C. Mits; What modern mathematics means to you. *The Technology Review* 46: 390, 392. [CW4], 1010.

[45a] La teoría de la extrapolacion estadistica. *Boletín de la Sociedad Matemática Mexicana* 2: 316–322. [CW3], 95–100.

[45b] (with Rosenblueth, Arturo) The role of models in science. *Philosophy of Science* 12: 316–321. [CW4], 445–451.

[46a] (with Heins, Albert E.) A generalization of Wiener-Hopf integral equation. *Proceedings of the National Academy of Sciences (USA)* 32: 98–101; *Mathematical Reviews* 8: 29. [CW3], 54–57.

[46b] (with Rosenblueth, A.) The mathematical formulation of the problem of conduction of impulses in a network of connected elements, specifically in cardiac muscle. *Archivos del Instituto de Cardiología de México* (July) 16: 205–265. [CW4], 511–571; *Boletín de la Sociedad Matemática Mexicana* 2 (1945): 37–42; *Mathematical Reviews* 9 (1948): 604.

[46c] Theory of statistical extrapolation. *Mathematical Reviews* 7: 416.

[46d] Self-correcting and goal-seeking devices and their breakdown. In The New York Academy of Sciences, *Preprinted Abstracts. Conference on "Teleological Mechanisms", Monday, October 21 and Tuesday, October 22, 1946*, typescript, in Box 0–14 of Margaret Mead Papers in the Division Manuscripts at the Library of Congress, Washington, D.C. [hereinafter referred to as the *Preprinted Abstracts*]. I thank the Library of Congress and particularly Bruce Kirby for kindly sending me copies of these documents.

[47a] (with Mandelbrojt, Szolem) Sur les fonctions indéfiniment dérivables sur une demi-droite. *Comptes Rendus de l'Académie des Sciences*. I. 225: 978–980; [CW2], 909–911. *Mathematical Reviews* 9 (1948): 230.

[47b] A scientist rebels. *The Atlantic Monthly* 179: 46. [CW4], 748.

[48a] Time, communication, and the nervous system. In *Teleological mechanisms*. Official proceedings of the "Conference on Teleological Mechanisms" from 21 to 22 October 1946. *Annals of the New York Academy of Sciences*, 50 (1948) [hereinafter referred to as the TM 1948], 197–220. [CW4], 220–243; *Mathematical Reviews* 10 (1949): 133.

[48b] Cybernetics. *Scientific American* 179: 14–18. [CW4], 784–788.

[48c] (with Rosenblueth, A., Pitts, W. and García Ramos, J.) An account of the spike potential of axons. *Journal of Comparative Physiology* 32-3 (December): 275–317. [CW4], 572–614.

[48d] A rebellious scientist after two years. *Bulletin of Atomic Scientist* 4: 338. [CW4], 749–750.

[48e] Review of: Infeld, L., Whom the Gods love. The story of Evariste Galois. *Scripta Mathematica* 14: 273–274. [CW4], 1015–1016.

[48f1] *Cybernetics. Or control and communication in the animal and the machine*, Paris, Hermann & Cie [...], 1948. (This work was commissioned by the French publisher Hermann & Cie, but it is in English)

[48f2] *Cybernetics. Or control and communication in the animal and the machine*, 1st ed. Cambridge (Mass.): The MIT Press, 1948.

[49a] Sur la théorie de la prévision statistique et du filtrage des ondes. *Analyse harmonique, Colloques Internationaux du CNRS,* No. 15 (1949): 67–74. Paris: Centre National de la Recherche Scientifique; [CW3], 101–108. *Mathematical Reviews* 11 (1950): 376.

[49b] (with Rosenblueth, A., Pitts, W. and García Ramos, J.) A statistical analysis of synaptic excitation. *Journal of Cellular and Comparative Physiology* 34-2, October: 173–205. [CW4], 615–647.

[49c] A new concept of communication engineering. *Electronics* 22: 74–77. [CW4], 197–199.

[49d] Sound communication with the deaf. *Philosophy of Science* 16-3: 260–262. [CW4], 409–411.

[49e] (with Levine, Leon) Some problems in sensory prosthesis. *Science* 110-11: 512. [CW4], 412.

[49f] Obituary: Godfrey Harold Hardy (1877–1947). *Bulletin of the American Mathematical Society* 55: 72–77. [CW4], 1033–1038.

[49g] *Extrapolation, interpolation, and smoothing of stationary time series, with engineering applications*, New York and London, Wiley, 1949; Cambridge (Mass.), MIT Press. First published during the war as a classified report to Section D2, National Defense Research Committee. 1 February 1942, nicknamed the "Yellow Peril".

[49h] Review of: Frank, P., Modern science and its philosophy, New York. *Book Review* 7: 3. [CW4], 996–997.

[50a] (with Geller, L.) Some prime-number consequences of the Ikehara theorem. *Acta Scientiarum Mathematicarum.* Szeged (HU). B, 12: 2528; [CW2], 754–757. *Mathematical Reviews* 11 (1950): 644; ibid. 12 (1951): 1002.

[50b] see [52d].

[50c] Some maxims for biologists and psychologists. *Dialectica* 4 September: 186–191. [CW4], 452–457.

[50d] (with Rosenblueth, Arturo) Purposeful and non-purposeful behavior. *Philosophy of Science* October: 318–326. [CW4], 187–195.

[50e] Cybernetics. *Bulletin of American Academy of Arts and Sciences* 3: 2–4. [CW4], 790–792.

[50f] Speech, language, and learning. *The Journal of Acoustical Society of America* 22: 696–697. [CW4], 200–201.

[50g] Entropy and information. *Proceedings of Symposia in Applied Mathematics* 2: 89. Providence (RI): American Mathematical Society. [CW4], 202. *Mathematical Reviews* 11 (1950): 305.

[50h] Too big for private enterprise. *Nation* 170: 496–497. [CW4], 702–703.

[50i] Too damn close. *The Atlantic Monthly.* 186: 50–52. [CW4], 704–706.

[50j] *The human use of human beings.* London: Eyre and Spottiswoode, 1950.

[50k] The brain. *Tech Engineering News* 31.

[50l] The thinking machine. *Time* (January) 55: 23.

[51a] Problems of sensory prosthesis. *Bulletin of the American Mathematical Society* 57: 27–35. [CW4], 413–421.

[51b] Homeostasis in the individual and society. *Journal of the Franklin Institute* 251-1: 65–68; reprinted in *Selected papers* (64f), 440–443 and in *God & Golem, Inc.* [64e]. [CW4], 380–383.

[51c] Two industrial revolutions. *Bulletin of the British Society for the History of Science* (October) 1-6: 159–160.

[52a] Cybernetics. *Scientia*, Italy 87: 233–235. [CW4], 203–205.

[52b] The miracle of the broom closet. *Tech Engineering News* 33-7.

[52c] Cybernetics. *Encyclopedia Americana*, ed 1952, 187–188. [CW4], 804.

[52d] Comprehensive view of prediction theory. In *Proceedings of the International Congress of Mathematicians, Cambridge, MA, August 30–September 6, 1950* (ICM 1950). Providence [RI]: American Mathematical Society, 1952: 308–321. [CW3], 109–122 as [50b].

[53a] Optics and the theory of stochastic processes. *Journal of the Optical Society of America* 43: 225–228; [CW3], 486–489. *Mathematical Reviews* 17 (1956): 33.

[53b] (with Siegel, Armand) A new form for the statistical postulate of quantum mechanics. *Physical Review* 9: 1551–1560. [CW3], 492–501. *Mathematical Reviews* 15 (1954).

[53c] Distributions quantiques dans l'espace différentiel pour les fonctions d'ondes dépendant du spin. *Comptes Rendus de l'Académie des Sciences* 237: 1640–1642. [CW3], 502–504. *Mathematical Reviews* 15 (1954): 490.

[53d] Les machines à calculer et la forme (Gestalt). In *Les machines à calculer et la pensée humaine, Colloques Internationaux du Centre National de la Recherche Scientifique, Paris, 1953*, 461–463. [CW4], 422–424. *Mathematical Reviews* 16 (1955): 529.

[53e] The concept of homeostasis in medicine. *Transactions and Studies of the College of Physicians of Philadelphia* (4) 20-3. [CW4], 384–390.

[53f] Problems of organization. *Bulletin of the Menninger Clinic* 17: 130–138. [CW4], 391–399.

[53g] The future of automatic machinery. *Mechanical Engineering* February: 130–132. [CW4], 663–665.

[53h] *Ex-prodigy. My childhood and youth*, New York: Simon and Schuster.

[53i] The electronic brain and the next industrial revolution. *Cleveland Athletic Club Journal*, January. [CW4], 666–672.

[53j] The machine as threat and promise. *St. Louis Post-Dispatch*, December 1953. [CW4], 673–678.

[53k] *This I believe: We can't attain truth without risk of error* (from "This I Believe radio show"), Minneapolis Tribune, November 1953. [CW4], 751.

[54a] Men, machines, and the world about. In *Medicine and Science, New York Academy of Medicine and Science*, 13–28, ed. I. Galderston. New York: International Universities Press, 1954. [CW4], 793–799.

[54b] Conspiracy of conformists. *Nation* 178: 375. [CW4], 752.

[54c] (with Campbell, D.) *Automatization, St. Louis post-dispatch* December. [CW4], 679–683.

[54d] *The human use of human beings*, 2nd ed. Garden City [NY]: Doubleday, 1954 (Completely revised, with several passages and chapters cut or rewritten).

[55a] Nonlinear prediction and dynamics. In *Proceedings Third Berkeley Symposium on Mathematical Statistics and Probability, University of California Press*, Berkeley [CA] 1954/5, 247–252. [CW3], 371–376. *Mathematical Reviews* 18 (1957): 949.

[55b] On the factorization of matrices. *Commentarii Mathematici Helvetici* 29: 97–111; *Mathematical Reviews* 16: 921. [CW3], 123–137.

[55c] (with Siegel, Armand) The differential space theory of quantum systems. *Nuovo Cimento* (10) 2: 982–1003, No. 4, Supplement.

[55d] Thermodynamics of the message. In *Neurochemistry. The chemical dynamics of brain and nerve*, ed. K.A.C. Elliott, Irvine H. Page, and J.H. Quastel. Springfield [IL]: Charles C. Thomas; Oxford: Blackwell Scientific Publications. [CW4], 206–211.

[55e] Time and organization. In *Second Fawley Foundation Lecture*. University of Southampton, 1–16. [CW4], 309–322.

[56a] (with Wintner, Aurel) *On a local L^2-variant of Ikeara's theorem*, 53–59. [CW2], 758–764.

[56b] The theory of prediction. *Modern mathematics for the engineer*, ed. E.F. Beckenbach. New York: McGraw-Hill. [CW3], 138–163.

[56c] (with Siegel, A.) The theory of measurement in differential space quantum theory. *Physical Review* 101 (1956): 429–432. [CW3], 527–530.

[56d] Pure patterns in a natural world. In *The new landscape in art and science*, ed. G. Kepes. Chicago: Paul Theobald and Co., 1956, 274–276. [CW4], 857–859.

[56e] Brain waves and the interferometer. *Journal of the Physical Society of Japan* 18-8. [CW4], 323–331.

[56f] Moral reflections of a mathematician. *Bulletin of the Atomic Scientists* 12: 53–57; taken from *I am a mathematician*. [CW4], 753–757.

[56g] *I am a mathematician. The later life of a prodigy; an autobiographical account of the mature years and career of childhood in Ex-prodigy*. Garden City [NY]: Doubleday; also in paperback the same year: Cambridge [MA]: The MIT Press, 1956.

[57a] (with Akutowicz, E.J.) The definition and Ergodic properties of stochastic adjoint of a unitary transformation. *Rendiconti del Circolo Matematico di Palermo* 6 (2): 205–217, Addendum 349. [CW1], 744–756; *Mathematical Reviews* 20 (1959), rev. no. 4328.

[57b] Notes on Pòlya's and Turan's hypotheses concerning Liouville's factor. *Rendiconti del Circolo Matematico di Palermo* 6 (2): 240–248. [CW2], 765–773; *Mathematical Reviews* 20 (1959), rev. no. 5759.

[57c] (with Wintner, Aurel) On the non-vanishing of Euler products. *American Journal of Mathematics* 79: 801–808. [CW2], 774–781.

[57d] (with Masani, Pesi R.) The prediction theory of multivariate stochastic processes, part I. *Acta Mathematica* 98: 111–150. [CW3], 164–203.

[57e] Rhythms in physiology with particular reference to encephalography. *Proceedings of Rudolf Virchow Medical Society in the City of New York.* 16: 109–124. [CW4], 332–347.

[57f] The role of the mathematician in a materialistic culture (A Scientist's dilemma in a materialistic world). In *Columbia Engineering Quarterly. Proceedings of the second Combined Plan Conference, Arden House,* October 6–9 (1957), 22–24. [CW4], 707–709.

[57g] *The role of the small cultural college in education of the scientists*; a speech given at Wabash College, Indiana, 10 October 1957. [CW4], 897–906.

[57h] Cybernetics. In *The Universal Standard Encyclopedia* [abridgement of *The New Funk and Wagnall's Encyclopedia*], New York: Standard Reference Works Publishing Co., 180. [CW4], 805.

[58a] Logique, probabilité et méthode des sciences physiques. In *La méthode dans les sciences modernes.* Paris: François Le Lionnais, 111–112. [CW3], 537–538.

[58b] (with Masani, Pesi R.) The prediction theory of multivariate stochastic processes, part II. *Acta Mathematica* 99: 93–137. [CW3], 204–248.

[58c] (with Wintner, Aurel) Random time. Nature 181 (1958): 561–562. [CW4], 254–256.

[58d] (with Masani, Pesi R.) Sur la prévision linéaire des processus stochastiques vectoriels à densité spectrale bornée, part I. *Comptes Rendus de l'Académie des Sciences.* I 246: 1492–1495. [CW3], 249–252. *Mathematical Reviews* 20 (1959), rev. no. 4324a.

[58e] (with Masani, Pesi R.) Sur la prévision linéaire des processus stochastiques vectoriels à densité spectrale bornée, parte II. *Comptes Rendus de l'Académie des Sciences. I. Paris* 246 (1958): 1655–1656. [CW3], 253–254; *Mathematical Reviews* 20 (1959), rev. no. 4324b.

[58f] My connection with cybernetics. Its origin and its future. *Cybernetica* 1–14. [CW4], 107–120.

[58g] Time and the science of organization. *Scientia* September. [CW4], 247–252.

[58h] Science: The megabuck era. *New Republic* 138: 10–11. [CW4], 710–711.

[58i] *Nonlinear problems in random theory.* Cambridge [MA]: The MIT press; New York: John Wiley; London: Chapman & Hall; *Mathematical Reviews* 20 (1959), rev. no. 7337.

[59a] (with Akutowicz, Edwin J.) A factorization of positive Hermitian matrices. *Journal of Mathematics and Mechanics* 8: 111–120. [CW3], 264–273.

[59b] (with Masani, Pesi R.) Non-linear prediction. In *Probability and statistics*, ed. U. Grenander. Stockholm: The Harald Cramer Volume, 190–212. [CW3], 377–399.

[59c] (with Masani, Pesi R.) On bivariate stationary processes and the factorization of matrix-valued functions. *Theory of Probability and its Applications* (Moscow) 4: 322–331; (English translation 300–308). [CW3], 255–263.

[59d] Man and the machine (Interview with N. Wiener). *Challenge: the Magazine of Economic Affairs* 7: 36–41. [CW4], 712–717.

[59e] *The tempter*. New York: Random House [in the form of a novel].

[60a] The application of physics to medicine. In *Medicine and other disciplines*, New York Academy of Medicine, ed. I. Galderston. Madison (CT): International Universities Press, 41–57. [CW4], 458–464.

[60b] The brain and the machine (summary of an address). In *Dimension of mind*, ed. S. Hook. Collier Books, 1960 (Proceedings of Third Annual New York University Institute of Philosophy held on May 15–16, 1959), 113–117. [CW4], 684–688.

[60c] Kybernetik. In *Das Soziologische Wörterbuch*. Stuttgart: F. Enke Verlag. [CW4], 806–807.

[60d] Some moral and technical consequences of automation. *Science* 131: 1355–1358. [CW4], 718–721.

[60e] The duty of the intellectual. *The Technology Review* 62 February: 26–27. [CW4], 758–759.

[60f] Preface. In *Cybernetics of natural systems*, ed. D. Stanley-Jones. London: Pergamon Press. [CW4], 1022–1025.

[60g] Possibilities of the use of the interferometer in investigating macromolecular interactions. In *Fast Fundamental Transfer Processes in Aqueous Biomolecular Systems*, 52–53, ed. F.O. Schmitt. Cambridge [MA]: Massachusetts Institute of Technology. [CW4], 358–359.

[60h] The grand privilege. *Saturday Review*. March 5.

[61a] Über Informationstheorie. *Naturwissen-chaften* 7: 174–176. [CW4], 212–214.

[61b] Science and society. *Voprosy Filosofii*. 7 and in *The Technology Review*, July: 49–52. [CW4], 773–776.

[61c] Cybernetics. *Or control and communication in the animal and the machine*, 2nd ed. New York: MIT Press; New York and London: Wiley (Revised edition of [48f] with two additional chapters: 9. "On learning and Self-Reproducing machines"; 10. "Brain Waves and Self-Organizing Systems").

[62a] A verbal contribution to *Proceedings of the International Symposium on the Application of Automatic Control in Prosthetics Design, Opatija, Yugoslavia, August 27–31, 1962*, 132–133. [CW4], 425–426.

[62b] The mathematics of self-organizing system. In *Recent developments in information and decision processes*. New York and London: Wiley. [CW4], 260–280.

[62c] *Short-time and long-time planning*, originally presented at 1954 IV ASPO National Planning Conference. Jersey Plans, An ASPO Anthology (1962), 29–36. [CW4], 808–815.

[62d] *On the history and prehistory of cybernetics*, Conferenza tenuta all'is-
       tituto di Fisica Teorica, Napoli, in *Cybernetics on neural processes*,
       Roma, CNR, 1965.

[63a] see [64j].

[63b1] Introduction to neurocybernetics. In *Progress in brain research*, 2:
       Nerve and memory models, ed. N. Wiener and J.P. Shade. Amsterdam:
       Elsevier Publishing Co., 1–7. [CW4], 400–406.

[63b2] Epilogue. In *Progress in brain research*, vol. 2: *Nerve and memory
       models*, ed. N. Wiener and J.P. Shade. Amsterdam: Elsevier Publishing
       Co., 1963, 264–268. [CW4], 427–431.

[63c] (with Deutsch, Karl) The lonely nationalism of Rudyard Kipling. *Yale
       Review* 52: 499–517. [CW4], 868–886.

[64a] On the oscillations of nonlinear systems. In *Proceedings of Symposium
       on Stochastic Models in Medicine and Biology, Mathematics Research
       Center, U.S. Army, 12–14 June, 1963*, ed. John Gurland. Madison (WI):
       University of Wisconsin Press, 167–177. [CW4], 292–302.

[64b] *Dynamical systems in physics and biology*, contribution to the series
       «Fundamental Science in 1984», London: The New Scientist. [CW4],
       659–660.

[64c] Machines smarter than men? *U.S. News & World Report* 56: 84–86.
       [CW4], 722–723 [An interview with Wiener]; summary in *Reader's
       Digest* 84 (1964): 121–124.

[64d] Intellectual honesty and the contemporary scientist (partial transcription
       of a conversation in the Hillel Group, MIT). *The Technology Review* 66:
       17–18, 44–45, 47. [CW4], 725–729.

[64e] *God & Golem, Inc. A comment on certain points where Cybernetics
       impinges on religion*. Cambridge [MA]: The MIT Press.

[64f] see [65e].

[64g] *I am a mathematician. The later life of a prodigy; an autobiographical
       account of the mature years and career of childhood in Ex-prodigy*, 2nd
       ed. Cambridge [MA]: The MIT Press, 1964.

[64h] *Ex-prodigy. My childhood and youth*, 2nd ed. Cambridge [MA]:
       The MIT Press, 1964.

[64i] *Time series*. Wiley paperbacks; Cambridge [MA]: The MIT Press, 1964.
       Reprint of [49a].

[64j] (with Della Riccia, Giacomo) Random theory in classical phase space
       and quantum mechanics. In *Proceedings of the International Conference
       on Functional Analysis. MIT, Cambridge [MA], 9–13 June, 1963*, 3–14.
       Cambridge [MA]: The MIT Press. [CW3], 540–551, as [63a].

[65a] *L'homme et la machine*. In *Le Concept d'Information dans la Science
       Contemporaine*. Colloques Philosophiques Internationaux de
       Royaumont, July 1962. 99–132. Paris: Gauthier-Villars. [CW4], 824 ff.

[65b] Perspective in cybernetics in Cybernetics of the nervous system.
       Progress in Brain Research, vol. 17, ed. N. Wiener and J.P. Schadè.
       Amsterdam, New York: Elsevier Pub. C., 399–408. [CW4], 360.

[65c] Cybernetics. In *Collier's encyclopedia*, ed. William D. Halsey, 598–599. New York: The Cornwell-Collier Publishing Co. [CW4], 816.

[65d] Progresso ed entropia. In *Profezie and realtà del nostro secolo. Testi e documenti per la storia di domain*, 603–621, ed. Franco Fortini. Bari: Laterza.

[65e] *Selected papers of Norbert Wiener. Including generalised harmonic analysis and Tauberian theorems*, with contributions from Y.W. Lee, N. Levinson, and W.T. Martin. Cambridge [MA]: The MIT Press, 1965.

[66a] (with Della Riccia, Giacomo) Wave mechanics in classical phase space, Brownian motion, and quantum theory. *Journal of Mathematics and Physics* 7: 1372–1383. [CW3], 552–563.

[66b] (with Siegel, A., Rankin, B. and Martin, W.T.) *Differential space, quantum systems and prediction*, ed. B. Rankin. Cambridge [MA]: The MIT Press.

[66c] *Generalized harmonic analysis and Tauberian theorems*, paperback ed of [30a] and [32a]. Cambridge [MA]: The MIT Press, 1966.

[68] Cybernetics in history. In *Modern systems research for the behavioral scientist: A sourcebook*, 31–36, ed. Walter F. Buckley. Chicago: Aldine Publishing Company (Reprinted from "Cybernetics in History," Chapter I of [54d]).

[75a] (with Landis, F.) Cybernetics. In *Funk and Wagnall's New Encyclopedia*. New York: Funk & Wagnall, 228. [CW4], 817.

[85a] Letter covering the memorandum on the scope [...] of a suggested computing machine, 21 September 1940 published for the first time in CW4, 122–124.

[85b] *Memorandum on mechanical solution of partial differential equations.* Published for the first time in [CW4], 125–134.

[85c] (with Rosenblueth, A., Pitts, W. and García Ramos, J.) *Muscular clonus: cybernetics and physiology*, published for the first time in [CW4], 466–510.

[89] *The human use of human beings*, introduction by S.J. Heims, London, Free Association, 1989. New publication of (54d), second edition of (50j).

[91] *Ergodic Theory* given at *The Statistics Seminar, MIT, 1942–1943*, ed. Lawrence Klein. In *Statistical Science*, 6-4 (November): 320–330.

[91] *Invention. The care and feeding of ideas.* Cambridge [MA]: The MIT Press, 1993 (Posthumous work, written about 1954).

Works of Other Authors

21CW. *Norbert Wiener in the 21st Century (21CW), 2014 IEEE Conference on 24–26 June 2014, Boston.* The conference papers here: http://ieeexplore.ieee. org/xpl/mostRecentIssue.jsp?punumber=6884877.

Abbagnano, Nicola. 1995. *Storia della filosofia. 6. La filosofia dei secoli XIX and XX. Dallo spiritualismo all'esistenzialismo.* Milano: Tea.

Abraham, Tara H. 2000. *"Microscopic cybernetics": Mathematical Logic, Automata Theory, and The Formalization of Biological Phenomena, 1936-1970.* Ph. D. Thesis, Graduate Department of the Institute for the History and Philosophy of Science and Technology. Toronto: University of Toronto.

Abraham, Tara H. 2002. "(Physio)logical circuits. The intellectual origins of the McCulloch–Pitts neural networks. *Journal of the History of the Behavioral Sciences* 38-1: 3–25.

Abraham, Tara H. 2004. Nicolas Rashevsky's Mathematical Biophysics. *Journal of the History of Biology* 37: 333–385.

Agar, J. 2001. Wiener, Norbert 1894-1964. In *Encyclopedia of computers and computer history*, ed. R. Rojas, 823–825. Chicago and London: Fitzroy Dearborn publishers.

Akera A., and Nebeker F. eds. 2002. *From 0 to 1. An authoritative history of modern computing.* New York: Oxford University Press.

Aldrich, John. 2007. But you have to remember P.J. Daniell of Sheffield. *Electronic Journ@l for History of Probability and Statistics*, 3-2 December. http://www.jehps.net/Decembre2007/Aldrich.pdf. Accessed March 21–27, 2017.

Allen, G.E. 1978. *Life science in the twentieth century.* Cambridge (UK)[UK]: Cambridge University Press.

Anon. 1945a. Radar. The Technique [pp. 139–146]. The industry. A clandestine business in the billions was built on the work of the physicists. [pp. 146–162]. *Fortune.* (October).

Anon. 1945b. Radar. Longhairs and short waves. *Fortune.* (November): 163–169.

Anon. 1949. Bialystock. In *Enciclopedia Italiana* [...], Roma, Istituto dell'Enciclopedia Italiana, vol 6, 857–858.

Anon. 2010. *50th Anniversary. Evolving from Calculators to Computers.* www. lanl.gov/history/atomicbomb/computers.shtml. Accessed July 21. 2010.

Anon. 1963. Can I.B.M. keep up the pace? *Business Week* 2 (February): 92–98.

Archibald, R.C. 1946. Conference on advanced computation techniques. *Mathematical Tables and Aids to Computation* (April) 2–14: 65–68.

Aspray, William. 1989. The transformation of numerical analysis by the computer. An example from the work of John von Neumann. In Rowe and McCleary 1988, 307–325.

Aspray, William. 1990a. *John von Neumann and the origins of modern computing.* Cambridge [MA] and London [UK]: The MIT Press, 1990.

Aspray, William. 1990b. *The origins of John von Neumann's theory of automata,* in Glimm, Impagliazzo, and Singer 1990, 289–309.

Atzema, E.J. 2003. Into the woods. Norbert Wiener in Maine. *The Mathematical Intelligencer* 25-2: 7–17.

BAMS. 1966. *Bulletin of the American Mathematical Society*, 72 (1). Part 2. 1966. Special edition: "Norbert Wiener 1894-1964".

Baracca, A., Ruffo, S., and Russo, A. 1979. *Scienza and industria 1848-1915. Gli sviluppi scientifici connessi alla seconda rivoluzione industrial*. Roma-Bari: Laterza.

Bates, J.A.V. 1947. Some characteristics of a human operator. *Journal of the Institution of Electrical Engineers*. Part IIA: Automatic Regulators and Servo Mechanisms. 94-2: 298–304.

Bateson, Gregory. 1946a. Circular causal systems in society. In *Preprinted abstracts*.

Bateson, Gregory. 1946b. *Circular causal systems in society*, final typewrited copy of the Bateson paper, with the same title one finds in the *Preprinted abstracts*. It has a cover sheet with the following words: "Will be published: Annals, N.Y. Acad. Sci. Vol. 49". I thank the Library of Congress and particularly Bruce Kirby for kindly sending me copies of these documents.

Bateson, Gregory. 1958 [1938]. *Naven*. 2nd edn. Stanford: Stanford University Press.

Bateson, Gregory. 1975 [1972]. *Steps to an ecology of mind. Collected essays in anthropology, psychiatry, evolution, and epistemology*. New York [NY]: Ballantine Books.

Bateston, G., and Mead, M. 1976. "For God's Sake, Margaret", interview with Stewart Brand. *CoEvolution Quarterly* Summer. http://www.wholeearth.com/issue/2010/article/361/for.god%27s.sake.margaret. Accessed March 22–27, 2017.

Battimelli, Giovanni. 1986. On the history of the statistical theory of turbulence. *Revista Mexicana de Fisica*. S1 32: 3–48.

Battimelli, Giovanni. 2002. *Illustri sconosciuti. Il gigante della meccanica*, in www.galileonet.it.

Bennett, Stuart. 1979. *History of control engineering. 1800-1930*. Stevenage [UK]: Peregrinus [for] the Institution of Electrical Engineers.

Bennett, Stuart. 1993. *History of control engineering. 1930-1955*. Stevenage [UK]: Peregrinus [for] the Institution of Electrical Engineers.

Bennett, Stuart. 1994. Norbert Wiener and control of anti-aircraft guns. *IEEE Control Systems Magazine* 14-6: 58–62.

Bergmann, Peter G. 1942. "Notes on: *The Extrapolation, Interpolation and Smoothing of Stationary Time Series, NorbertWiener*, Research Project DIC-6037, Report to the Services 19, The Massachusetts Institute of Technology, 14 December. Div. 7-313.1-M3 (Cit. by Summary Technical Report of the Division 7 vol. 1 Gunfire Control, 165).

Bergson, Henri. 1911 [1896]. *Matter and memory*. London: George Allen & Unwin LTD; New York: The Macmillan Company.

Bergson, Henri. 1911 [1907]. *Creative evolution*. New York: Henry Holt and Company.

Bergson, Henri. 1934. *La pensée et le mouvant*. Paris: Alcan.

Bernard, Claude. 1878. *La science expérimentale*. Paris: Librarie J.-B. Baillière & Fils.

Bernstein, Richard J. 1992. The resurgence of pragmatism. *Social research*. 59-4: 813–840.

Bernstein, Richard J. 2007 [1991]. Appendix: Pragmatism, pluralism, and the healing of wounds. In *The new constellation. The ethical-political horizons of modernity/postmodernity*, 324–339. Cambridge [UK]: Polity Press.

Bigelow, Julian. 1971. Interview by Richard R. Mertz on 20 January 1971. In *Computer Oral History Collection*. Archives Center. Smithsonian National Museum of American History. http://invention.smithsonian.org/. Accessed January 1, 2010.

Bode, H.W. 1945. *Network analysis and feedback amplifiers design*. New York: Nostrand.

Bode, H.W. 1960. Feedback. The history of an idea. In *Proceedings of the Symposium on Active Networks and Feedback Systems—New York, N.Y. April 19, 20, 21, 1960*. New York: Polytechnic Institute of Brooklyn.

Bode, H.W., and Shannon, C. A simplified derivation of the method of linear least square smoothing and prediction theory. *Proceedings of the IRE* (April) 38-4: 417–425.

Bosco, N. 1987. *Invito al pensiero di George Santayana*. Milan: Mursia.

Bottazzini, Umberto. 2016. *Hilbert's Flute. The history of modern mathematics*. New York: Springer.

Boulton, Marjorie. 1960. *Zamenhof, creator of Esperanto*. London: Routledge and Paul.

Boutroux, Émile. 1911. *William James*. Paris: Colin.

Brainerd, John G. 1976. Genesis of the ENIAC. *Technology and Culture*. 17/3 (July): 482–488.

Brandt, F. 1967. Harald Høffding. In *The encyclopedia of philosophy*, ed. P. Edwards, vol. IV.

Breton, Philippe. 1987. *Histoire de l'informatique*. Paris: La Découverte.

Breton, Philippe. 1992. *L'utopie de la communication. L'emergence de l'homme sans interieur*. Paris: La Decouverte.

Bridgman, Williams Percy. 1927. *The logic of modern physics*. New York: The Macmillan Company.

Bruce, H. Addington. 1911. New ideas in child training. *American Magazin* (July): 291–292.

Buchanan, R. Angus. 1996. The history of technology. In *EB 1996*.

Bulmer, Martin. 1984. *The Chicago school of sociology. Institutionalization, diversity and the rise of sociological research*. Chicago: The University of Chicago Press.

Burks, Arthur W. 1947. Electronic computing circuits of the ENIAC. *Proceedings of the IRE* 35: 756–767.

Bush, Vannevar, 1941. *Report of the National Defense Research Committee for the First Year of Operation*, June 27, 1940 to June 28, 1941, 1–36; attachments:

Declassified paper sent to president Roosevelt with a covering letter on 16 July 1941 (FDRPL), Box 2.

Bush, Vannevar. 1945. As we may think. *The Atlantic Monthly* July.

Bush, Vannevar. 1970. *Pieces of the action*. New York: William Morrow.

Bush, Vannevar. 1982 [1940]. *Arithmetical Machine*, 2 March 1940, Arithmetical Machine, Washington, D.C., 2 March 1940. In Randell 1982a, 337–343. The original is in Vannevar Bush Papers, Container 18, Folder: Caldwell, Samuel, 1939-1940. Library of Congress, Washington, D.C.

Calimani R., and Lepschy A. 1990. *Feedback. Guida ai cicli a retroazione: dal controllo automatico al controllo biologico*. Milan: Garzanti.

Calimani, Dario. 1998. *T. S. Eliot. Le geometrie del disordine* [T. S. Eliot. Geometries of Disorder]. Napoli: Liguori.

Calimani, Riccardo and Lepschy, Antonio. 1990. *Feedback. Guida ai cicli a retroazione: dal controllo automatico al controllo biologico*. Milano: Garzanti.

Campbell, Donald T. 2005. Etnocentrism of discipline and the fish-scale model of omniscience. In *Interdisciplinary collaboration: An emerging cognitive science*, ed. S.J. Derry, C.D. Schunn, and M.A. Gernsbacher, 3–21. Mahwah, NJ: Lawrence Erlbaum, 2005.

Cannon, Walter B. 1932. *The wisdom of the body*. New York: Norton and Co.

Capra, Fritjof. 1996. *The web of life. A new scientific understanding of living systems*. New York: Anchor Books.

Carnap, Rudolf. 1937. *The logical syntax of language*. London: K. Paul, Trench, Trubner & co. ltd.

Carnegie Institution of Washington. 1946. *Year Book 1946 of Carnegie Institution of Washington* (1 July, 1946–June 30, 1947), 77–78.

Carroll, Lewis. 1901 [1865]. *Alice's adventures in Wonderland*. New York and London: Harper & Brothers publishers.

Ceruzzi, Paul E. 1998. *A history of modern computing history of computing*. Cambridge [MA]: MIT Press.

Chandrasekhar S., Gamow, W.G., and Tuve, M.A. 1938. The problem of stellar energy. *Nature* 28 May, 141: 882.

Chandrasekhar, Subrahmanyan 1943. Stochastic problems in physics and astronomy. *Reviews of Modern Physics* 15: 1–89.

Chandrasekhar, Subrahmanyan. 1977. Interview of S. Chandrasekhar by S. Weart on 17 May 1977, Niels Bohr Library & Archives, American Institute of Physics, College Park, MD USA.

Chandrasekhar, Subrahmanyan. 1939. *An introduction to the study of stellar structure*. Chicago: University of Chicago Press.

Cini, Marcello. 1985. The context of discovery and the context of validation. The proposal of Von Neumann and Wiener in the development of 20th century physics. *Rivista di storia della scienza*. 2-1: 99–122.

Cini, Marcello. 1994. *Un paradiso perduto. Dall'universo delle leggi naturali al mondo dei processi evolutivi* [A lost paradise]. Milano: Feltrinelli.

Cohen, Bernard I. 2003. Mark I, Harvard. In *Harvard encyclopedia of computer science*, 4th ed, 1078–1080. Chichester [UK]: John Wiley and Sons.

Comrie, L.J. 1944. Recent progress in scientific computation. *Journal of Scientific Instruments*. 21 August 1944: 129–135.

Comrie, Leslie J. 1943. Mechanical computing. In David Clark. 1943. *Plane and geodetic surveying*, vol. 2, 3rd ed, revised by James Glendenning. Appendix I, 462–473. London: Constable, 1943.

Comstock D.F., and Troland, L.T. 1917. *The nature of matter and electricity. An outline of modern views*. New York: D. Nostrand Co.

Continenza B., and Gagliasso, E. eds. 1998. *L'informazione nelle scienze della vita*. Milan: FrancoAngeli.

Conway, F., and Siegelman, J. 2005. *Dark hero of the information age. In search of Norbert Wiener, the father of Cybernetics*. New York: Basic Books.

Cooper, Steven J. 2008. From Claude Bernard to Walter Cannon. Emergence of the concept of homeostasis. *Appetite*. (November) 51-3: 419–427.

Copleston, Frederick. 1996 [1966]. *A history of philosophy. VIII. Modern philosophy: Modern philosophy: Empiricism, idealism, and pragmatism in Britain and America*. New York etc.: Doubleday.

Cordeschi, Roberto 2002. *The discovery of the artificial. Behavior, mind and machines before and beyond cybernetics*. Dordrecht: Kluwer Academic Publishers.

Craik, K.J.W. 1944. Physiological and Psychological Aspects of Gun Control Mechanisms. Pt. II, B.P.C. (March) 43–254.

Craik, K.J.W. 1947. Theory of the human operator in control systems. I. The operator as an engineering system. *British Journal of Psychology*. General Section, 38-2: 56–61.

Craik, K.J.W. 1966. *The nature of psychology*, ed. S.L. Sherwood. Cambridge [UK]: Cambridge University Press.

Craik, K.J.W. 1967 [1943]. *The nature of explanation*. Cambridge [UK]: Cambridge University Press.

Craik, K.J.W., and M.A. Vince. 1945a. The design and manipulation of instrument knobs. M.R.C. Report 46/272. Applied Psychology Unit. 14th January.

Craik, K. J. W. and Vince, M. A., 1945b. Physiological and psychological aspects of gun control mechanisms. Pt. III, B.P.C. 45/405, February 1945.

Crawford, Perry O. Jr. 1942. *Automatic control by arithmetical operations*. M.S. Thesis. MIT.

Creager, A., and Santesmases, M. 2006. Radiobiology in the atomic age. Changing research practices and policies in comparative perspective. *Journal of the History of Biology* 39-4: 637–647.

Cull, Paul. 2007. The mathematical biophysics of Nicolas Rashevsky. *BioSystems* 88: 178–184.

Cunliffe, Marcus. 1963 [1954]. *The Literature of the United States*. Baltimore [MD], etc.: Penguin Books.

D'Agostini, Franca. 1999. *Breve storia della filosofia del Novecento. L'anomalia paradigmatica*. Turin: Einaudi.

Dahan Dalmedico, Amy. L'essor des mathématiques appliquées aux Etats-Unis: l'impact de la seconde guerre mondiale. *Revue d'histoire des mathématiques* 2: 149–213.

Daniell, P.J. 1917. The modular difference of classes. *Bulletin of the American Mathematical Society* 23: 446–450.

Daniell, P.J. 1918a. A general form of integral. *Annals of Mathematics*, 19, 279–294.

Daniell, P.J. 1918b. Differentiation with respect to a function of limited variation. *Transactions of the American Mathematical Society* 19: 353–362.

Daniell, P. J. 1918c. Integrals around General Boundaries. *Bulletin of the American Mathematical Society* 25: 65–68.

Daniell, P.J. 1919a. A general form of Green's Theorem. *Bulletin of the American Mathematical Society* 25: 353–357.

Daniell, P.J. 1919b. The derivative of a functional. *Bulletin of the American Mathematical Society* 25: 414–416.

Daniell, P.J. 1919c. Integrals in an infinite number of dimensions. *Annals of Mathematics* 20: 281–288.

Daniell, P.J. 1919d. Functions of limited variation in an infinite number of dimensions. *Annals of Mathematics* 21: 30–38.

Daniell, P.J. 1919e. Solution to a problem posed by W.D. Cairns. *American Mathematical Monthly* 26: 321.

Daniell, P.J. 1920a. Further properties of the general integral. *Annals of Mathematics* 21, 203–220.

Daniell, P.J. 1920b. Stieltjes derivatives. *Bulletin of the American Mathematical Society* 26, 444–448.

Daniell, P.J. 1920c. Observations weighted according to order. *American Journal of Mathematics* 42: 222–236.

Daniell, P.J. 1921a. Stieltjes-Volterra Products. *Comptes rendus du Congrès international des mathématiciens. Strasbourg, 22–30 Septembre 1920* [Proceedings of the International Congress of Mathematicians (ICM 1920)], ed. H. Villat, 130–136. Toulouse: Imprimerie É. Privat.*Comptes Rendus du Congrès International des Mathematiciens* 22–30 Septembre 1920. Paris: Villat.

Daniell, P.J. 1921b. The integral and its generalizations. *The Rice Institute Pamphlet* 8-1: 34–62.

Daniell, P.J. 1921c. Integral products and probability. *American Journal of Mathematics* 43: 143–162.

Daniell, P.J. 1921d. Two generalizations of the Stieltjes integral. *Annals of Mathematics* 23: 168–182.

Daniell, P.J. 1943. *The Extrapolation, Interpolation and Smoothing of Stationary Time Series with Engineering Applications by Nobert Wiener. Digest of Manual.* OSRD Report 370, OSRD Liaison Office W-386-1. Div. 7-313.1-M5 (Cit. by *Summary Technical Report of the Division 7 vol. 1 Gunfire Control*, p. 165).

Darwin, Charles. 1860 [1959]. *On the origin of species by means of natural selection, or, the preservation of favoured races in the struggle for life.* London: J. Murray.

Darwin, Charles. 1875 [1871]. *The descent of man, and selection in relation to sex.* New York: D. Appleton and company.

De Luca A., and Ricciardi, L.M. 1986 [1971]. *Introduzione alla cibernetica.* Milan: Franco Angeli.

De Luca, Aldo. 2006. Some reflections on cybernetics and its scientific heritage. *Scientiae Mathematicae Japonicae.* 64-2: 243–253.

De Ruggiero, Guido. 1921 [1912]. *Modern philosophy.* New York: Macmillan.

Dechert, Charles R. ed. 1966. *The social impact of cybernetics.* New York, Simon and Schuster. [Proceedings of a conference on "Cybernetics and society" held in Washington in November 1964, jointly organized by Georgetown University, the American University, George Washington University, and the American Society for Cybernetics.

Den Hartog, J.P., and Peters, H. eds. 1939. *Proceedings of the Fifth International Congress for Applied Mechanics, held at Harvard University and the Massachusetts Institute of Technology.* Cambridge [MA], Massachusetts, 12–16 September 1938. New York: Wiley.

Desch, Joseph R. 1942a. Report to H.M. Williams, Vice-President of National Cash Register, in Charge of Engineering & Research Department. Propriety of National Cash Register, 21 January 1942. 9 pages. Confidential, then declassified. http://www.daytoncodebreakers.org. Accessed 1 January 2010. Report without a title, but with explicit purpose: "A brief history of the art, the objectives and accomplishments of our electrical research laboratory" (p. 1), in particular on research concerning counters and electronical calculators.

Desch, Joseph R. 1942b. *Memo of Present Plans for an Electro-Mechanical Analytical Machine,* 15 September 1942. Propriety of National Cash Register. "Secret" , then declassified. (NARA) Record Group 38, Crane Library, File: CNSG 5750/441; Published by Ralph Erskine, Philip Marks and Frode Weiemd on 29 November 2000, http://cryptocellar.web.cern.ch/cryptocellar/USBombe. Accessed January 1, 2010.

Dewey, John. 1896. The reflex arc concept. *Psychological Review* 3: 357–370.

Douch, E.J.H. 1947. The use of servos in the army during the past war. *Journal of the Institution of Electrical Engineering* Part IIA, 94: 177–189.

Du Bois-Reymond, Emil. 1891 [1880]. Die sieben Welträthsel [Seven riddles of the Universe]. In *Über die Grenzen des Naturerkennens - Die sieben Welträthsel.* Leipzig: Verlag von Veit & Comp.

Dupuy, Jean-Pierre. 1999. *Aux origine des sciences cognitives,* new edn Paris: La Découverte.

EB. 1996. *Encyclopaedia BritannicaCD97,* edition on CD-ROM copyright 1996.

Eckert, J. Presper, et al. 1945. *Description of the ENIAC and comments on electronic digital computing machines.* In J.P. Eckert, J.W. Mauchly, H. Goldstine, and J.G. Brainerd. Restricted, then declassified. Prepared by the Moore School of Electrical Engineering, University of Pennsylvania, NDRC AMP Memo 171.2R, 30 November.

Eckert, J. Presper. 1980. The ENIAC. In Metropolis, Howlett and Rota, 525–539.

Edwards, Paul N. 1997. *The closed world. Computers and the politics of discourse in Cold War America.* Cambridge [MA]: The MIT Press.

Einstein, Albert. 1938. Why do they hate the Jews? A great scientist analyzes the sorrows of his people. *Collier's Weekly,* 26 November: 9–10, 38.

Einstein, Albert. 1963. *Einstein on peace,* ed. Otto Nathan and Heinz Norden, preface by Bertrand Russell. London: Methuen & Co Ltd.

Einstein, Albert. 1998 [1905]. On the motion of small particles suspended in liquids at rest required by the molecular-kinetic theory of heat. In *Einstein's Miraculous Year: Five Papers that Changed the Face of Physics.* Edited and introduced by John Stachel [...]; and with a foreword by Roger Penrose, 85–98. Princeton: Princeton University Press.

Elias, Norbert. 1987 [1983] *Involvement and detachment.* Oxford and New York: Blackwell.

Elias, Peter 1994. *The rise and fall of cybernetics in the US and the USSR.* In Jerison, Singer, and Strook 1997, 21–29.

Eliot, Thomas S. 1930 [1922]. *The waste land.* New York: Horace Liveright.

Eliot, Thomas S. 1964. *Knowledge and experience in the philosophy of F. H. Bradley.* London: Faber and Faber.

Ferrarotti, Franco. 1989. *La sociologia alla riscoperta della qualità.* Rome-Bari: Laterza.

Ferrarotti. Franco. 1977. *Introduzione* to *La sociologia del potere,* ed Franco Ferrarotti. Rome-Bari: Laterza.

Ferry, D.K., and Saeks, R.E. 1985. Comments on [85a], [85b]. In [CW4], 137–140.

Feynman, Richard. 1992 [1985]. *Surely you're joking, Mr. Feynman! Adventure of a curious character, as told to Ralph Leighton.* London: Vintage Books.

Fisch, M.H. 1996. General introduction. In *Classic American Philosophers. Peirce. James. Royce. Santayana. Dewey. Whitehead,* ed M.H. Fisch, 1–40. New York: Fordham University Press.

Fishburn, P., and Monjardet, B. 1992. Norbert Wiener on the theory of measurement (1914, 1915, 1921). *Journal of mathematical psychology* (June) 36-2: 165–184.

Fitzpatrick, Anne. 1999. *Igniting the light elements: The Los Alamos thermonuclear weapon. Project, 1942–1952.* Thesis. LA-13577-T. Issued: July 1999. "Approved for public release; distribution is unlimited."

Flugel, J.C. ed. 1950. *International Congress on Mental Health, London, 1948. International Committee for Mental Hygiene.* London: H. K. Lewis, and New York: Columbia University Press.

Fontana. M. 1991. I 'luoghi alti' della ricerca: un ritratto del MIT. In *Scienza & Tecnica 91/92. Annuario della EST. Enciclopedia della Scienza and della Tecnica.* Milan: Mondadori, 311–325.

Ford, J.J. 1966. Cibernetica sovietica and sviluppo internazionale. In Dechert 1966, 108–127.

Frank, Lawrence K. 1948. Foreword. In *TM* 1948, 189–196.

Frank, Lawrence K. 1948 *Society as the patient.* New Brunswick, N.J.: Rutgers University Press.

Frank, W.J., Cochran, T.B. and Norris, R.S. 1996. The Technology of War. Nuclear weapons. In EB97.

Gale, G. 1997. *The role of Leibniz and Haldane in Wiener's cybernetics.* In Mandrekar and Masani 1997, 247–262.

Galimberti, Umberto. 1992. *Dizionario di psicologia.* Turin: UTET.

Galison, Peter. 1994. The ontology of the enemy: Norbert Wiener and the cybernetic vision. *Critical Inquiry.* 21: 228–265.

Gamow G. and Tuve, M.A. 1938 "The Fourth Washington Conference on Theoretical Physics," March 28, 1938, Office of Public Relations, The George Washington University. http://encyclopedia.gwu.edu. Accessed January 1, 2010.

Gamow, G. and Abelson, P.H. 1946. The ninth Washington Conference on Theoretical Physics. *Science* 104: 574.

Gardner, Howard. 1985. *The mind's new science.* New York: Basic Book.

Gâteaux, René. 1913. Sur les fonctionnelles continues et les fonctionnelles analytiques. *Comptes rendus hebdomadaires des séances de l'Académie des sciences* 157: 325–327.

Gâteaux, René. 1919. Sur la notion d'intégrale dans le domaine fonctionnel et sur la théorie du potentiel. *Bulletin de la Société Mathématique de France* 47: 47–70.

Gâteaux, René. 1922. Sur diverses questions du calcul fonctionnel. *Bulletin de la Société Mathématique de France.* 50: 1–37.

Gerovitch, Slava. 2001. 'Mathematical Machines' of the Cold War: Soviet computing, American cybernetics and ideological disputes in the Early 1950s. *Social Studies of Science* (April) 31-2: 253–287.

Gerovitch, Slava. 2002. *From newspeak to cyberspeak. A history of Soviet cybernetics.* Cambridge [MA]: The MIT Press.

Geyer F., and Van Der Zowen, J. 1994. Norbert Wiener and the social sciences *Kybernetes* 23-6/7: 46–61.

Gladwin, Lee. 2004. Alan M. Turing's contribution to co-operation between the UK and the US. In *Alan Turing. Life and legacy of a great thinker*, ed. Christof Teuscher, 463–474. Berlin and New York: Springer.

Gleick, James. 2011 [1992]. *Genius. The life and science of Richard Feynman.* New York: Integrated media.

Glimm, J., Impagliazzo, J., and Singer, I. eds. 1990. The legacy of John von Neumann. In *"Proceedings of Symposia in Pure Mathematics"* 50. Providence [RI], American Mathematical Society.

Gödel, Kurt. 1967 [1931]. On formally undecidable propositions of Principia mathematica and related systems, In Van Heijenoort, 596–616.

Goldstine, Hermann H. 1980 [1973]. *The computer from Pascal to von Neumann.* Princeton [NJ]: Princeton University Press.

Gomez, Victor. 2016. *Multivariate time series with linear state space structure.* New York, Berlin and Heidelberg: Springer.

Gosling, Francis G. 1999. *The Manhattan project. Making the atomic bomb.* Washington D.C.: U. S. Dept. of Energy; Oak Ridge [TN]: Office of Scientific and Technical Information.

Grattan-Guiness, Ivor. 1975. Wiener on the logics of Russell and Schröder. An account of his doctoral thesis, and his discussion of it with Russell. *Annals of Science* 32: 103–132.

Greenberg, J.R., and Mitchell, S.A. 1983. *Object relations in psychoanalytic theory.* Cambridge [MA]: Harvard University Press.

Grinevald, Jacques. 1988. Sketch for the history of the idea of the biosphere. In *Gaia. The Thesis, the Mechanisms, and the Implications,* ed. P.B. and E. Goldsmith, 1–34. Camelford [UK]: Wadebridge Ecological Centre.

Groves, Leslie R. 1962. *Now it can be told. The story of the Manhattan project.* New York: Harper.

Gunfire Control. Summary Technical Report of the Division 7. Part 1: Gunfire Control. Vannevar Bush, James Conant, and Harold Hazen, Summary Technical Report of the Division 7, NDRC. Vol. 1: Gunfire Control. Washington 1946. Part II: R. B. Blackman, H. W. Bode, and C. E. Shannon, Data Smoothing and Prediction in Fire-Control Systems (= Data Smoothing and Prediction) with affiliation to BTL. Confidential. Declassified number AD200795, DTIC http://www.dtic.mil/dtic/. There are two other volumes in the series: Vol. 2: Optical range finders and Vol. 3: Airborne fire control systems.

Haeckel, Ernst. 1900 [1899]. *Riddle of the universe at the close of the Nineteenth century.* New York and London: Harper & Brothers, 1900.

Hagemeyer, F.W. 1979. *Die Entstehung von Informationskonzepten in der Nachrichtentechnik* [The emergence of the concept of information in communication technology]. Doctoral thesis at Freie Universität Berlin, FB 11: Philosophie und Sozialwissenshaften, November 8.

Haldane, J.B.S. 1924. *Daedalus, or, science and the future,* London: Kegan Paul, Trench, Treubner & Co.

Haldane, J.B.S. 1934. Quantum mechanics as a basis for philosophy. *Philosophy of Science* 1: 78–98.

Hall, Rupert. 1959. Scholar and the craftman in the scientific revolution. In *Critical Problems in the History of Science. Proceedings Institute for the History of Science, University of Wisconsin, 1957,* ed. Marshall Clagett, 3–23. Madison: University of Wisconsin Press.

Hardy, Godfrey H. 1994 [1941]. *A mathematician's apology.* Cambridge [UK]: Cambridge University Press.

Hawkins, David. 1946. *Manhattan district history project Y. The Los Alamos Project.* Vol. I.: *Inception until August 1945.* Los Alamos, New Mexico, Los Alamos Scientific Laboratory of the University of California, 1946. Report written 1946. Distributed December 1, 1961. Source: US Department of Energy. http://www.cfo.doe.gov. Accessed January 1, 2010.

Hazen, Harold. 1934. Design and test of high performance servomechanism, *Journal of the Franklin Institute,* 218: 543–580.

Heims, Steve J. 1975. Encounter of behavioral sciences with new machine-organism analogies in the 1940s. *Journal of the History of the Behavioral Sciences* 11: 368–373.

Heims, Steve J. 1977. Gregory Bateson and the mathematicians: From interdisciplinary interaction to societal functions. *Journal of the History of the Behavioral Sciences*, 13: 141–159.

Heims, Steve J. 1984 [1980]. *John von Neumann and Norbert Wiener. From mathematics to the technologies of life and death.* Cambridge [MA]: The MIT Press.

Heims, Steve J. 1991. *The cybernetics group.* Cambridge [MA]—London [UK]: The MIT Press.

Heims, Steve J. 1993. *Introduction* to [93], IX-XXI.

Heine, C.J.H. [1851] Romanzero. It includes: Hebräische Melodien, Historien and Lamentationen.

Heisenberg, Werner. 1981 [1977]. *Tradition in science.* New York: Continuum.

Hellman, Walter D. 1981. *Norbert Wiener and the growth of negative Feedback in scientific explanation; with a proposed research program of "cybernetic analysis".* Ph.D. Thesis. Oregon State University. Completed December 16, 1981. Commencement June 1982.

Hilbert, David. 1996 [1918]. Axiomatic thought. In *From Kant to Hilbert. A source book in the foundations of mathematics,* ed. William Bragg Ewald, vol. 2, 1105–1115. Oxford: Oxford University Press.

Høffding, H. 1900 [1894–1895]. *A brief history of modern philosophy.* The Macmillan Company.

Høffding, H. 1905 [1903]. *The problems of philosophy.* New York and London: The Macmillan company.

Hölldobler, B., and Wilson, E.O. 1990. *The ants.* Cambridge [MA]: Belknap Press of Harvard University Press.

Holmes, Oliver Wendell. 1858. *The Autocrat of the Breakfast-Table.* Houghton: Mifflin and Company.

Holt, E.B., et al. 1910. *The program and first platform of six realists.* By Edwin B. Holt, Walter T. Marvin, W. P. Montague, Ralph Barton Perry, Walter B. Pitkin, and Edward Gleason Spaulding. *The Journal of Philosophy, Psychology and Scientific Methods* (July) 7-15: 393–401.

Holt, Edwin B., et al. 1912. *The new realism: Cooperative studies in philosophy.* New York: The Macmillan Company.

Hongsen, W. 1996. Norbert Wiener at Qinghua University. *Boston Studies in the Philosophy of Science.* 179: 447–452.

Hook, D.H., Norman, J.M. and Williams, M.R. 2002. *Origins of cyberspace. A library on the history of computing, networking, and telecommunications.* Novato [CA], History of science.com.

Hunsaker, J.C., and von Karman, T. 1939. Report of the Secretaries. In Den Hartog and Peters 1939, XVII–XXII.

Husserl, Edmund. 1970. *Crisis of European sciences and transcendental phenomenology; an introduction to phenomenological philosophy.* Translated, with an introd. by David Carr. Evanston [IL]: Northwestern University Press.

Husserl, Edmund. 1970. *The crisis of European sciences and transcendental phenomenology. An introduction to phenomenological philosophy.* Translated with an introduction by David Carr. Evanston [IL]: Northwestern University Press.

Hutchinson, George Evelyn. 1948. Circular causal systems in ecology. In *TM* 1948, 221–246.

Hutchinson, George Evelyn. 1946. Circular causal mechanisms in ecology. In *Preprinted abstracts.*

Hutchinson, George Evelyn. 1978. *Introduction to population ecology.* New Haven [CT]: Yale University Press.

Hutchinson, George Evelyn. 1993 [1957]. *A treatise on limnology.* New York: Wiley.

Hutton W., and Giddens, A. eds. 2000. *Global capitalism.* New York: The New Press.

Huxley, Thomas Henry in *EB 1996.*

Introductory Statement. 1946. In *Preprinted abstracts.*

Israel, G., and Millan Gasca, A. 2009. *The world as a mathematical game. John von Neumann and twentieth century science.* Basel and Boston: Birkhäuser.

Ito, K. 1976. Commentary on [20f], [21c, d], [23d], [24d]. In [CW1], 513–519.

Jacobs, William. W. 1902. *The monkey's paw.* In W.W. Jacobs, *The lady of the barge.* 27–53. London and New York: Harper & Brothers publishers.

James, William. *Pragmatism. 1907. A new name for some old ways of thinking.* New York, Longman Green and Co, 1907.

James, William 1958 [*1909*]. *A pluralistic universe;* Hibbert Lectures to Manchester College on the present situation in philosophy. New York: Longmans, Green, and Co.

Jaspers, Karl. 1919. *Psychologie der Weltanschauungen* [Psychology of the worldviews]. Berlin: Springer.

Jdanko, A. 1994. Norbert Wiener et l'étude cybernetique de la religion. *Cybernetica.* 3/4: 291–314.

Jerison, D., Singer, I.M. and Strook, D.W. eds. 1997. *The legacy of Norbert Wiener. A centennial symposium.* In Honor of the 100th Anniversary of Norbert Wiener's Birth October 8–14, 1994 Massachusetts Institute of Technology, Cambridge [MA]. Providence [RI]: AMS.

JNMD. 1965. *The Journal of Nervous and Mental Disease.* 140–141. January 1965, serial no. 986. Issue largely dedicated to Wiener.

Jonas, Hans. 1953. A critique of cybernetics. *Social Research* (Summer) 20-2: 172–192.

Jungk, Robert. 1958 [1956]. *Brighter than a thousand suns. A personal history of the atomic scientists.* New York: Harcourt Brace.

Kac, Mark. 1966. Wiener and integration in function spaces. In BAMS, 52–68.

Kant, Immanuel. 2007 [1790]. *Critique of judgement*. Oxford: Oxford University press.

Kay, Lily E. 1997. Cybernetics, information, life. The emergency of scriptural representations of heredity. *Configurations*, 5: 23–91.

Kay, Lily E. 2000. *Who wrote the book of life? A history of the genetic code*. Stanford: Stanford University press.

King, Robert. 1923. Thermionic vacuum tubes and their application. *Bell System Technical Journal*. October. 2–4: 31–100.

Kinsey, A.C., Pomeroy, W.B. and Martin, C.E. 1948. *Sexual Behavior in the human male*. Philadelphia and London: W. B. Saunders Co.

Kleene, Stephen C. 1951. *Representation of events in nerve nets and finite automata*. U.S. Air Force Project Rand, Research Memorandum, MR-704, 15 December 1951. 98 pages.

Kline, Morris. 1972. *Mathematical thought from ancient to modern times*. New York: Oxford University Press.

Kline, Ronald R. 2015. *The cybernetics moment. Or why we call our age the information age*. Baltimore [MD]: Johns Hopkins University Press.

Knapp, V. 1978 [1963]. *L'applicabilità della cibernetica al diritto*, Introduzione di Mario G. Losano. Torino: Einaudi (originally published in Czechoslovakian 1963).

Köhler, W. 1966. *The task of Gestalt psychology*. Princeton: Princeton University Press.

Kolmogorov, Andrey. 1941. Interpolation und Extrapolation von stationären zufälligen Folgen [Interpolation and extrapolation of stationary random sequences]. Bulletin de l'Académie des sciences de U.R.S.S., Series on Mathematics. 5: 3–14 (Russian).

Kosulajeff, P.A. 1941. Sur les problèmes d'interpolation et d'extrapolation des suites stationaires. *Comptes rendus de l'académie des sciences de U.R.S.S.* 30: 13–17 (French).

Kyburg, H.E. 1976. Comments on [14a]. In [CW1], 33.

La Mettrie, Julien Offray de. 1943 [1748]. *Man and machine*, La Salle [IL]: The Open Court Publishing Company.

Landes, David S. 1969. *The Unbound Prometheus. Technological change and industrial development in Western Europe from 1750 to the present*. Cambridge [UK]: Cambridge University Press, 1969.

Henderson, Lawrence J. 1913. *The fitness of the environment. An inquiry into the biological significance of the properties of matter*. In part delivered as lectures in the Lowell Institute, February, 1913. New York: The MacMillan Company.

Lazarsfeld, Paul. 1946. Circular processes in public opinion. In *Preprinted abstracts*.

Le Concept d'Information dans la Science Contemporaine. Colloques Philosophiques Internationaux de Royaumont, July 1962. Paris: Gauthier-Villars, 1965.

Lepschy, Antonio. 1998. Interdisciplinarità and metadisciplinarità dai punti di vista dell'ingegneria dell'informazione and della cibernetica. In *Accademia and interdiplinarità. I: Saggi.* Padova: Accademia Galileiana di Scienze Lettere ed Arti, 169–190.

Lettvin, Jerome Y. 2000. Interview with Jerome Y. Lettvin, 2 June 1994. With new material added in 1997. In Anderson and Rosenfeld eds 2000. *Talking nets. An oral history of neural networks.* Cambridge [MA]. The MIT Press.

Levinson, Norman 1942 *Prediction of Stationary Time Series by a Least Squares Procedure,* Report under a US Army Air Corps Meteorological contract. About March 1942 (Report cited by Bennett 1993, note 51, 184).

Levinson, Norman. 1945. *An Exposition of Wiener's Theory of Prediction.* OSRD 5328, OEMsr-1384, Note 20, AMG-Harvard, June 1945. AMP-13-M21 (Cit. by *Summary Technical Report of the Division 7 vol. 1 Gunfire Control,* 165).

Levinson, Norman. 1966. Wiener's Life. In BAMS 1966, 1–32.

Lewin, Kurt. 1948. *Resolving social conflicts. Selected papers on group dynamics.* New York: Harper & Brothers.

Lewis, W. Bennet. 1948 [1942]. *Electrical counting: With special reference to counting alpha and beta particles,* 2nd ed. Cambridge [UK]: The University press.

Lipset, David. 1982 [1980]. *Gregory Bateson: The legacy of a scientist.* Boston: Beacon Press.

Litvinoff, Barnet. 1988. *The burning bush. Antisemitism and world history.* London: William Collins Sons & C. Ltd..

Livingston, William K. 1946. The vicious circle in causalgia. In *Preprinted abstracts.*

Livingston, William K. 1948. The vicious circle in causalgia. In *TM* 1948, 247–258.

Lo, A.W. 1997. A non-random walk down wall street. In Jerison, Singer, and Strook, 149–183.

Locke, William N., and Booth, A. Donald. eds. 1955. *Machine translation of languages: Fourteen essays.* Cambridge [MA]: Technology Press; and New York: Wiley.

Lorente de Nó, Rafael. 1938. Analysis of the activity of the chains of internuncial neurons. *Journal of Neurophysiology* 1: 207–244.

Losano, M.G. 1978. *Introduzione* to Knapp.

Lotka, A. 1925. *Elements of physical biology.* Baltimore [MD]: Williams and Wilkins.

Lowe, A. s.d. *T. S. Eliot and F. H. Bradley.* S. Pietro in Lariano [Verona].

MacColl, Leroy A. 1945. *Fundamental theory of servomechanisms.* New York: Van Nostrand.

Madge, John. 1962. *The origins of scientific sociology.* New York: The Free Press of Glencoe.

Mahowald, Mary Briody. 1972. *An idealistic pragmatism. The development of the pragmatic element in the philosophy of Josiah Royce.* The Hague: Martinus Nijhoff.

Mandelbrot, Benoît B. 1963. The variation of certain speculative prices. *The Journal of Business* 36-4: 394–419.

Mandelbrot, Benoît B. 1967. How long is the coast of britain? Statistical self-similarity and fractional dimension. *Science.* 3775-156: 636–638.

Mandelbrot, Benoît B. 1975. *Les objets fractals. Forme, hazard et dimension.* Paris: Flammarion.

Mandelbrot, Benoît B. 1983. *The fractal geometry of nature.* San Francisco: W. H. Freeman.

Mandrekar, V., and Masani P. eds. 1997. *Proceedings of Norbert Wiener Centenary Congress Michigan State University November 27–December 3, 1994.* Providence [RI]: AMS.

Mangione C., and Bozzi, S. 1995. *Storia della logica. Da Boole ai nostri giorni.* Milano: Garzanti.

Masani P.R., and Phillips, R.S. 1985. Anti-aircraft fire-control and the emergence of cybernetics. In [CW4], 141–179.

Masani, Pesi R. 1985. Leibnitz, quantum mechanics, and cybernetics. Comments on [32c], [34c], [36g]. In [CW4], 98–104.

Masani, Pesi R. 1990. *Norbert Wiener, 1894-1964.* Basel. Boston and Berlin: Birkhäuser Verlag.

Mason, Samuel J. 1953. Feedback theory. Some properties of signal flow graphs. *Proceeding of the I.R.E.* 41: 1144–1156.

Maurizi, Stefania. 2004. *Una bomba, dieci Storie. Gli Scienziati e l'atomica* [A bomb Ten Stories. Scientists and Atomic]. Milan: Bruno Mondadori.

Maxwell, J. Clerk. 1867–1868. On governors. *Proceedings of the Royal Society* 16: 270–283.

Mayr, Otto. 1970. *The origins of feedback control.* Cambridge [MA]: M.I.T. Press.

McCulloch Summary. McCulloch, Warren *To the members of the conference on teleological mechanisms—Oct. 23 & 24, 1947.* Typewritten summary of the first three Conferences Macy, prepared by Warren McCulloch, distributed to participants to the fourth Macy Conference, held on days 23 to 24 October 1947] In (MCAPS). Series II. Conferences and publications, 1931-1992. Macy Meeting I, VII 1946-; Box 19 - Library of the American Philosophical Society di Philadelphia. I thank the APS and particularly Robert S. Cox for having kindly sent the document photocopies.

McCulloch, W.S., and Pitts, W. 1943. A logical calculus of the ideas immanent in nervous activity. *The Bulletin of Mathematical Biophysics* 5: 115–137.

McCulloch, Warren S. 1941. What is a number, that a man may know it, and a man, that he may know a number? *General Semantics Bulletin* 26 and 27: 7–18.

McCulloch, Warren S. 1948. A recapitulation of the theory, with a forecast of several extensions. In *TM* 1948, 259–277.

McCulloch, Warren S. 1965a. *Norbert Wiener and the art of theory.* In [JNMD 1965], 16.

McCulloch, Warren S. 1974. Recollections of the many sources of cybernetics. *ASC Forum* 6-2: 5–16.

McCulloch, Warren. 1946. Summary of theory and extension to Gestalt psychology. In *Preprinted Abstract.*

McK.Rioch, D. 1948. Intervention by D. McK.Rioch. In *General discussion* following Wiener's report. *TM* 1948, 219–220.

McMillan, B., and Deem, G.S. 1976. The Wiener program in statistical physics. Commentary on [38a], [39b, h], [40d], [43a]. In [CW1], 654-671.

Mead, Margaret. 1968. *Cybernetics of cybernetics*. In *Purpositive systems*, ed. Von Förster et al., 1–11. New York: Spartan Books.

Meetings 1940. Meetings of the Association and its sections. *American Mathematical Monthly*. 46: 502.

Merton, Robert C. 1997. On the role of the Wiener process in finance theory and practice. The case of replicating portfolios. In Jerison, Singer and Strook 1997, 209–221.

Merton, Robert K. 1968. The Matthew effect in science. *Science* (5 January) 159-3810: 56–63.

Metropolis, N., Howlett, J. and Rota, Gian-Carlo. eds. 1980. A history of computing in the twentieth century. In *Proceeding of the International Research Conference on the History of Computing, Los Alamos Scientific Laboratory*, 1976. New York: Academic Press.

Metropolis, N., and Nelson, E.C. 1982. Early computing at Los Alamos. *Annals of the History of Computing* (October) 4-4: 348–357.

Metropolis, Nicholas C. 1987. "An Interview with," OH 135, interview by William Aspray on 29 may 1987, Los Alamos (NM) Charles Babbage Institute. The Center for the History of Information Processing, University of Minnesota, Minneapolis, Charles Babbage Institute. Oral history database http://special.lib. umn.edu/cbi/oh. Accessed January 1, 2010.

Mikulak, M.W. 1966. *Cibernetica and marxismo-leninismo*. In Dechert 1966, 90–107.

Miller, George A. 1951. *Language and communication*. New York: McGraw-Hill.

Miller, Marc. 2007. *Representing the immigrant experience. Morris Rosenfeld and the emergence of Yiddish literature in America*. Syracuse [NY]: Syracuse University Press.

Mindell, D. A. 1996. *Datum for its annihilation. Feedback, control, and computing. 1916-1945*. PhD thesis, MIT Press.

Mindell, D.A. 2002. *Between human and machine. Feedback, control and computing before cybernetics*, Baltimore [MD]: The Johns Hopkins University Press.

Mindell, David A. 1995. Automation's finest hour. Bell Labs and automatic control in World War II. *IEEE Control Systems Magazine* 15-6: 72–80.

Monk, Ray. 1990. *Ludwig Wittgenstein. The duty of genius*. New York: Free Press. Maxwell Macmillan International.

Monod, Jacques. 1971. *Chance and necessity. An essay on the natural philosophy of modern biology*. New York: Alfred A. Knopf.

Montagnini, L., Tabacchi, M.E. and Termini, S. 2016. Out of a creative jumble of ideas in the middle of last Century. Wiener, interdisciplinarity, and all that. *Biophysical Chemistry* 208: 84–91.

Montagnini, Leone, 2015b. The mathematical art of Norbert Wiener. *Lettera Matematica*. International edition. (September) 3-3: 129–134.

Montagnini, Leone. 1900–2000. Bit & Plutonium, inc. Le relazioni tra Norbert Wiener and John von Neumann alle origini della cibernetica. *Atti dell'Istituto Veneto di Scienze Lettere ed Arti* II-158: 361–390.

Montagnini, Leone. 1996. *Norbert Wiener e le scienze sociali*. M. A. Thesis. Supervisor Franco Ferrarotti. Faculty of Sociology. University of Rome "La Sapienza", A.Y. 1994/95, discussed on 23 March 1996.

Montagnini, Leone. 2000. *Cibernetica e guerra fredda* [Cybernetics and the Cold War]. *Ácoma. Rivista Internazionale di Studi Nordamericani* (Spring-Summer) 19: 76–84.

Montagnini, Leone. 2000–2001. Norbert Wiener e le scienze sociali. Il qualitativismo metodologico di un matematico. *Atti dell'Istituto Veneto di Scienze Lettere ed Arti* II-159: 469–501.

Montagnini, Leone. 2001. Il bello della scienza. Aspetti qualitativi nel pensiero matematico. *Oikos* 12: 9–52.

Montagnini, Leone. 2001–2002a. Gli occhiali nuovi di un matematico. Il periodo filosofico di Norbert Wiener. *Atti e memorie dell'Accademia Galileiana di Scienze, Lettere ed Arti già dei Recovrati and Patavina* II: 114: 55–86.

Montagnini, Leone. 2001–2002b. La rivoluzione cibernetica. L'evoluzione delle idee di Norbert Wiener sulla scienza and la tecnica. *Atti and memorie dell'Accademia Galileiana di Scienze, Lettere ed Arti*. II-114: 109–135.

Montagnini, Leone. 2005. *Le Armonie del Disordine. Norbert Wiener matematico-filosofo del Novecento*. Venice [IT]: Istituto Veneto di Scienze, Lettere ed Arti.

Montagnini, Leone. 2007. Looking for "scientific" social science: the Macy Conferences on Cybernetics in Bateson's itinerary. *Kybernetes* 36-7/8: 1012–1021.

Montagnini, Leone. 2008. Philosophical Approaches towards Sciences of Life in Early Cybernetics. In L. M. Ricciardi, A. Buonocore; E. Pirozzi, eds. *Collective Dynamics on Competition and Cooperation in Biosciences. A Selection of Papers in the Proceedings of BIOCOMP2007 International Conference. Vietri Sul Mare, Italy, 24–28 September 2007. AIP Conference Proceedings* 1028: 11–17.

Montagnini, Leone. 2010a. L'interdisciplinarità per Norbert Wiener e per Eduardo Caianiello. In: Greco, P. and Termini, S. eds *Memoria e progetto. Un modello per il Mezzogiorno che serva a tutto il Paese.*, ed. P. Greco, and S. Termini, 47–68. Monte San Pietro (Bologna): GEM.

Montagnini, Leone. 2010b. Identities and Differences. A stimulating aspect of Early Cybernetics. In: *Cybernetics and Systems 2010*, ed. R. Trappl, 157–162. Vienna: Austrian Society for Cybernetic Studies.

Montagnini, Leone. 2012. La Cibernetica alle origini delle Scienze dell'Informazione. Storia e problemi attuali. PhD dissertation in computational and computer science. Dept. Mathematics and Applications, "E. Caccioppoli". University of Naples "Federico II", discussed on 13 February 2012.

Montagnini, Leone. 2013. Interdisciplinary issues in Early Cybernetics. In: Lilia Gurova, László Ropolyi, and Csaba Pléh, eds *New Perspectives on the history of cognitive science*, ed. Lilia Gurova, László Ropolyi, and Csaba Pléh, 81–89. Budapest: Akadémiai Kiadò.

Montagnini, Leone. 2016. Quando Wiener era di casa a Napoli. Come la scienza può volare alto. Il caso del gruppo di Cibernetica di E. R. Caianiello. *Rivista del Centro Studi Città della Scienza*. Naples, 21 marzo 2016.

Montagnini, Leone. 2017a. Leibniz's deep Footprints in Wiener's Scientific Path. In *Leibniz and the Dialogue between Sciences, Philosophy and Engineering, 1646-2016. New Historical and Epistemological Insights*, ed. R. Pisano, M. Fichant, P. Bussotti, A.R.E. Oliveira, 253–284. London: The College Publications.

Montagnini, Leone. 2017b. Interdisciplinarity in Norbert Wiener, a mathematician-philosopher of our time. *Biophysical Chemistry*. Online: 28 June 2017. DOI:10.1016/j.bpc.2017.06.009

Montagnini, Leone. 2017c, W for Wiener. Consistency is no longer a virtue. *Lettera Matematica International edition*. Online: 10 July 2017. DOI:10.1007/s40329-017-0186-0

Montessori, Maria. 1949. *The absorbent mind*. Madras (India): The Theosophical Publishing House.

Moore, George E. 1922. *Philosophical studies*. London: K. Paul, Trench, Trubner & co., ltd. and New York: Harcourt, Brace & co. inc.

Morelli, Marcello. 2001. *Dalle calcolatrici ai computer degli anni Cinquanta. I protagonisti e le macchine della storia dell'informatica*. Milan: Franco Angeli.

Morison, S.E., and Commager, H.S. 1950. *The growth of the American Republic*. New York: Oxford University Press.

Morse, M. 1940. Report of the war preparedness committee. *American Mathematical Monthly* 46: 500–502.

Morse, M., and Hart, W. 1941. Mathematics in the defence program. *American Mathematical Monthly* 48: 293–302.

Nacci, Michela. 1991. *Introduzione* to Haldane J. B. S. and Russell, B. *Dedalo, o La scienza and il futuro / Icaro, o Il futuro della scienza*, VII-XXXIX. Turin: Bollati Boringhieri.

Nagel, Ernest. 1985. Comments on [14c], [14d]. In [CW4], 67.

Narasimhan, T.N. 1999. Fourier's heat conduction equations' history, influence, and connection. *Review of Geophysics* 37, 1: 151–171.

Neher, André. 1987. *Faust et le Maharal de Prague. Le mythe et le réel*. Paris: Presses Universitaires de France.

Nietzsche, Friedrich W. 2005 [1878] *Human, all too human I*. Cambridge [UK]: Cambridge University Press.

Nora, S., and Minc, A. 1978. *L'informatisation de la société*. Paris: La documentation française.

Nyquist, Harry. 1932. Regeneration theory. *Bell System Technical Journal*. 11: 126–147.

Office of Public Relations *News Release*. 1946. *News Release from the George Washington University and the Carnegie Institution of Washington Theoretical Physics meet with biologists for study of living processes*. Washington, D. C., 2 November. Office of Public Relations, The George Washington University. University Archives http://encyclopedia.gwu.edu/gwencyclopedia. Accessed 24 May 2011.

Owens, Larry. 1986. Vannevar bush and the differential analyzer: The text and context of an early computer. *Technology and Culture* January. 27-1: 63–95.

Owens, Larry. 1989. Mathematicians at war: Warren Weaver and the Applied Mathematics Panel, 1942-1945. In Rowe and McCleary 1989, 286–305.

Owens, Larry. 1996. Where are we going, Phil Morse? Changing Agendas and the Rhetoric of Obviousness in the Transformation of Computing at MIT, 1939-1957. *IEEE Annals of the History of Computing archive* (December) 18-4: 34–41.

Park, R.E., and Miller, H.A. 1921. *Old world traits transplanted*. New York and London: Harper & Brothers Publishers.

Peirce, Charles S. 1878. How to make our ideas clear. *Popular Science Monthly* January, 12: 286–302.

Peirce, Charles S. 1891. The architecture of theories. *The Monist*. January 1: 161–176.

Peirce, Charles S. 1892a. The doctrine of necessity examined. *The Monist*. April 2: 321–337.

Peirce, Charles S. 1892b. The law of mind. *The Monist* July 4: 533–559.

Percival, W.S. 1953. The solution of passive electrical networks by means of mathematical trees. In *Proceedings of the IEE*. Part III. 100-65: 143–150.

Percival, W.S. 1955. The graphs of active networks. *Proceedings of the IEE*. Part C: Monographs 102: 270–278.

Perrin, Jean. 1910 [1909]. *Brownian movement and molecular reality*. Annales de Chimie et de Physique. September 1909. London, Taylor and Francis.

Perrin, Jean. 1916 [1913]. *Atoms*. New York: D. Van Nostrand Company.

Perry, Helen Swick. 1982. *Psychiatrist of America*. Cambridge [MA]: Harvard University Press.

Peterson, Erik L. 2010. *Finding Mind, Form, Organism, and Person in a Reductionist Age: The Challenge of Gregory Bateson and C. H. Waddiington to Biological and Anthropological Orthodoxy, 1924–1980*. 2 volumes. A Dissertation Submitted to the Graduate School of the University of Notre Dame in Partial Fulfillment of the Requirements for the Degree of Doctor of Philosophy. Phillip R. Sloan, Director Graduate Program in History and Philosophy of Science Notre Dame, Indiana, April 2010.

Phillips, R.S. 1943. Servomechanism. 11 May 1943, Rad. Lab. Report n. 372.

Piccinini, Gualtiero. 2003. *Computations and computers in the sciences of mind and brain*. PhD Thesis, University of Pittsburgh. Faculty of Arts and Sciences. 20 June 2003.

Pierce, John R. 1961. *An introduction to information theory. Symbols, Signal & Noise*. NewYork, Harper.

Pietro Greco and Ilenia Picardi. 2005. *Hiroshima. La fisica riconosce il peccato.* Roma: Nuova Iniziativa Editoriale, 2005.

Pope, Alexander 1717 [1709]. *An essay on criticism.* In Pope, Alexander. *The works of Mr Alexander Pope*, 75–14. London: W. Bowyer, for Bernard Lintot between the Temple-Gates.

Popper, Karl. 1972, *Objective knowledge: An evolutionary approach.* Oxford: Oxford University Press.

Porter, Arthur. 1965. The servo panel. A unique contribution to control-systems engineering. *Electronics & Power* (October) 11-10: 330–333.

Poundstone, W. 1993. *Prisoner's dilemma.* Oxford: Oxford University Press.

Pratt, V. 1987. *Thinking machines. The evolution of artificial intelligence.* Oxford: Basil Blackwell.

Preprinted *abstracts.* 1946. The New York Academy of Sciences, *Preprinted Abstracts. Conference on "Teleological Mechanisms".* Monday, October 21 and Tuesday, October 22, 1946, typescript, in Box 0-14 of Margaret Mead Papers in the Division Manuscripts at the Library of Congress, Washington, D.C. I thank the Library of Congress and particularly Bruce Kirby for kindly sending me copies of these documents.

Quintanilla, S. 2002. Arturo Rosenblueth y Norbert Wiener: dos científicos en la historiografía de la educacíon contemporanea. *Revista Mexicana de Investigación Educativa.* May-August, 7–15: 303–329.

Randell, Brian ed. 1982a. *The origins of digital computers. Selected papers*, 3. Ed. Berlin, Heidelberg, New York: Springer.

Randell, Brian. 1982b. From analytical engine to electronic digital computer: The contributions of Ludgate, Torres, and Bush. *Annals of the History of Computing.* October 4-4: 327–341.

Randell, Brian. 1985. Comments on [85a], [85b]. In [CW4], 135–136.

Rapoport, A., and Landau, H.G. 1951. Mathematical Biology. *Science* 27 July.

Rapoport, Anatol. 1965. Norbert Wiener as the prophet of the cybernetic revolution, in [JNMD 1965], 8–9.

Rashevsky, Nicholas. 1933. Outline of a physico-mathematical theory of excitation and inhibition. *Protoplasma* October. 20-1, 42–56.

Rashevsky, Nicholas. 1934. Foundations of mathematical biophysics. *Philosophy of Science*, 1-2: 176–196.

Rashevsky, Nicholas. 1936, Mathematical biophysics and psychology. *Psychometrika*, 1: 1–26.

Rashevsky, Nicholas. 1938. *Mathematical biophysics. Physico-mathematical foundations of biology.* Chicago: University of Chicago Press.

Restaino, F. 1978. Gli inizi della filosofia analitica inglese: Moore, Russell and Wittgenstein. In *Storia della filosofia*, vol 10, 81–106. Milan: Vallardi.

Rhodes, Richard, 1986. *The making of the atomic bomb.* New York: Simon & Schuster.

Ribot, Théodule Armand. 1882 [1881]. The diseases of memory. New York: New York, D. Appleton and Company.

Ribot, Théodule Armand. 1906 [1900]. *Essay on the creative imagination.* Chicago: The Open Court publishing company.

Richardson, Lewis Fry. 1960. *Arms and insecurity. A mathematical study of the causesand origins of war.* London: Stevens.

Rockefeller Foundation. 1946. Massachussetts Institute of Technology Electronic Computation. In *The Rockefeller Foundation Annual Report 1946*, New York, 168–169.

Rockefeller Foundation. 1947. Massachussetts Institute of Technology. Mathematical Biology. In *The Rockefeller Foundation Annual Report 1947*, New York, 111–112.

Rockefeller Foundation. 1947. The Cross-Breeding of Biology. In *Annual Report 1947.* New York, 31–34.

Rojas, R., and Hashagen, U. eds. 2000. *THE FIRST computers. History and architectures*, ed. Raul Rojas and Ulf Hashagen. Cambridge: MIT Press.

Rosenblith, W., and Wiesner, J. 1965. From philosophy to mathematics to biology. In [JNMD 1965], 3–8.

Rosenblueth A., and Cannon, W.B. 1942. Cortical responses to electric stimulation. *American Journal of Physiology* 31 January 135: 690–741.

Rosenblueth, A., Bond, D.D. and Cannon, W.B. 1942. The control of clonic responses of the cerebral cortex. *American Journal of Physiology* 1 November 137: 681–694.

Rosenblueth, Arturo. 1946. The control of movements of animals organisms. In *Preprinted abstracts.*

Rosenblueth, Arturo. 1950. *The transmission of nerve impulses at neuroeffector junctions and peripheral synapses.* Cambridge [MA]: The MIT press and New York: Wiley and Sons.

Rosenblueth, Arturo. 1970. *Mind and brain. A philosophy of science.* Cambridge [MA]: The MIT Press.

Rosenfeld, Morris. 1897. *Lieder-buch* [Lider Bukh]. New York: s.n.

Rosenfeld, Morris. 1898. *Songs from the Ghetto.* With prose translation, glossary, and introduction, ed. Leo Wiener. Boston: Copeland and Day.

Rosenfeld, Morris. s.d. *Encyclopedia Judaica*, vol. 14, col. 286. Jerusalem: Keter Publishing House.

Rossi, Paolo. 1970 [1962]. *Philosophy, technology, and the arts in the early modern era.* New York: Harper & Row.

Rossi, S. 1971. *Evoluzione dei calcolatori elettronici. Natura and prospettive dell'informatica.* Milan: Hoepli.

Routh, Edward John. 1877. *A treatise on the stability of a given state of motion, particularly steady motion.* London: Macmillan and Co.

Rowe, D., and McCleary, J. 1989. The history of modern mathematics. Vol. II: Institutions and applications. In *Proceeding of the Symposium on the History of Modern Mathematics*, Vassar College, Poughkeepsie, New York, June 20–24, 1988, ed. D. Rowe and J. McCleary. Boston [...]: Academic Press.

Rowe, D. 1986. "Jewish mathematics" at Göttingen in the era of Felix Klein. *Isis* 77: 422–449.

Rowe, D. 1989. Klein, Hilbert, and the Göttingen mathematical tradition. *Osiris* S 2. 5: 186–213.

Royce, Josiah. 1885. *The religious aspect of philosophy; a critique of the bases of conduct and of faith*. Boston and New York: Houghton, Mifflin and company.

Royce, Josiah. 1905. *The relations of the principles of logic to the foundations of geometry*. In Royce, Josiah. *Logical essays. Collected logical essays of Josiah Royce*, 379–441, ed. Daniel S. Robinson. Dunbuque [IA]: W.C. Brown, 1951.

Royce, Josiah. 1913. *The problem of Christianity*. Lectures delivered at the Lowell institute in Boston, and at Manchester College. New York: Macmillan.

Royce, Josiah. 1914. The mechanical, the historical and the statistical. *Science* (April) 29: 551–566.

Royce, Josiah. 1923 [1901]. *The world and the individual*. Second Series. Nature, Man and the Moral order. New York and London: The Macmillan Company.

Royce, Josiah. 1959 [1899]. Supplementary essay. The one, the many, and the Infinite. In Royce 1959 [1899], 471–588.

Royce, Josiah. 1959 [1899]. *The world and the individual*. First Series. The Four historical conception of God. New York: Dover publication.

Royce, Josiah. 1963. *Josiah Royce's seminar, 1913–1914: as recorded in the notebooks of Harry T. Costello, ed. by Grover Smith*. With an essay on the philosophy of Royce by Richard Hocking. New Brunswick [NJ]: Rutgers University Press.

Rubin, M.D. 1968. History of technological feedback. In *Positive feedback. A general systems approach to positive/negative feedback and mutual causality*, ed J.H. Milsum, 9–22. Oxford [...]: Pergamon Press.

Russell, Bertrand. 1900. *The philosophy of Leibniz*. London: Allen & Unwin.

Russell, Bertrand. 1912. *The problems of philosophy*. New York: H. Holt and Company.

Russell, Bertrand. 1914. *Our knowledge of the external world as a field for scientific method in philosophy*. Chicago: The Open Court Publishing Co.

Russell, Bertrand. 1918–1919. *The philosophy of logical atomism*. Monist 28: 495–527; 29: 32–63, 190–222, 345–380.

Russell, Bertrand. 1967–1969. *The autobiography*. [v. 1: 1872–1914; v. 2: 1914–1944; v. 3. 1944–1967]. London: George Allen and Unwin.

Russell, Bertrand. 2009. *My mental development*. In *The Basic Writings of Bertrand Russell*, ed Robert E. Egner and Lester E. Denonn, London and New York: Routledge.

Sacerdote, Gino. 1953. *Prefazione* to Wiener, Norbert, *Introduzione alla cibernetica* (tr. it. of [50j]). Turin: Edizioni Scientifiche Einaudi.

Santayana, George. 1896. *The sense of beauty; being the outlines of aesthetic theory*. New York: C. Scribner's sons.

Santayana, George. 1900 *Interpretations of poetry and religion*. New York: C. Scribner's Sons.

Santayana, George. 1905–1922. *The life of reason; or, The phases of human progress*. [v. 1: Introduction, and Reason in common sense; v. 2: Reason in

society; v. 3: Reason in religion; v. 4: Reason in art. v. 5: Reason in science].
New York: C. Scribner's sons, and London: Archibald Constable & Co.

Santayana, George. 1913a. *The intellectual temper of the age.* In Santayana 1913c: 1–24.

Santayana, George. 1913b. *The philosophy of Henri Bergson.* In Santayana 1913c: 58–109.

Santayana, Georgec. 1913. *Winds of doctrine. Studies in contemporary opinion,* London: J. M. Dent & Sons ltd.; New York: Charles Scriber & Sons.

Segal, Jérôme. 1998 *Théorie de l'information. Sciences, techniques et société de la seconde guerre mondiale à l'aube du XXIe siècle.* Doctoral thesis presented at the Université Lumière Lyon 2, 1998 http://theses.univ-lyon2.fr/Theses/jsegal/tdm.html. Accessed in 2001.

Schwartz, Stephen P. 2012. *A Brief History of Analytic Philosophy: From Russell to Rawls.* Chichester: Wiley-Blackwell.

Segal, Jérôme. 2003. *Le Zéro et le Un. Histoire de la notion scientifique d'information au 20ᵉ siècle.* Paris: Syllepse.

Shannon, C.E., and Weaver, W. 1963 [1949]. *The mathematical theory of communication.* Urbana and Chicago: University of Illinois Press, Urbana, Illinois.

Shannon, C.E., and McCarthy, J. 1956. *Automata studies.* Princeton [NJ]: Princeton University Press.

Shannon, Claude E. 1938. A symbolic analysis of relay and switching circuits. *Transactions of the American Institute of Electrical Engineers* 57: 713–723.

Shannon, Claude E. 1940. An algebra for theoretical genetics. PhD thesis. Massachusetts Institute of Technology, Department of Mathematics, April 15, 1940. http://dspace.mit.edu. Accessed in 2010.

Shannon, Claude E. 1941. Mathematical theory of the differential analyzer. *Journal of Mathematics and Physics* 20: 337–354.

Shannon, Claude E. 1945. *A mathematical theory of cryptography,* 1 September 1945. Memorandum MM 45-110-02 written by the Signal Security Agency. "Declassified" in 1957. A copy is held at the MIT Library.

Shannon, Claude E. 1948. A mathematical theory of communication. *The Bell System Technical Journal* 27-3: 379–423 and 27-4: 623–656.

Skaff, W. 1986 *The philosophy of T.S. Eliot. From skepticism to a surrealist poetic (1909-1927).* Philadelphia: University of Pennsylvania Press.

Slack, Nancy G. 2010. *G. Evelyn Hutchinson and the invention of modern ecology.* Foreword by Edward O. Wilson. New Haven [CT] and London: Yale University Press.

Sloane, N.J.A., and Wyner, A.D. 1993. *Claude Elwood Shannon,* in Shannon, C. E. 1993 *Collected Papers,* New York: IEEE Press.

Smalheiser, Neil R. 2000. Walter Pitts. *Perspectives in Biology and Medicine* (Winter) 43-2: 217–226.

Smith, Grover. 1963. *Editor's Introduction* to Royce 1963, 1–15.

Snell, J. Laurie. 1997. A conversation with Joe Doob. *Statistical Science* 12-4: 301–311.

Spinoza, Benedictus. 1883 [1677]. *Ethic demonstrated in geometrical order and divided into five parts*. New York: Macmillan & co.

Stapp, H.P. 1997. Quantum mechanical coherence, resonance, and mind. In Mandrekar and Masani, 263–300.

Stern, Nancy. 1981. *From ENIAC to UNIVAC. An appraisal of the Eckert-Mauchly computers*. Bedford [MA]: Digital Press.

Stewart, Irvin. 1948. *Organizing Scientific Research for War. The administrative history of the Office of Scientific Research and Development*. Boston: Little, Brown and Company.

Stibitz, George R. 1945a. *Relay Computers*. Restricted. NDRC-RC AMP, Memo 171.1 R, February 1945, v. 70; detailed summary in *Mathematical Tables and Other Aids to Computation* 2–20 (October 1947): 364–365.

Stibitz, George R. 1945b. A talk on relay computers. NDRC-RC: AMP. Memo 171.1 M 1945.

Stibitz, George R. 1980. Early computers. In Metropolis, Howlett and Rota, 479–483.

Stibitz, George R. 1986 "Curriculum Vitae," updated 1986, available at http://stibitz.denison.edu/info.html. Accessed in 2010.

Struik, Dirk. 1997. *Reminiscences of Norbert Wiener*. In Jerison, Singer, and Strook 1997, 31–32.

Sullivan, Harry Stack. 1964. *The fusion of psychiatry and social science*. With introd. and commentaries by Helen Swick Perry. New York: Norton.

Summary Technical Report of the Division 7. Vol. 1 Gunfire Control: Vannevar Bush, James Conant, and Harold Hazen, *Summary Technical Report of the Division 7, NDRC*. Vol. 1: *Gunfire Control*. Washington 1946. In two parts: Part I: *Gunfire Controll*, containing the description of the various projects during the 5 war years. Part II: *Data Smoothing and Prediction*, containing R. B. Blackman, H. W. Bode, and C. E. Shannon, *Data Smoothing and Prediction in Fire-Control Systems* written with the affiliation of BTL. Unclassified number AD200795, DTIC http://www.dtic.mil/dtic/.

Talcott Parsons. 1966. *Societies. evolutionary and comparative perspectives*. Englewood Cliffs [NJ]: Practice-Hall.

Telecommunications Systems: Network Milestones. in *EB* 1996.

Termini, Settimo, ed. 2006. *Imagination and Rigor. Essays on Eduardo R. Caianiello's Scientific Heritage*. Milan: Springer.

Termini, Settimo. 2006. Remarks on the development of Cybernetics. *Scientiae Mathematicae Japonicae* 64-2: 461–468.

TM. 1948. Teleological mechanisms. In *Proceedings of the "Conference on Teleological Mechanisms"*, 21 and 22 October 1946. *Annals of the New York Academy of Sciences*. 50: 259–277.

Tolle, M. 1905. *Die Regelung der Kraftmaschinen* [Control of Power Machines]. Berlin: Springer.

Tolstoy, Leo. 1904 [1899]. *Resurrection*. In *The complete works of Count Tolstoy*, vol. 21.

Tolstoy, Leo. 1904-1905. *The complete works of Count Tolstoy*. Translated from the original Russian and edited by Leo Wiener [24 vols]. Boston: D. Estes & company.

Tolstoy, Leo. 1905 [1893]. *The kingdom of god is within you*. In *The complete works of Count Tolstoy*, vol. 20, 1–380.

Troland, Leonard T. 1914. Adaptation and the chemical theory of sensory response. *The American Journal of Psychology* 25: 500–527.

Troland, Leonard T. 1916. Philosophy and the world peace. *The Journal of Philosophy Psychology and Scientific Methods* (3 August) XIII-16.

Troland, Leonard T. 1914. The chemical origin and regulation of life. *The Monist* 24: 92–133.

Tropp, Henry S. 1980. The Smithsonian Computer History Project and some personal recollections. In *Metropolis, Howlett and Rota*, 114–122.

Tudico, Christopher. 2012. *The History of the Josiah Macy Jr. Foundation*; New York: Josiah Macy Jr. Foundation.

Tukey, John W. 1952. Review of: Wiener, Norbert. 1949 [1942]. *The extrapolation, interpolation and smoothing of stationary time series with engineering applications*. In *Journal of the American Statistical Association*. (June) 47–258 (June): 319–321.

Turing, Alan. 1936. On computable numbers, with an application to the entscheidungs problem. In *Proceedings of the London Mathematical Society*. 42-2: 230–265.

Tustin, Arnold. 1947b. A method of analysing the behaviour of linear systems in terms of time series. *Journal of the Institution of Electrical Engineers*. Part IIA. Automatic Regulators and Servo Mechanisms. 94-1: 130–142.

Tustin, Arnold. 1947a. The nature of the operator's response in manual control, and its implications for controller design. *Journal of the Institution of Electrical Engineers*—Part IIA: Automatic Regulators and Servo Mechanisms. 94-2: 190–206.

Ulam, Stanislaw M. 1980. Von Neumann. The interaction of mathematics and computing. In Metropolis, Howlett and Rota, 93–99.

Ulam, Stanislaw M. 1969(?). The role of the Los Alamos Laboratory work in the history of the modern computing machines. [Cit. by Aspray 1990a, note 21, p. 260. Paper in (SUAP)].

Uttley, J.A. 1944. "The human operator as an intermittent servo", in report of the 5th meeting of the Manual Tracking Panel of the Servo-Panel, 17/8/1944.

Van Heijenoort, Jean, ed. 1967. *From Frege to Gödel: A source book in Mathematical Logic (1979-1931)*. Cambridge [MA]: Harvard University Press.

Verdict 1973. US District Court of Minnesota Fourth Division Civil Action, File No. 4-67 CIV. 138 Honeywell Inc., Plaintiff, vs. Sperry Rand Corporation and Illinois Scientific Developments, Inc., Defendants, Findings of Fact, Conclusions of Law and Order For Judgment. Verdict trial Honeywell *vs* Sperry Rand, October 1973. Text available at www.ushistory.org/more/eniac/index.htm. Accessed in 2010.

Vernadsky, Vladimir I. 1944. *Problems of biogeochemistry*, ed. G. Evelyn Hutchinson. New Haven [CT]: Connecticut academy of arts and sciences.

Vishniac, Roman 1983. *A vanished world*, with a foreword by Elie Wiesel. New York: Farrar, Straus, and Giroux.

Volterra, Vito. 1901. Sui tentativi di applicazione delle matematiche alle scienze biologiche e sociali [On attempts to apply mathematics to the biological and social sciences]. *Giornale Degli Economisti e Annali di Economia*. (November): 436–458.

Volterra, Vito. 1926a. Variazioni and fluttuazioni del numero d'individui in specie animali conviventi. *Memorie della Regia Accademia Nazionale dei Lincei* 2: 31–113.

Volterra, Vito. 1926b. Fluctuations in the abundance of a species considered mathematically. *Nature* 118: 558–560.

Volterra, Vito. 1931. *Leçons sur la théorie mathématique de la lutte pour la vie*. Paris: Gauthier-Villars.

Von Förster, Heinz. 1968. *Purposive systems*, ed. H. Von Förster et al. New York: Spartan Books.

Von Förster, Heinz. 1991. *Cibernetica ed epistemologia: storia e prospettive*, in [Bocchi-Ceruti 1991], 112–140.

Von Förster, H., Mead, M. and Teuber, H.L. eds. 1950. Cybernetics. Circular causal and feedback mechanisms in biological and social systems. In *Transactions of 6th Conference* [New York] March 24–25, 1949. New York: Josiah Macy, Jr. Foundation.

Von Förster, H., Mead, M. and Teuber, H.L. eds. 1951. Cybernetics. Circular causal and feedback mechanisms in biological and social systems.In *Transactions of 7th Conference* [New York] March 23–24, 1950. New York: Josiah Macy, Jr. Foundation.

Von Förster, H., Mead, M. and Teuber, H.L. eds. 1952. Cybernetics. Circular causal and feedback mechanisms in biological and social systems. In *Transactions of the 8th Conference* [New York] March 15–16, 1951. New York: Josiah Macy, Jr. Foundation.

Von Förster, H., Mead, M. and Teuber, H.L. eds. 1953. Cybernetics. Circular causal and feedback mechanisms in biological and social systems. *Transactions of the 9th Conference* [New York] March 20–21, 1952. New York: Josiah Macy, Jr. Foundation.

Von Förster, H., Mead, M. and Teuber, H.L. eds. 1955. Cybernetics. Circular causal and feedback mechanisms in biological and social systems. *Transactions of the 10th Conference* [Princeton, NJ], April 22, 23 and 24, 1953. New York: Josiah Macy, Jr. Foundation.

Von Förster, Heinz. 1991. Cibernetica ed epistemologia: storia e prospettive. In *La sfida della complessità*, ed. G. Bocchi and M. Ceruti, 112–140. Milan: Feltrinelli.

Von Neumann, J. and Goldstine, H.H. 1961. On the principles of large scale computing machines. In Von Neumann, J. *Collected Works*, vol. V, 1–32. London: Pergamon Press.

Von Neumann, J. and Morgenstern, O. 1953 [1944]. *Theory of games and economic behavior*. Princeton [NJ]: Princeton University Press.

Von Neumann, J. and Richtmyer, R.D. 1950. A method for the numerical calculation of hydrodynamic shocks. *Journal of Applied Physics* 21: 232–237.

Von Neumann, John. 1932a. Proof of the quasi-ergodic hypothesis. *Proceedings National Academy of Science* 18: 70–82.

Von Neumann, John. 1932b. Physical applications of the ergodic hypothesis. *Proceedings National Academy of Science* 18: 263–266.

Von Neumann, John. 1944. Report by John von Neumann to J. Robert Oppenheimer, NDRC, 1 August 1944, 8 pages report, typescript on letterhead NDRC, without title, but we read at the very beginning: "The object of this letter is to report on the calculating machine". www.lanl.gov/history. Accessed July 21, 2010.

Von Neumann, John. 1945. *First draft of a report on the EDVAC*. Contract No. W-670-ORD-4926 between the United States Army Ordnance Department and the University of Pennsylvania. Moore School of Electrical Engineering, University of Pennsylvania, 30 June 1945. Mimeographed paper, 101 pages, printed only on one side. PDF scan of the original document in the Smithsonian Institution owned by the Smithsonian Institution Libraries. https://archive.org/details/firstdraftofrepo00vonn. Accessed March 27, 2017.

Von Neumann, John. 1949. *Governed*. Review of Wiener, Norbert *Cybernetics*. In *Physics Today* (May) 2–5: 33–34.

Von Neumann, John. 1951. The general and logical theory of automata. In *Cerebral Mechanisms in Behavior. The Hixon Symposium*, ed. Lloyd A. Jeffress, 1–31. New York and London: Wiley and Sons Chapman & Hall Jeffress.

Von Neumann, John. 1955 [1932c]. *Mathematical foundations of quantum mechanics*. Princeton [NJ]: Princeton University Press.

Von Neumann, John. 1956. Probabilistic logics and the synthesis of reliable organisms from unreliable components. In *Conference held at the California Institute of Technology in January 1952*. In Shannon and McCarthy 1956, 43–98.

Von Neumann, John. 1963. Refraction, intersection, and reflection of shock waves. In Von Neumann 1963, 300–308.

Von Neumann, John. 1963. *Use of variational methods in hydrodynamics*. Memorandum from J. von Neumann to O. Veblen, 26 March 1945. In von Neumann, John. 1963, *Collected works*, vol. VI, 1–32. London: Pergamon Press, 357–359.

Von Neumann, John. 1966. *Theory of self-reproducing automata*, edited and completed by Arthur W. Burks. Urbana [IL]: University of Illinois Press.

Von Neumann, John. 1967 [1922]. An axiomatisation of set theory. In *From Frege to Gödel: A source Book in Mathematical Logic (1979-1931)*, ed. Jean van Heijenoort, 393–413.

Von Neumann, John. 1981 [1946]. The principles of large-scale computing machines. *Annals of the History of Computing*, vol. 3, No. 3, July 1981

[Conference held at the meeting of the Mathematical Computing Advisory Panel of the Office of Research and Inventions of the Navy, Department in Washington, D.C., 15 May, 1946].

Von Neumann, John. 1986 [1958]. *The computer and the brain.* New Haven [CT]: Yale University Press.

Von Neumann, John. 2005. *Selected letters,* ed. Miklos Redei; Providence [RI]: AMS; London: London Mathematical Society.

War Preparedness Committee. 1940. War Preparedness Committee of the AMS and MAA. *American Mathematical Monthly* 47: 500–502.

Weaver, Warren 1963 [1949]. *Recent Contribution to the Mathematical Theory of Communication. Introductory Note on the General Setting of the Analytical Communication Studies.* In Shannon and Weaver 1963 [1949]. [Curiously, the first title is given by the book as the title of the Weaver contribution. After a blank page a second title follows showing the more appropriate nature of the text as introduction to the Shannon contribution, reproducing Shannon 1948].

Weaver, Warren. 1946. Foreword to summary technical report of the applied mathematics panel, vol. 3. VII–VIII.

Weaver, Warren. 1949, "Translation", 15 July 1949. In *Machine translation of languages: fourteen essays,* ed. William N. Locke and A. Donald Booth, 15–23. Cambridge [MA]: The Technology press, and New York: Wiley.

Werskey, Gary. 1988 [1978]. *The visible college. A collective biography of British scientists and socialists of the 1930s*; foreword by Robert M. Young. London: Free Association Books.

Whitehead, A.N., and Russell, B. 1910–1913. *Principia Mathematica.* 3 vols. Cambridge [UK]: Cambridge University Press.

Wiener, Leo. 1898. The popular poetry of the Russian Jews. *Americana Germanica,* 2–2.

Wiener, Leo. 1899. *The history of Yiddish literature in the nineteenth century.* New York: C. Scribner's Sons.

Wiener, Leo. 1902–1903. *Anthology of Russian literature from the earliest period to the present time,* ed. Leo Wiener. New York and London: G.P. Putnam's Sons.

Wiener, Leo. 1915. *Commentary to the Germanic laws and mediaeval documents.* Cambridge [MA]: Harvard University Press; London: Humphrey Milford Oxford University Press.

Wiener, Leo. 1917. *Contributions toward a history of Arabico-Gothic culture.* New York: The Neale publishing company.

Wiener, Leo. 1922. *Africa and the discovery of America.* Philadelphia: Innes and Sons.

Wiener, Leo. s.d. *Encyclopedia Judaica.* Jerusalem: Keter Publishing House. vol. 16, Col. 499.

Wilder, R.L. 1985. Comments on [29h]. In [CW4], 860.

Wildes, K.L., and A. Lindgren, Nilo. 1985. *A century of electrical engineering and computer science at MIT, 1882-1982.* Cambridge [MA]: The MIT Press.

Williams, Michael R. 1997. *History of computing technology*, 2nd ed. Los Alamitos [CA]: IEEE Computer Society Press.

Wilson, Edward O. 1971. *The insect societies.* Cambridge [MA]: Belknap Press of Harvard University Press.

Wilson, Edward O. 1975. *Sociobiology. The new synthesis.* Cambridge [MA]: Harvard University Press.

Witte, Karl H.G. 1914 [1819]. *The education of Karl Witte; or, The training of the child,* with an introduction by H. Addington Bruce; tr. from the German by Leo Wiener. New York: Thomas Y. Crowell Company.

Wittgenstein, Ludwig 1963 [1921]. *Tractatus logico-philosophicus.* London: Routledge & Kegan Paul; New York: The Humanities Press.

Woodger, Joseph Henry. 1937. *The axiomatic method in biology.* With appendices by Alfred Tarski and W.F. Floyd. Cambridge [UK]: The University Press.

Yockey, H.P., Platzman, R.L. and Quastler, H. eds. 1958. *Symposium on information theory in biology.* Gatlinburg [TN], October 29–31, 1956, New York [...]: Pergamon Press.

Young, David M. Jr. 1950. *Iterative methods for solving partial difference equations of elliptic type,* PhD thesis. Department of Mathematics, Harvard University, 1 May.

Young, David M. Jr. 1987. A historical review of iterative methods. In *HSNC '87.* In *Proceedings of the ACM Conference on History of Scientific and Numeric Computation,* ed. G.E. Crane, 117–124. New York: ACM Press.

Zellini, Paolo. 1996. Il calcolo e la scoperta della complessità. In SISSA. *Caos and complessità.* Napoli: Cuen.

Zilsel, Edgar. 1942. The sociological roots of science. *American Journal of Sociology* (January) 47-4: 544–562.

Index

A

Abel, Niels Henrik, 75
Abelson, P.H., 202
Abraham, Tara H., 135, 137, 202
Agar, J, 103
Aiken, Howard H, 144, 145, 149, 150, 161–163, 165, 203
Akera, A., 103
Aldrich, John, 52
Allen, G. E., 84
Archibald, R. C., 234
Archimedes of Syracuse, 99
Aristotle, 41
Aspray, William, 86, 142, 144, 145, 146, 149, 154, 156, 159, 162, 166, 197, 216, 250, 251
Atanasoff, John Vincent, 151
Atzema, Eisso J., 39

B

Babbitt, Irving, 33
Baker, H. F., 48
Banach, Stefan, 51, 52, 65
Barnett, Isaac Albert, 52
Bates, J. A. V., 125
Bateson, Gregory, 125, 185, 186, 189, 193–195, 198, 223, 225, 235, 236, 238
Behrend, B. A., 250
Bell, Alexander Graham, 50
Bennett, Stuart, 114, 119, 120, 125, 126, 129, 132, 134
Bergmann, Peter G., 134
Bergson, Henri, 12, 13, 17, 18, 24, 30, 31, 33, 35, 41, 42, 209, 210, 214, 217
Berkeley, Edmund C., 143
Berkeley, George, 19, 75
Bernal, John Desmond, 94, 195, 219
Bernard, Claude, 22, 84
Bernstein, Richard J., 8, 12, 29, 48

Berry, Clifford E., 151
Bigelow, Julian, 114, 125, 126, 128–131, 137, 150, 158, 159, 166, 180, 184, 207, 210
Birkhoff, George David, 37, 38, 72, 86, 121
Blackman, R. B., 124
Blaschke, Wilhelm, 93
Bode, Hendrik Wade, 118, 124, 126
Boltzmann, Ludwig, 58, 72, 209
Bolyai, János, 75
Bond, D. D., 181
Born, Max, 70, 71, 85
Bosco, N., 17
Bottazzini, Umberto, 87
Boulton, Marjorie, 4
Boutroux Émile, 12
Boyce, J.C., 131, 134
Bozzi, S., 34
Bradley, Francis Herbert, 19, 23, 24, 30, 33, 42, 61
Brainerd, John G., 151, 152, 154, 156, 157
Brandt, F., 12
Bremer, Frédéric (1892–1982), 192
Bridgman, Percy Williams, 21, 22, 87, 89, 190
Brouwer, L. E. J., 85, 90, 199
Brown, Gordon, 131
Brown, Robert, 52, 142
Buchanan, R. Angus, 96
Bulmer, Martin, 187
Burks, Arthur W., 152, 154
Bush, Vannevar, 68, 69, 103, 105, 107, 132, 133, 146, 151–154, 158, 228, 248

C

Caianiello, Eduardo Renato, 251
Caldwell, Samuel H., 108, 113, 115, 152, 154, 179, 182–184
Calimani, Dario, 56
Calkin, John Williams, 144
Campbell, Donald T., 65

© Springer International Publishing AG 2017
L. Montagnini, *Harmonies of Disorder*, Springer Biographies,
DOI 10.1007/978-3-319-50657-9

Cannon, Walter Bradford, 17, 82, 84, 126, 140,
     168, 172, 181, 188
Cantor, Georg, 20, 29, 38, 61, 190
Carnap, Rudolf, 85, 137
Carroll, Lewis, 245
Cauchy, Augustin-Louis, 48, 75, 118
Ceruzzi, Paul E., 103, 152
Chafee, Emory L., 144
Chandrasekhar, Subrahmanyan, 167, 170
Christopherson, Derman, 105
Cini, Marcello, 55, 56
Cochran, T B., 141, 142, 169
Cockcroft, John, 81, 96, 97
Columbus, Christopher, 25
Compton, Arthur H., 141, 175, 184
Comrie, Leslie J., 145
Conant, James B., 133
Continenza, Barbara, 211
Conway, F., 251
Cooper, Steven J., 84
Copleston, Frederick, 11, 15
Costello, Harry T., 57
Courant, Richard, 70, 71
Cousins, Norman, 233
Craig, Wallace, 40
Craik, Kenneth James William, 125
Crawford, Perry, 152
Creager, A., 202
Cull, Paul, 137
Cunliffe, Marcus, 50
Cunningham, Leland E., 162, 165, 167
Cushing, Harvey Williams, 79
Cushman, Herbert Ernest, 11

D
Daedalus, 77, 95, 97
Daniell, Percy John, 52, 53
Darwin, Charles, 10, 13, 81, 91
Dean, Gordon, 233
De Broglie, Louis, 81
Deem, G. S., 60
Delbrück, Max, 201
Della Riccia, Giacomo, 251
De Luca, Aldo, 193, 211
Deming, W. Edwards, 162, 163, 165, 167
Den Hartog, J. P., 98, 105
De Ruggiero, Guido, 19
Descartes, René, 213, 218, 242
Desch, Joseph R., 108, 154
Dewey, John, 8, 13, 24, 35, 37, 187, 188
Doob, Joseph, 122
Dreyfus, Philippe, 208

Driesch, Hans, 42
Du Bois-Reymond, Emil, 10, 11
Dupuy, Jean-Pierre, 187, 228

E
Eckert, J. Presper, 152, 154–157, 167, 180
Eckert, Wallace J., 144
Edwards, Paul N., 234
Eger, Akiba, 4
Einstein, Albert, 7, 21, 26, 27, 53, 57, 175
Eisenhower, Dwight D., 236
Elias, Peter, 228
Eliot, Thomas Stearns, 19, 24, 32, 33, 37, 50,
     56, 57
Engelmann, Margaret, 66
Engels, Friedrich, 228, 229
Engstrom, Howard T., 106
Epstein, Jason, 237, 249
Euler, Leonhard, 75, 90, 142

F
Feller, William, 116, 135
Fermi, Enrico, 141
Ferrarotti, Franco, 225
Ferry, D. K., 104
Feynman, Richard, 232
Fishburn, Peter, 37
Fitzpatrick, Anne, 141, 143, 144, 171
Fontana, M., 50
Ford, J. J., 228, 229
Forrester, Jay, 184, 234
Fourier, Joseph, 38, 68, 70
Frankel, Stanley, 143, 170, 171
Frank, Lawrence K., 169, 185–188, 194
Frank, Philipp, 89, 93, 133, 141, 185
Frank, W. J., 141, 142
Fréchet, Maurice René, 49, 51
Frederick the Great, King of Prussia, 4
Freeman, Harold A., 135
Frege, Gottlob, 25, 30, 85
Fremont-Smith, Frank, 185, 186, 189
Freymann, Enrique, 205
Fry, T. C., 106, 129

G
Gagliasso, Elena, 211
Galilei, Galileo, 7, 67
Galison, Peter, 124, 125, 129, 130, 132, 134,
     175
Galois, Évariste, 75
Gamow, George, 81, 170, 202
Gandhi, Mahatma Mohandas Karamchand, 5

Gardner, Howard, 228
Gâteaux, René, 52, 53
George, William H., 94
Gerard, Ralph W., 186, 192
Gerovitch, Slava, 228
Geulincx, Arnold, 213
Geyer F., 239
Gibbs, Josiah Willard, 57, 60, 64, 71, 72, 121,
    209, 211
Giddens, Anthony, 227
Gillon, Paul N., 153, 155, 156
Gladwin, Lee, 121
Gleick, James, 147
Gödel, Kurt, 28, 32, 85, 87, 88, 90, 93, 191,
    210
Goldschmidt, Viktor M., 197
Goldsmith, Hyman, 233
Goldstein, Irving R., 163
Goldstine, Adele, 155
Goldstine, Herman, 145, 149–152, 154–157,
    159, 161, 162, 165, 169–171, 205, 234
Grattan-Guiness, Ivor, 10, 11, 14–16, 23, 26,
    27, 37, 47
Greco, Pietro, 141
Greenberg, J. R., 187
Green, Thomas, 19, 49
Grinevald, Jacques, 188
Groves, Leslie Richard Jr., 141, 159, 179

H
Haeckel, Ernst, 10, 11, 13, 216
Hagemeyer, F. W., 124
Haldane, John Burdon Sanderson, 80–82, 88,
    91, 92, 94, 95, 97, 98, 124, 125, 214,
    218, 219, 225, 251
Hall, Rupert, 247
Hardy, G. H., 27, 38, 48, 68, 74, 76–78
Harrison, George R., 168, 174
Harrower, Molly, 186, 192
Hart, James Norris, 39, 40, 148
Hazen, Harold L., 119, 124, 133, 152, 153, 179
Heaviside, Oliver, 68, 75, 118, 119,
    247, 249, 250
Heck, Cloyd, 171
Hegel, Georg Wilhelm Friedrich, 21, 24
Heims, Steve J., xx, 6, 7, 10, 28, 54, 55, 57, 59,
    66, 86, 87, 125, 149, 164, 168, 173, 185,
    187, 188, 191, 196, 203, 231, 233, 235,
    236, 249
Heine, Heinrich, 6
Hellman, Walter D., 112, 114, 149, 159,
    162–164, 166, 168, 172, 174
Henderson, Lawrence Joseph, 21, 22, 84
Hermite, Charles, 54

Heyting, Arend, 85
Hilbert, David, 28, 29, 48, 70, 85, 87, 191, 199
Hitler, Adolf, 85, 112, 172
Høffding, Harald, 11
Hölldobler, B., 241
Holmes, Oliver Wendell, 50
Holt, Edwin B., 15, 41
Hopf, Eberhard, 71, 114, 115, 147
Hotelling, Harold, 135, 167
Huntington, Edward Vermilye, 35,
    38–40, 47, 48
Hurwitz, Adolf, 118, 119
Husserl, Edmund, 24, 29, 67
Hutchinson, George Evelyn, 188, 197, 198
Huxley, Thomas H., 10, 11
Huyghens, Christiaan, 212

I
Ingham, Albert Edward, 68
Irving, Washington, 50
Israel, Giorgio, 86

J
Jackson, Dugald Caleb, 68
Jacobs, W. W., 243
James, William, 5, 8, 11–13, 15, 17, 21, 24, 30,
    37, 39, 41, 50, 117
Jaspers, Karl, xvii
Jdanko, A., 242
Jerison, D., xvii
Jevons, William Stanley, 226
Johnson, Lyndon B., 252
Johnson, T. H., 152

K
Kac, Mark, 64
Kahn, Bertha, 3
Kant, Immanuel, 4, 20, 21, 29, 36, 92, 128
Kay, Lily E., 202, 215
Kennedy, John F., 252
Kettering, Charles F., 76
Khrushchev, Nikita Sergeyevich, 228
King, Robert, 118
Kinsey, A. C., 231
Kline, Morris, 54
Klüver, Heinrich, 186
Knapp, Viktor, 229
Koch, Helge von, 53, 61
Kolmogorov, Andrey, 116, 132
Konopiski, Emil, 171
Kosulajeff, P. A., 116
Kruif, Paul de, 76
Kubie, Lawrence S., 192
Kuhn, Thomas Samuel, 131

Kyburg, H. E., 34

L
Lagrange, Joseph-Louis, 38
Laguerre, Edmond, 113
Lakatos, Imre, 219
La Mettrie, Julien Offray de, 213
Landau, H. G., 29, 48, 137
Laplace, Pierre-Simon marquis de, 38
Lawrence, Ernest Orlando, 141
Lazarsfeld, Paul, 186, 194–196, 198
Lebesgue, Henri, 49, 52–55, 64, 71, 72
Lee, Yuk-Wing, 70, 86, 105, 113
Leibniz, Gottfried Wilhelm, xvii, 4, 13, 25, 62,
        63, 64, 81, 82, 91, 145, 154, 212, 213,
        214, 216, 218
Lepschy, Antonio, 83, 116
Lettvin, Jerome Y., 137, 139
Levinson, Norman, 14, 47, 106, 111, 114, 134
Lewin, Kurt, 186, 188
Lewis, Bennet, 152
Lewis, Clarence Irving, 32
Lindbergh, Charles Augustus, 96
Lindgren, Nilo A., 109, 134, 153, 182–184
Lipset, David, 193
Littlewood, John Edensor, 48, 68, 77
Livingston, William K. (1892–1966), 197
Locke, John, 4, 182
Loomis, Alfred Lee, 133
Lorente de Nó, Rafael, 163, 164, 186, 189,
        191, 192
Lorenz, Edward, 244
Losano, Mario G., 227, 228
Lotka, Alfred J., 137
Lovell, C. A., 124

M
MacColl, Leroy A., 119, 134, 197
Mach, Ernst, 21, 85, 190
Madge, John, 188
Mahowald, Mary Briody, 20
Maimonides, Moses, 4
Malinowski, Bronislaw, 193
Mandelbrot, Benoît, 54, 62, 227
Mandrekar, V., xvii
Mangione, C., 23, 34
Martin, C. E., 231
Marwin, Walter T., 15
Marx, Karl, 227, 228
Masani, Pesi R., 24, 32, 34, 35, 56, 72, 76, 82,
        87, 105, 106, 113, 114, 129, 175, 191,
        197, 199, 201, 206, 225

Mason, Samuel J, 120
Mauchly, John W, 106, 151, 152, 154, 155,
        157
Maupertuis, Pierre Louis, 63
Maxwell, James Clerk, 58, 64, 117, 118, 195,
        209, 211
McCulloch, Warren Sturgis, 65, 127, 135–140,
        157, 160, 164–166, 173, 181, 185, 186,
        189–192, 194, 198–200, 211
McMillan, B., 60
McTaggart, John M. E., 19, 24
Mead, George Herbert, 187
Mead, Margaret, 125, 185, 186, 188, 193, 194
Mendel, Gregor, 81
Mendelssohn, Moses, 4
Menger, Karl (1902–1985), 89, 93
Mercer, James, 48
Merton, Robert, 249
Metropolis, Nicholas C., 142–145, 147, 164,
        171
Mikulak, M. W., 227–229
Millán Gasca, Ana, 86
Miller, George A, 223
Miller, H. A, 7
Miller, Marc, 5
Mills, Charles Wright, 8
Minc, Alain, 208
Mindell, David A., 113, 116, 119, 124, 130,
        132, 152, 155
Mitchell, Dana, 143, 187
Mitchell, S. A., 187
Moe, Henry A., 165, 173, 181
Monjardet, Bernard, 37
Monk, Ray, 26, 27
Monod, Jacques, 202
Montagnini, Leone, 9, 62, 65, 74, 94, 116, 128,
        193, 225, 234, 251
Montague, William Pepperell, 15
Mooney, Paul, 129
Moore, George E., 15, 25, 30
Morgenstern, Oskar, 86, 186
Morison, Robert S., 181
Morrell, Ottoline, 27, 28
Morse, H. C. Marston, 106, 148
Münsterberg, Hugo, 37, 38
Mussolini, Benito, 172

N
Nacci, Michela, 94
Nagel, Ernst, 32
Narasimhan, T. N., 38
Nebeker, F., 103

Neddermeyer, Seth, 142
Nelson, Eldred C., 142–145, 147, 164, 167, 170
Newton, Sir Isaac, 63, 75, 81, 212, 213, 218
Nietzsche, Friedrich, 46
Nora, Simon, 208
Norris, R. S, 39, 142, 148, 169
Northrop, F. S. C, 136, 189
Nyquist, Harry, 118, 119

O

Oppenheimer, Robert, 142, 144, 146, 169, 171, 235, 236
Osgood, William Fogg, 39, 49
Overbeck, Wilcox, 180
Owens, Larry, 134, 144, 150, 151, 183

P

Pantaleoni, Maffeo, 226
Pareto, Vilfredo, 226
Park, R. E, 7, 187
Parsons, Talcott, 239
Pascal, Blaise, 154
Peano, Giuseppe, 23–25, 61
Peirce, Charles Sanders;, 8, 13, 23, 37, 57, 58, 60, 61, 209, 210
Peretz, Isaac Leib, 5
Pericles, 78
Perrin, Jean Baptiste, 52–54, 61
Perry, Ralph Barton, 15, 27, 38, 41
Perry, Swick, 187
Peters, H., 98, 105
Peterson, Erik L., 193
Phillips, Henry Bayard, 57, 105, 113, 168, 174
Picardi, Ilenia, 141
Piccinini, Gualtiero, 108, 124, 125, 136, 139, 140, 149, 159, 161, 162, 173, 181
Pitkin, Walter B., 15
Pitts, Walter, 65, 135, 136, 137–140, 157, 158, 159, 162, 164, 165, 172, 181, 186, 192, 199
Plato, 14, 17, 41, 78, 81, 91
Platzman, R. L., 202
Poe, Edgar Allan, 50, 238
Poitras, Edward, 113, 129
Pomeroy, W. B., 231
Pope, Alexander, 238
Popper, Karl, 56, 209
Potter, Humphrey, 191
Pound, Ezra, 33
Pratt, V., 106
Pupin, Idvorsky Michael, 250

Q

Quastler, H., 202
Quintanilla, S., 82

R

Radcliffe-Brown, Alfred, 193
Ramón y Cajal, Santiago, 164
Ramos, Juan García, 141, 181, 196
Randell, Brian, 105, 145
Rapoport, Anatol, 137
Rashevsky, Nicolas, 137, 139
Restaino, F., 26
Reynolds, Osborne, 55
Ribot, Théodule Armand, 10
Ricciardi, Luigi Maria, 193
Rioch, David McKenzie, 188
Rogers, William Barton, 50
Roosevelt, Franklin Delano, 98, 101, 111, 133, 173
Rosenblith, W., 252
Rosenblueth, Arturo, 21, 82–85, 88, 125, 126, 128, 135–137, 140, 160, 161, 163, 164, 168, 172–174, 181, 182, 185, 186, 189, 191, 196, 199–201, 206, 207, 224
Rosenfeld, Morris, 5
Rossi, Paolo, 220
Routh, Edward John, 118, 119
Royce, Josiah, 8, 15–17, 19–21, 23, 24, 26, 31, 33, 35, 37, 42, 43, 47, 57–60, 62, 66, 84, 207, 210, 217
Rubin, M. D., 118
Russell, Bertrand, 15, 18, 23–30, 32–38, 41, 43, 48, 51, 87, 90, 136, 137, 191, 216

S

Saccheri, Giovanni Girolamo, 248
Sacerdote, Gino; 259, 206
Saeks, R. E., 104
Sage, Nathaniel, 9, 14
Samuel, Arthur, 243
Samuelson, Paul, 135
Santayana, George, 8, 15–18, 31, 33, 34, 37
Santesmases, M., 202
Santillana, Giorgio Diaz de, 17, 159, 172, 175
Savage, Leonard J., 186
Schickard, Wilhelm, 154
Schilt, Jan, 144
Schmidt, Karl, 15, 23
Schröder, Ernst, 23
Schrödinger, Erwin, 71, 81
Schwartz, Stephen P., 28
Segal, Jérôme, 62, 119, 124, 186, 211, 214

Shannon, Claude, 121, 122, 124, 134, 140, 216
Siegelman, J., 166, 251
Singer, I. M., xvii
Skaff, W., 19, 33, 57
Slack, Nancy G., 188
Smith, Adam, 225, 228
Smith, Grover, 21, 57
Smoluchowski, Marian, 53
Snell, J. Laurie, 122
Southard, Elmer Ernest, 21
Southwell, Richard, 104, 148
Spaulding, Edward Gleason, 15
Spencer, Herbert, 10, 13, 17, 19
Sperry, Roger Wolcott, 192
Spiegelman, Sol, 202
Spinoza, Baruch, 13, 62, 214
Stalin, Joseph Vissarionovich, 227
Stewart, Irvin, 134, 143
Stibitz, George R., 106, 108, 124, 129,
    144–146, 150, 152, 156,
    167, 190, 220
Stieltjes, Thomas Joannes, 54
Stone, Marshall Harvey, 20, 144
Stratton, Julius A., 182
Strauss, Lewis Lichtenstein, 168, 235
Strook, D. W., xvii
Struik, Dirk Jan, 71, 135, 251
Sullivan, Harry Stak, 187, 188
Szász, Otto, 29
Szilard, Leo, 202

T
Tait, Peter, 132
Taylor, Richard, 168, 183
Taylor, Sir Geoffrey Ingram, 52, 99
Teller, Edward, 142, 170, 171,
    179, 202, 235
Termini, Settimo, xx, 211
Tolstoy, Leo, 5, 6
Troland, Leonard Thompson, 61
Tropp, Henry S., 106
Truman, Harry S, 173, 235
Tukey, John W., 134
Turing, Alan Mathison, 138, 160, 189, 199,
    200, 216
Tustin, Arnold, 120, 125
Tuve, Merle Anthony, 170, 202
Tyler, Harry W., 51

U
Ulam, Stanisław, 144, 148
Urey, Harold, 141, 146

V
Vallarta, Manuel Sandoval, 82
Van De Graaff, Robert J., 97
Van Der Zowen, J., 239
Veblen, Oswald, 48, 49, 85, 167
Vernadsky, Vladimir, 197
Vestine, Ernest H., 162, 163, 165
Vince, M. A., 125
Vishniac, Roman, 4
Vogt, Karl (also Carl), 216
Volterra, Vito, 49, 137, 197, 226
Von Bonin, Gerhardt, 186
Von Förster, Heinz, 186, 189
Von Koch, Helge, 53
Von Mises, Richard, 135
Von Neumann, John, 65, 72, 85–87, 89, 92,
    106, 121, 135, 138, 139, 141, 142,
    144–151, 153–159, 161, 164–174, 179,
    180, 183, 184, 186, 189–191, 198,
    200–203, 211, 214, 215, 231–238, 247

W
Wald, Abraham, 135
Walras, Léon, 226
Walton, Sir John, 81, 96, 97
Watson, Thomas J., 143, 144
Watt, James, 78, 94, 95, 97, 117, 191
Weaver, Warren, 108, 115, 119, 125, 129–131,
    134, 141, 144, 145, 150–152, 155, 165,
    182, 183
Weller, Lawrence, 174
Wente, E. C., 124
Werskey, Gary, 94
Weyl, Hermann, 87, 202
Whitehead, Alfred North, 20, 23, 25, 26, 34,
    35, 38, 51, 87, 136
Wiener, Barbara, 66
Wiener, Leo, 3–8, 10, 12, 24, 26, 65
Wiener, Margaret (Peggy), 66
Wiener, Solomon, 5
Wiesner, Jerome Bert, 238, 252
Wilder, Raymond Louis, 76
Wildes, K. L., 109, 134, 153, 182–184
Wilks, Samuel S., 106, 162, 163, 165
Williams, Michael R., 21, 103, 106, 150, 154,
    156, 216, 250, 251
Williams, Samuel B., 156
Wilson, Edward O., 241
Wintner, Aurel, 172
Witte, Karl H. G., 8
Witte, Karl Junior, 8
Wittgenstein, Ludwig, 26, 27, 32

Wolff, Christian, 4
Woodger, Joseph Henry, 136
Woods, Friedrich, 21

Y
Yockey, H. P., 202
Young, David M., 105, 148

Z
Zamenhof, Ludwik Lejzer, 4
Zellini, Paolo, 190
Zermelo, Ernst, 90
Zilsel, Edgar, 247

Printed in the United States
By Bookmasters